MOS/LSI
Design and Application

TEXAS INSTRUMENTS ELECTRONICS SERIES

Carr and Mize ▪ MOS/LSI DESIGN AND APPLICATION

Crawford ▪ MOSFET IN CIRCUIT DESIGN

Delhom ▪ DESIGN AND APPLICATION OF TRANSISTOR SWITCHING CIRCUITS

The Engineering Staff of Texas Instruments Incorporated ▪ CIRCUIT DESIGN FOR AUDIO, AM/FM, AND TV

The Engineering Staff of Texas Instruments Incorporated ▪ SOLID-STATE COMMUNICATIONS

The Engineering Staff of Texas Instruments Incorporated ▪ TRANSISTOR CIRCUIT DESIGN

The IC Applications Staff of Texas Instruments Incorporated ▪ DESIGNING WITH TTL INTEGRATED CIRCUITS

Hibberd ▪ INTEGRATED CIRCUITS

Hibberd ▪ SOLID-STATE ELECTRONICS

Kane and Larrabee ▪ CHARACTERIZATION OF SEMICONDUCTOR MATERIALS

Runyan ▪ SILICON SEMICONDUCTOR TECHNOLOGY

Sevin ▪ FIELD-EFFECT TRANSISTORS

MOS/LSI Design and Application

DR. WILLIAM N. CARR

Professor of Electronic Sciences and Electrical Engineering, SMU

DR. JACK P. MIZE

Visiting Industrial Professor of Electronic Sciences and Electrical Engineering, SMU

Edited by

Robert E. Sawyer and John R. Miller

McGRAW-HILL BOOK COMPANY

New York St. Louis San Francisco Düsseldorf Johannesburg
Kuala Lumpur London Mexico Montreal New Delhi
Panama Rio de Janeiro Singapore Sydney Toronto

Library of Congress Cataloging in Publication Data

Carr, William N

MOS/LSI design and application.

(Texas instruments electronics series)
1. Integrated circuits. 2. Digital electronics.
I. Mize, Jack P., joint author. II. Title.
TK7874.C38 621.381'73 72-7407
ISBN 0-07-010081-0

234567890 HDBP 76543

The editors for this book were Tyler G. Hicks and Lydia Maiorca and its production was supervised by Teresa F. Leaden. It was set in Times Roman by York Graphic Services, Inc.

It was printed by Halliday Lithograph Corporation and bound by The Book Press.

Contents

Preface

This book is written with the intended purpose of presenting a unified treatment of MOS/LSI for the systems designer. Details of digital circuitry are therefore stressed in Chapter 4—Inverters, Static Logic, and Flip-flops; Chapter 5—Shift Register For Data Delay, Logic, and Memory; Chapter 6—The MOS/Bipolar Interface; Chapter 7—Memory Applications; and Chapter 8—Programable Logic Arrays. For completeness of circuit discussion, Chapter 9 on MOS Analog Circuitry is included. Methods of achieving reliability in MOS integrated circuits and economic aspects of MOS/LSI (probably the most important feature of these new circuit methods) are presented in Chapters 3 and 10, respectively. Since MOS fabrication techniques contribute to system performance, a discussion of the existing MOS arsenal of technology is given in Chapter 2. The authors believe that without a basic understanding of semiconductor device physics as it directly pertains to MOS/LSI, the effectiveness of the systems designer will be impaired and therefore Chapter 1, on this topic, has been included.

In a field which is advancing as rapidly as is MOS/LSI, it is difficult to choose the propitious moment for documenting the status and methods provided by this new and powerful form of electronics. The writer immediately finds that he is involved in a struggle attempting to "tell it like it is" rather than "like it was." To cope with the situation, this book has been written with a conscientious effort to present an up-to-date rendition of MOS/LSI methods and techniques—yet at the same time it is tempered with the approach of presenting fundamental concepts which the systems designer can utilize in accommodating his analyses to future technology advances. In our presentation of fundamentals, we have purposely omitted discussion of second-order effects and have thereby sacrificed rigor in order that basic principles will be made evident to the reader.

We are indebted to our colleague Herman Van Beek who has aided us in assembling a major portion of the text. Keith Lovelace and John Hodge provided many of the details pertaining to reliability aspects of MOS/LSI. We received helpful comments with respect to the economics of MOS/LSI from Charles Phipps and Daniel Baudouin. M. Ramanathan assisted in the computer calculations of Chapter 4 and his special contribution to the programable logic array presentation has been very helpful. We acknowledge our disciplinarian Donald Scharringhausen whose overt influence contributed to bringing the book to successful completion. Our editors John Miller and Bob Sawyer have most importantly contributed to ensuring that the text has been made readable.

The authors are fortunate to have the continued assistance of J. S. Kilby. This acknowledgment falls far short of adequately expressing our appreciation for his helpful suggestions and encouragement pertaining to this book.

Finally, although Texas Instruments, Incorporated has provided the means by which this book is brought to press, the authors assume full responsibility for the interpretations and forecasts which appear herein.

William N. Carr
Jack P. Mize

Dallas, Texas

MOS/LSI
Design and Application

1

MOS Device Physics

1.1 INTRODUCTION

1.1.1 Metal-Oxide-Semiconductor (MOS) Device Physics

In view of the formidable array of new technologies and methods presently facing the electronic engineer, the question arises as to the necessity for devoting attention to the particular discipline of MOS device physics where silicon is primarily used as the semiconductor element. Justification can best be made for the required effort by separately considering the situation relevant to:

1. The system designer
2. The circuit designer
3. The device technologist

The system designer should understand that MOS technology has not, as yet, settled into a routine prescription which is completely specified, understood, and defined. Fortunately, MOS technology and circuit innovations continue to unfold and generate dramatic improvements in circuit performance. It presently appears that MOS ranks as a serious contender for utilization in large blocks of electronic functions formerly reserved for bipolar transistors. The present-day employment of MOS and bipolar devices in electronic systems can be likened to the structure of a building. In the analogy, MOS/LSI (metal-oxide-semiconductor large-scale integration) serves as large "building blocks," and bipolar circuits (TTL gates, for example) serve as the "mortar or concrete" which interconnects the building blocks to one another and to external appendages such as meters, visual displays, motors, or printers. The system design engineer must select from the rich spectrum of MOS technology that method which will optimumly provide the building blocks for his particular electronic system. A fundamental knowledge of MOS device physics will make his task easier and his judgment more effective.

The MOS circuit designer operates in an arena of ever-increasing complexity and design sophistication. To work effectively he must fully exploit the metal-oxide-silicon field-effect transistor (MOSFET) in circuit design. A basic understanding of the device and its operation is an essential requirement for the circuit designer who expects to accomplish his tasks successfully.

The device technologist involved with MOS processing and technical innovation must understand how his fabrication methods affect the device performance in the integrated circuit. A lack of understanding of MOS device physics will reduce the performance of the technologist to that of an automaton who merely passes semiconductor silicon slices through the process subject to the whims of the circuit designer. This is a situation to be avoided, since many of the advances in MOS technology during the past several years have been brought about through a strong interaction between circuit designer and technologist—an interaction which must be maintained to ensure future evolution of MOS in electronic systems.

1.1.2 Basic Structure of the MOSFET

A cross-sectional view of a p-channel MOSFET is shown in Fig. 1.1. Note that the device consists of p-type source and drain regions diffused into an n-type silicon substrate. A very thin layer of insulating SiO_2 is positioned on top of the silicon between the source and drain regions. This insulator is called the gate oxide and is nominally 1000 Å thick. Positioned on top of the gate oxide is a metal field plate (usually aluminum) which is referred to as the gate. When a voltage is applied to the gate, negative with respect to the substrate, positive charge is brought to the surface of the silicon between source and drain. If the negative gate voltage is large enough (a few volts), the channel between the source and drain will become p-type (inverted). Under these conditions, electrical conduction will take place between the source and drain when a potential difference is applied between these two terminals. In the absence of the inversion layer, the source and drain in Fig. 1.1 are effectively isolated from each other. It is to be noted that the device exhibits bilateral symmetry; i.e., the source and drain regions are electrically interchangeable and are not defined until the device is connected to voltage or current nodes within a circuit.

The inversion layer which forms the conducting bridge between source and drain is the seat of the electrical activity exhibited by the device. The analytical description of the inversion layer basically involves extension of concepts pertinent to that of bulk silicon. Other bulk silicon properties such as junction capacitance, junction avalanche breakdown, and junction leakage current play important secondary roles in MOS device action. In view of the importance of bulk-silicon concepts and

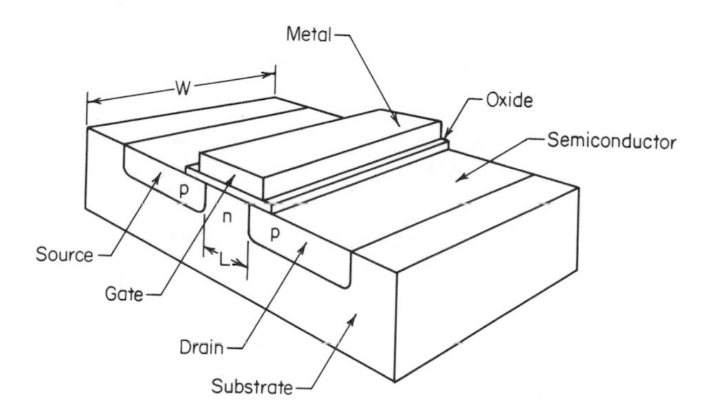

Fig. 1.1. MOSFET structure.

properties, a summary of those relevant to MOSFET performance will first be presented.

1.2 THE SILICON BULK

1.2.1 Definition of the Bulk

By the term *bulk silicon* we mean that body of silicon infinitely removed from the surface, the surface thereby having no influence on the bulk properties. The bulk silicon referred to is, of course, single crystalline in theory and in practice.

1.2.2 Crystalline Structure of Bulk Silicon

(a) The Cubic Lattice. Silicon crystalline structure is characterized by periodic repetition of silicon atoms. The crystallographer classifies single-crystal silicon as a diamond-lattice structure which consists of two interpenetrating face-centered cubic sublattices. In practical terms this means that silicon can be thought of in certain considerations as a simple cubic crystal. Although the crystal structure is repetitive, let us consider for analysis a single "cube" of silicon positioned on the x, y, z axes as shown in Fig. 1.2.

In MOS technology it is frequently necessary to specify various planes in a silicon crystal. *Miller indices* are employed for this purpose. To specify the Miller indices of a plane:

- Determine the intercepts a, b, c of the plane with the three axes x, y, z, respectively. The plane must be chosen so that no intercept is at the origin.
- Form the reciprocals of these intercepts.
- Express the reciprocal terms with the smallest set of integers that can be obtained by multiplying each of the fractions by the same number (i.e., reduce the terms to the smallest set of integer values possible).

examples

1. The (111) plane (Fig. 1.3). In this example the plane intercepts the axes at unit axial lengths 1, 1, 1. The reciprocals of these unit values is unity, and since

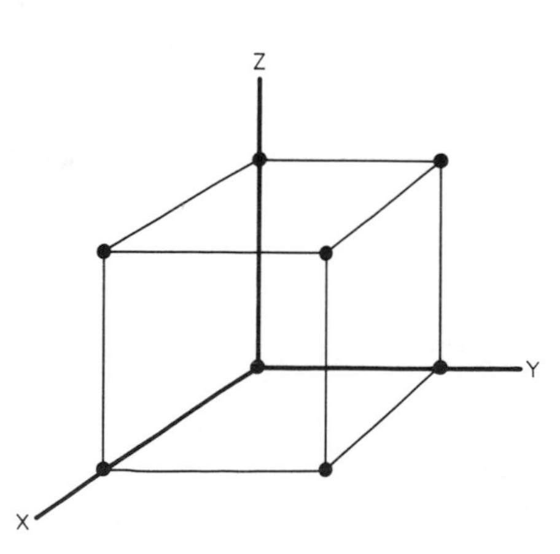

Fig. 1.2. Simple cubic crystal structure. $a = b = c =$ unity, and axes are mutually perpendicular.

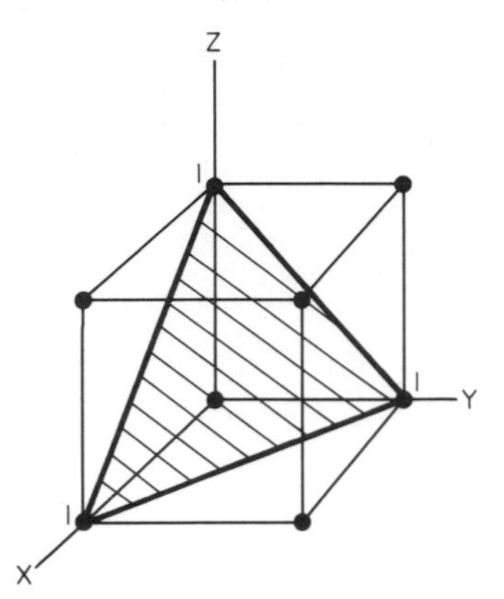

Fig. 1.3. The (111) plane.

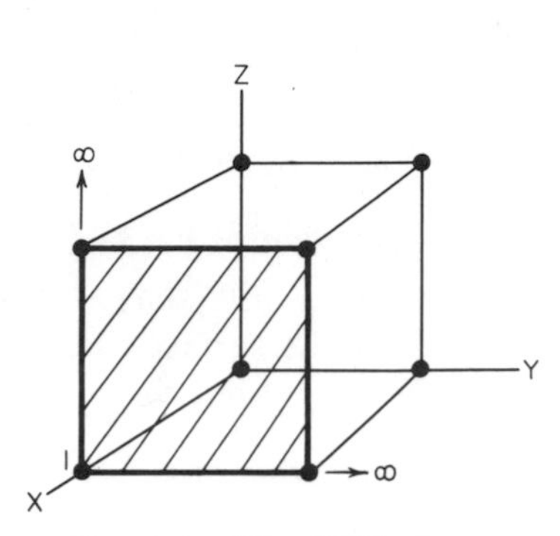

Fig. 1.4. The (100) plane.

no common factors other than 1 are present, the Miller indices are (111) and this plane is called the (111) plane.

2. The (100) plane (Fig. 1.4). In this example the plane intercepts the three axes at axial lengths of 1, ∞, ∞. The reciprocals of these values yield the Miller indices of (100) and the plane is called the (100) plane.

3. The (110) plane (Fig. 1.5). In this example the plane intercepts the three axes at axial lengths of 1, 1, ∞. The reciprocals of these values yield Miller indices of (110) and the plane is called the (110) plane.

It is of importance to observe that, excluding signs, the Miller indices are to be reduced to the smallest set of integers. Thus, for example, the plane with intercepts 2, 2, ∞ has Miller indices of (110) and is parallel to the plane shown in Fig. 1.5. Also note that a bar over a number indicates a negative intercept. Thus a plane having intercepts of $\bar{1}$, $\bar{1}$, ∞ with the x, y, z axes, respectively, has Miller indices of $(\bar{1}\bar{1}0)$ and is a plane parallel to the (110) plane.

(b) Physical Consequences of Crystal Symmetry. The fact that bulk silicon is characterized by a cubic-crystal structure places certain symmetry constraints on the physical properties of the material. The constraints obey a physical law given by Neumann,[1] which can be stated in elementary terms: *The physical properties of a crystalline material must exhibit the same symmetry properties as those of the crystalline structure of the material.* Since the cubic structure of silicon exhibits complete rotational symmetry, the physical parameters such as electrical and thermal conductivity, mobility, dielectric constant, and diffusivity are *always* isotropic in *bulk* silicon.

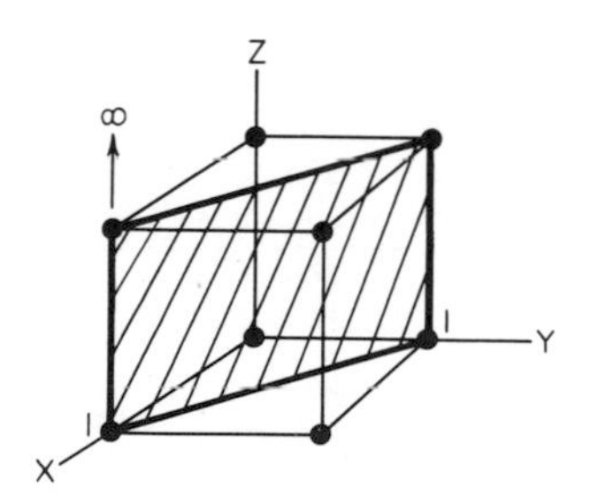

Fig. 1.5. The (110) plane.

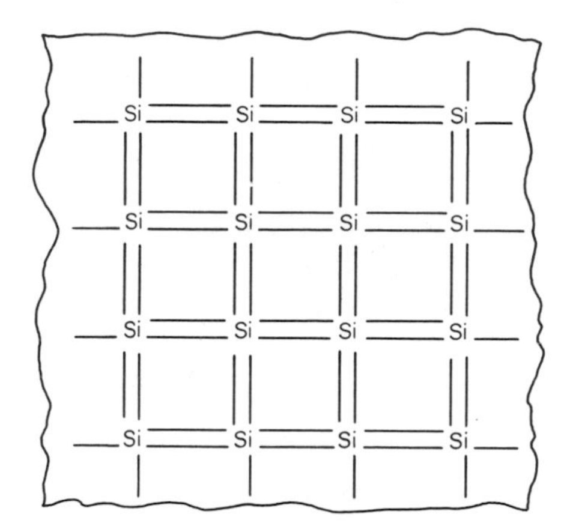

Fig. 1.6. Covalent bonding of silicon atoms.

(c) Covalent Bonding. The silicon crystal is held together by covalent bonding forces between silicon atoms. The covalent bonding forces arise from sharing valence electrons between neighboring atoms in an effort to complete and close an outer shell of electrons, thereby forming a configuration which is energetically most favorable. In a pure silicon crystal each silicon atom shares its four valence electrons with four neighboring silicon atoms, forming four covalent bonds. The situation is summarized in Fig. 1.6.

1.2.3 Intrinsic Silicon

Semiconductors are characterized by valence, forbidden, and conduction regions depicted for silicon as shown in Fig. 1.7. An energy level E_i, called the intrinsic level, is drawn midway between the valence and conduction band edges E_v and E_c, respectively. The width of the silicon forbidden gap, E_g, is 1.1 eV at 300°K. The drawing of Fig. 1.7 has physical meaning in the vertical direction only (it is a one-dimensional energy-level diagram).

For finite temperatures a probability exists that electrons from the top of the valence band will be thermally excited across the forbidden band into the conduction band. It can be shown that for pure, single crystalline silicon at 300°K, the number of electrons n_i residing in the conduction band as a result of thermal excitation from the valence band is[2]

$$n_i = 3.9 \cdot 10^{16} T^{3/2} e^{-1.21/2KT}/\text{cm}^3 \tag{1-1}$$

where $n_i = 1.4 \cdot 10^{10}/\text{cm}^3$ at $T = 300$°K

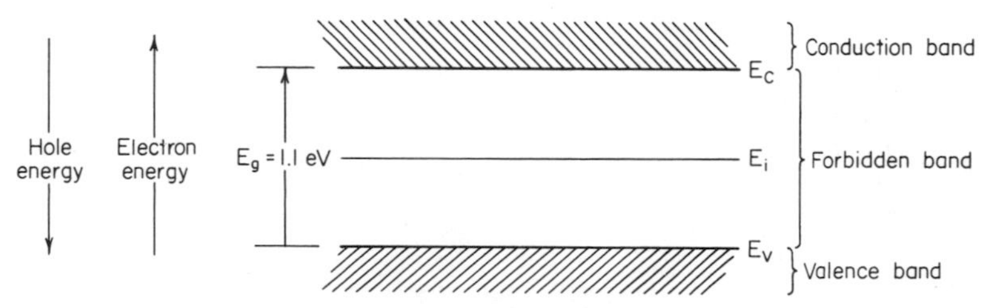

Fig. 1.7. Energy-level structure of silicon.

These n_i electrons are available for conduction, as are the holes (positive charges) left behind in the valence band. The number of electrons is equal to the number of holes

$$n = p = n_i$$

in the absence of impurities, and the silicon under these conditions is designated as *intrinsic*. For the intrinsic condition, the Fermi level[3] E_f can be shown to lie at the midgap position E_i, if the mass of the electron is taken equal to that of the hole.[4] It is to be noted that although pure silicon does not exist in practice, the concepts of E_i, n_i, and *intrinsicity* will be found to be of great importance in the analysis of MOSFET electrical characteristics.

1.2.4 Extrinsic Silicon

(a) Doping of Silicon with Phosphorus. All the silicon regions of the MOSFET contain dopant impurities that are purposely placed in the semiconductor body. The dopants most often used in MOS technology are boron and phosphorus. The effect of boron doping or phosphorus doping on the electrical conductivity of silicon is profound. We shall first consider the effect of phosphorus doping on silicon.

When phosphorus is added to silicon in minute quantities—in practice only a few parts per million—excess electrons become available for conduction. The phosphorus impurity atom is called a *donor atom* since it contributes a negatively charged mobile carrier to the crystal lattice. The excess electrons contributed by the phosphorus donors arise from two basic mechanisms:

1. The valence of phosphorus is five, and therefore when a phosphorus atom is substituted into a silicon lattice site, only four valence electrons are involved in covalent bonding, and the fifth electron can become available for conduction. The situation is summarized in Fig. 1.8.
2. There exists a small but finite binding energy for the fifth electron to the donor atom phosphorus. The fifth electron resides in a shallow energy state below the lowest electron state in the conduction band. This energy state for phosphorus lies 0.039 eV below the conduction band as shown in

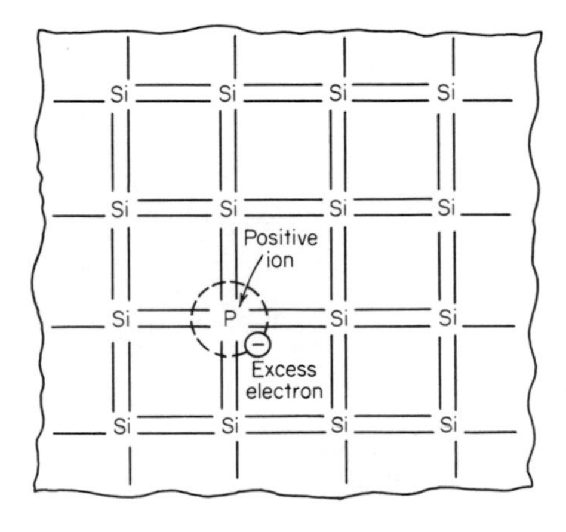

Fig. 1.8. Covalent bonding of phosphorus-doped silicon.

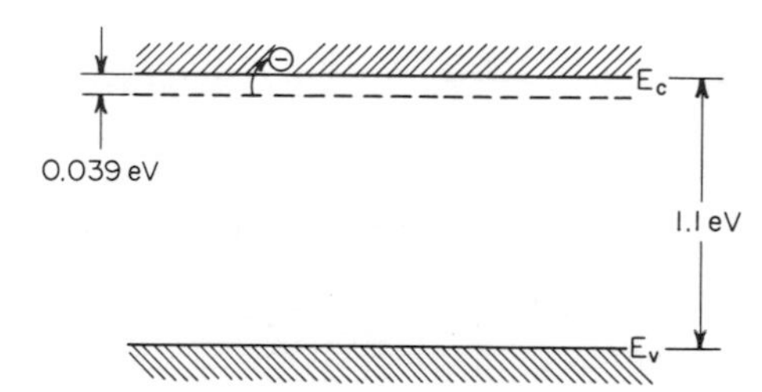

Fig. 1.9. Phosphorus donor level in silicon.

Fig. 1.9. At room temperature essentially all the donor electrons become thermally excited to the conduction band, where they can contribute to electrical conduction. With the availability of excess electrons, the semiconductor is classified as being *extrinsic* and is called *n-type* since conduction proceeds by negative carriers.

Because of the binding energy involved, the number of electrons excited to the conduction band from the donor level will be a function of temperature. Figure 1.10 pictorially indicates the number of electrons excited to the conduction band from a $10^{15}/\text{cm}^3$ donor impurity concentration as a function of temperature.[5] Note that carrier freeze-out begins to play an important role below $\approx 100\,°\text{K}$. For temperatures above $550\,°\text{K}$, n_i contributes appreciably to the electron population in the conduction band. As temperature continues to increase above $550\,°\text{K}$, the equal number of holes and electrons resulting from n_i generation brings the Fermi level for the n-doped extrinsic material back toward midband. At sufficiently high temperatures the Fermi level resides essentially at midband, and the material is said to have "gone intrinsic" (even though it had been extrinsic in the region of $300\,°\text{K} \pm \approx 200\,°\text{K}$).

For n-type extrinsic semiconductors, the Fermi level E_f lies above the intrinsic level E_i by an amount given by the expression[G]

$$n = n_i e^{(E_f - E_i)/kT} = n_i e^u \tag{1-2}$$

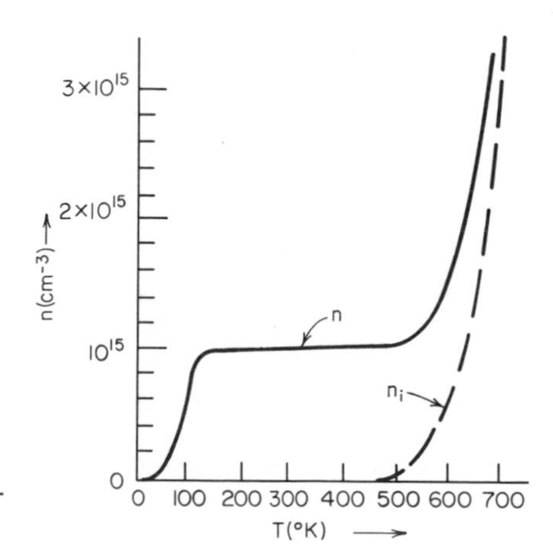

Fig. 1.10. Temperature dependence of electron concentration in n-type silicon.[5]

where $u = \dfrac{E_f - E_i}{kT}$ (dimensionless)

$$kT = 0.026 \text{ eV at } T = 300°\text{K}$$

$$n_i = 1.4 \cdot 10^{10}/\text{cm}^3 \text{ at } T = 300°\text{K}$$

Thus at 300°K for an n-type donor doping density of $10^{15}/\text{cm}^3$ (typical for silicon on which p-channel MOSFETs are fabricated), $E_f - E_i$ is, from Eq. (1-2):

$$E_f - E_i = kT \ln \frac{n}{n_i} = 0.026 \ln \frac{10^{15}}{1.4 \cdot 10^{10}}$$

and

$$E_f - E_i = +0.29 \text{ eV}$$

$$u = \frac{0.29}{0.026} = +11.1$$

The energy-level diagram for phosphorus-doped bulk silicon with a donor doping level of $10^{15}/\text{cm}^3$ and all donors ionized (implying that all donor electrons are thermally excited to the conduction band, as they would be at 300°K) is shown in Fig. 1.11. A plot of $E_f - E_i$ versus impurity doping level is shown in Fig. 1.12.

(b) Doping of Silicon with Boron. When silicon is doped with boron, a situation develops which is essentially the inverse of that for the phosphorus-doped silicon case. The presence of boron in the silicon lattice provides for a "negative" excess of electrons, resulting in the existence of positively charged carriers called *holes*. The boron impurity atoms are called *acceptor atoms* since they can be thought of as accepting an electron from the valence band (or equivalently exciting a hole into the valence band), thereby completing a covalent bond and leaving a mobile positively charged hole in the valence band which can contribute to electrical conduction. The excess holes contributed by the boron acceptors arise from two basic mechanisms:

1. The valence of boron is three, and therefore when boron is substituted into a silicon lattice site, only three valence electrons are available for covalent

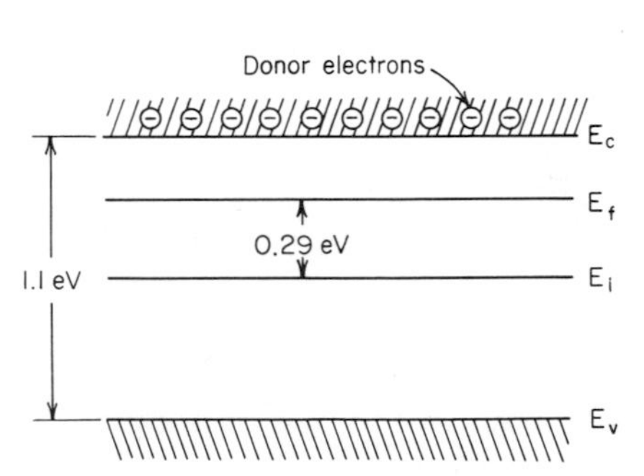

Fig. 1.11. Energy-level diagram for silicon with phosphorus doping concentration of $10^{15}/\text{cm}^3$.

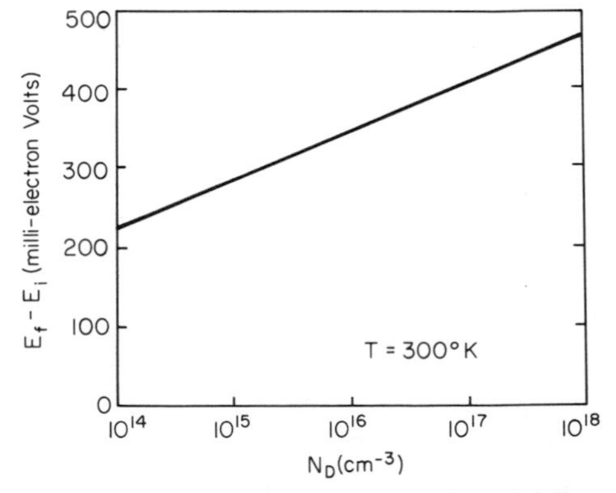

Fig. 1.12. $E_f - E_i$ *vs.* impurity doping level N_D.

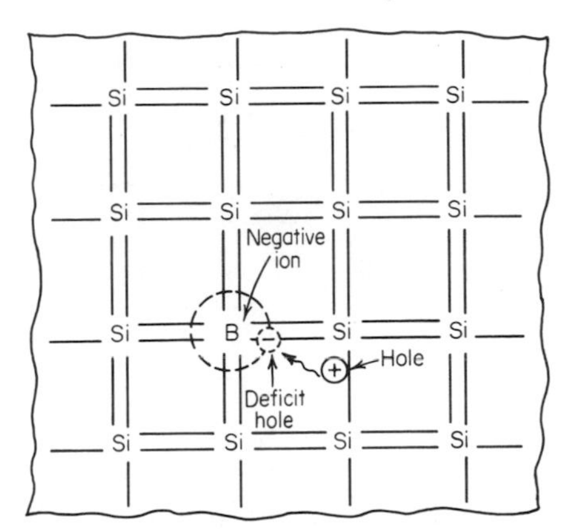

Fig. 1.13. Covalent bonding of boron-doped silicon.

bonding, and a hole can become available for conduction. The situation is summarized in Fig. 1.13.

2. Since there exists a small but finite binding energy for the hole to cling to the boron atom, the hole lies in a shallow energy-level state above the highest electron state in the valence band. The resulting hole energy state lies 0.045 eV above the valence band as shown in Fig. 1.14. At 300°K, essentially all the acceptor holes are thermally excited into the valence band, where they can contribute to electrical conduction.

In analogy with the phosphorus-doped silicon example, in the presence of excess holes the semiconductor is extrinsic and is called *p-type* because conduction proceeds by positive carriers. The same considerations summarized in Fig. 1.10 hold for p-type material. For p-type extrinsic semiconductors the Fermi level lies below the intrinsic level E_i by an amount given by Eq. (1-3), and for a p-type donor doping density of $10^{15}/\text{cm}^3$ we find

$$p = n_i e^{(E_i - E_f)/kT} = n_i e^{-u} \tag{1-3}$$

where $u = \dfrac{E_f - E_i}{kT}$ (dimensionless)

$$E_f - E_i = -0.29 \text{ eV}$$
$$u = -11.1$$
$$T = 300°\text{K}$$

The energy-level diagram for boron-doped bulk silicon with a doping level of $10^{15}/\text{cm}^3$ and all acceptors ionized (implying that all holes are thermally excited

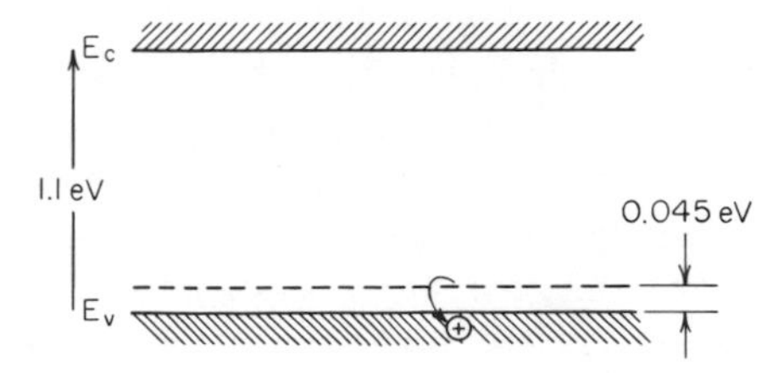

Fig. 1.14. Boron acceptor level in silicon.

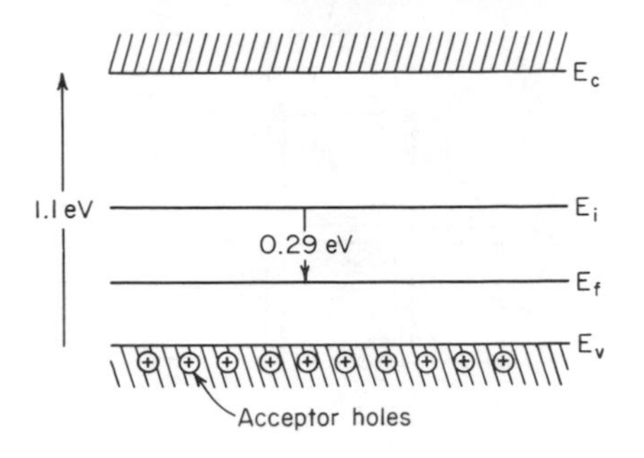

Fig. 1.15. Energy-level diagram for silicon with boron doping concentration of $10^{15}/\text{cm}^3$.

to the valence band) is shown in Fig. 1.15. A plot of $E_f - E_i$ versus impurity doping level would be identical to that of Fig. 1.12, with the exception that the quantities $E_f - E_i$ and u are negative for the boron-doped silicon case.

1.2.5 Electrical Conductivity of Extrinsic Bulk Silicon

(a) Carrier Mobility. When an electric field $\mathcal{E}$ is applied to an extrinsic semiconductor, a drift velocity v is imparted to electron charge carriers if the material is purely n type, or to hole carriers if the material is purely p type. The drift velocity of the charge carrier is directly proportional to the applied electric field and the proportionality factor is defined as the carrier mobility μ. The relationship can be summarized as:

$$v = \mu \mathcal{E} \tag{1-4}$$

Although the path of the charge carriers is fraught with collisions within the crystal lattice (lattice scattering) or by encounters with ionized impurities (impurity scattering), the concept of drift velocity (although it be an averaged velocity) is of great importance in describing semiconductor device phenomena.

Carrier mobility in bulk silicon is a function of doping concentration, as would be expected from consideration of the impurity scattering process of Coulombic origin (recall that phosphorus donor atoms or boron acceptor atoms are fully ionized in their lattice sites at room temperature). Measurements employing the Hall effect coupled with electrical conductivity on silicon samples reveal that the majority carrier electron mobility value (electron mobility in n-type silicon) is greater by a factor of ≈ 3 than that for majority carrier hole mobility values (hole mobility in p-type silicon). The data are summarized in Fig. 1.16.[7] Mobility values exhibited by minority carriers (electrons in p-type silicon or holes in n-type silicon) are quite similar to the values they exhibit as majority carriers.[8]

It is to be noted that the data of Fig. 1.16 are applicable when the value of the electric field is less than $\approx 10^3$ V per cm. For higher field regions the mobility values actually decrease, resulting eventually in the saturation of drift velocity. The mobility is then referred to as being *field dependent*. The reduction of carrier mobility in high-field regions is caused by a fundamental change in the interaction of carriers with the lattice. The change is brought about by the carriers becoming "hot"; i.e., the carriers acquire velocity components that are large in comparison to the velocity associated with thermal motion.

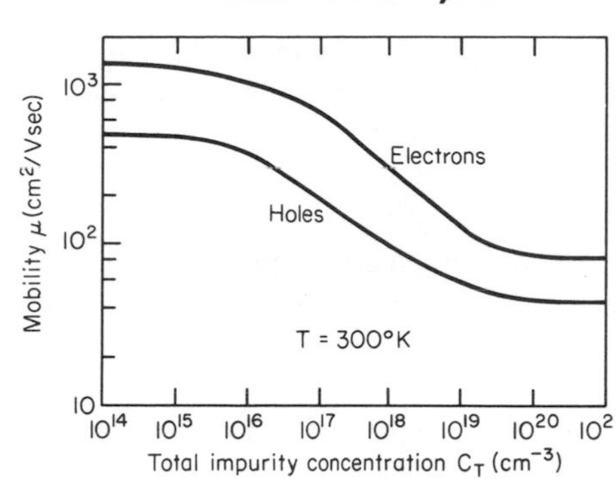

Fig. 1.16. Majority-carrier mobility values as a function of impurity concentration.[7] $T = 300°K$.

Bulk carrier mobility values exhibit a strong and somewhat complex dependence on temperature. The functional dependence on temperature arises from the nature of the scattering mechanisms, i.e., whether lattice scattering or impurity scattering dominates the particular situation. The temperature dependence of carrier mobility in bulk silicon for various impurity concentration values is summarized in Fig. 1.17.[9]

Since carrier drift velocity is directly proportional to carrier mobility, it would be expected that the inherent frequency response of a semiconductor device should be a direct function of mobility. It is thus observed that bulk-silicon devices which rely on drift or diffusion of electrons as their basic conduction mechanism (NPN bipolar silicon transistors) are "faster" in circuit performance than hole-carrier devices (PNP bipolar transistors)—all other parameters being equivalent, of course.

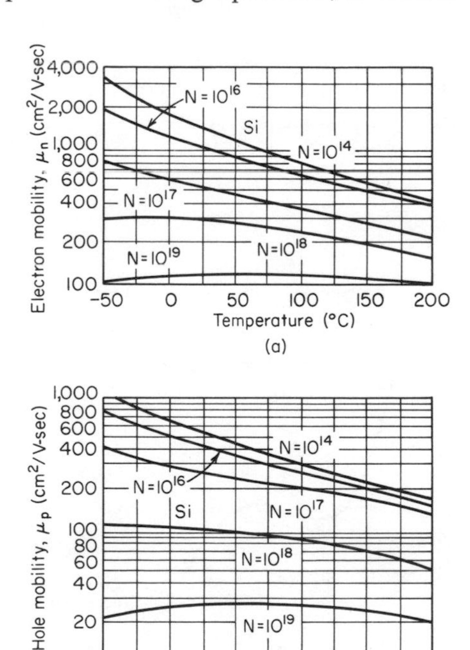

Fig. 1.17. Electron and hole mobility values in silicon as a function of temperature for various impurity concentrations.[9]

(b) Resistivity–Conductivity Equations. From the fundamental definition of current density in terms of electric field ϵ and conductivity σ_n, for the case of an n-type semiconductor:

$$j_n = \sigma_n \mathcal{E} \tag{1-5}$$

In terms of electron charge ($q = 1.6 \cdot 10^{-19}$ Coul), carrier density n, and carrier velocity v, current density can also be expressed as

$$j_n = nqv \tag{1-6}$$

Then from Eqs. (1-4), (1-5), and (1-6)

$$j_n = nqv = nq\mu_n\epsilon = \sigma_n \mathcal{E} \tag{1-7}$$

and therefore

$$\sigma_n = nq\mu_n \tag{1-8}$$

Similarly for a p-type semiconductor, conductivity σ_p can be expressed as

$$\sigma_p = pq\mu_p \tag{1-9}$$

Since resistivity ρ is the reciprocal of conductivity, the expressions for resistivity readily follow and are:

$$\rho_n = \frac{1}{nq\mu_n} \tag{1-10}$$

$$\rho_p = \frac{1}{pq\mu_p} \tag{1-11}$$

If an extrinsic bulk semiconductor is doped with both donors and acceptors, the resistivity is expressed as

$$\rho = \frac{1}{nq\mu_n + pq\mu_p} \tag{1-12}$$

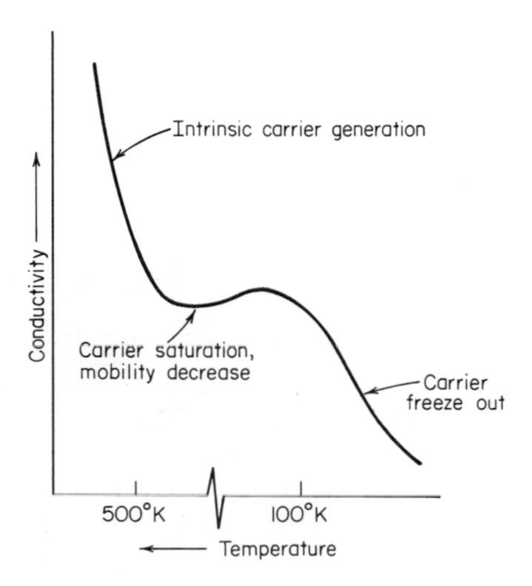

Fig. 1.18. Generalized dependence of silicon conductivity with respect to temperature.

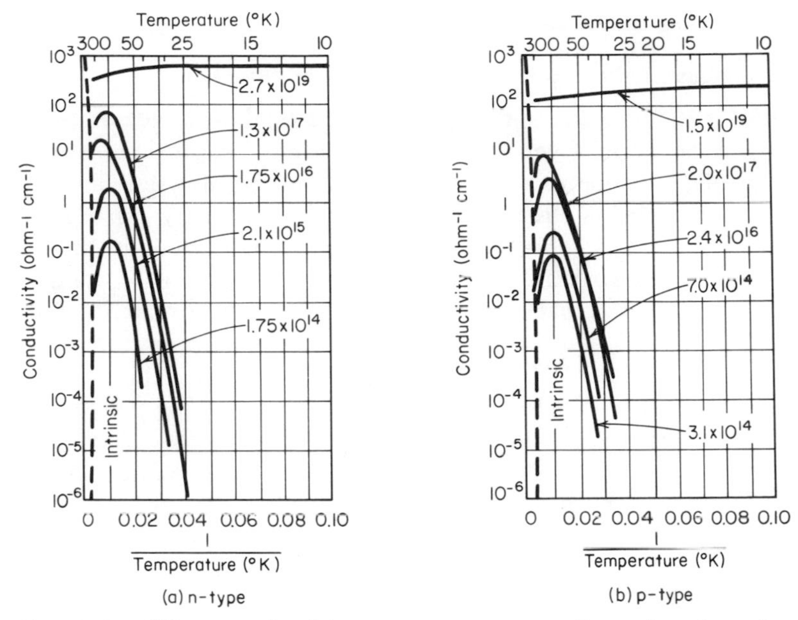

Fig. 1.19. Silicon conductivity versus temperature for various impurity concentrations.[10]

Since conductivity and resistivity are functions of carrier concentration and mobility [Eq. (1-12)] with both of the latter being temperature dependent, it is to be expected that conductivity and resistivity will also be temperature dependent. The general behavior of bulk-silicon conductivity as a function of temperature is summarized in Fig. 1.18. The actual data are shown in Fig. 1.19.[10]

A graph showing the relationship between resistivity and doping level (Fig. 1.20) has been prepared for silicon by Irvin.[11] The data of Fig. 1.20 have been obtained from a compilation of a large number of measurements on silicon samples containing either donor or acceptor impurities. Figure 1.20 is probably the most frequently used graph in silicon device design.

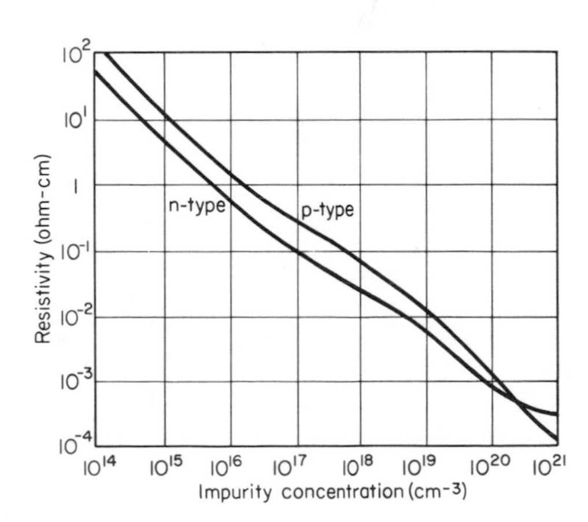

Fig. 1.20. Silicon resistivity at 300°K as a function of impurity concentration.[11]

1.3 THE SILICON p-n JUNCTION

1.3.1 The p-n Junction at Equilibrium—the Built-in Potential

In Sec. 1.2.4 it was pointed out that for n-type material the large number of excess electrons move the Fermi level E_f above the center of the forbidden energy gap E_i. Similarly, the large number of excess holes in p-type material move the Fermi level downward and below the center of the forbidden energy gap E_i. When an interface of p- and n-type silicon is formed (for example, by impurity doping of boron into n-type material), a carrier gradient is produced which results from the net transfer (by diffusion) of electrons from n material to p material and the net transfer of holes (by diffusion) in the opposite direction. Equilibrium occurs when a built-in (contact) potential V_B, with accompanying electric field, is developed to counteract carrier diffusion across the interface of the resulting p-n junction.

To derive an expression for the built-in potential V_B, assume that a "step" p-n junction has been formed as shown in Fig. 1.21. A large gradient of holes and electrons will appear in the vicinity of $x = 0$ as a result of the abrupt doping profile. Since the electron and holes are mobile, we would expect them to diffuse from the step profile region of impurity dopants N_D and N_A, respectively. The diffusion of mobile carriers is described by Fick's law:

$$j = -qD \text{ (gradient of carrier concentration)}$$

$$j = -qD\frac{\Delta C}{\Delta x} \quad \text{(in one dimension where } C \text{ defines carrier concentration and } D \text{ is the diffusion coefficient)} \tag{1-13}$$

The resulting diffusion of mobile hole and electron carriers uncovers N_A and N_D stationary dopant impurities, respectively, which are ionized. The stationary dopant impurities set up an electric field which is generated in a direction to oppose the further diffusion of electrons and holes from their respective gradients. The interface region now possessing an electric field is called the *transition* or *depletion-layer*

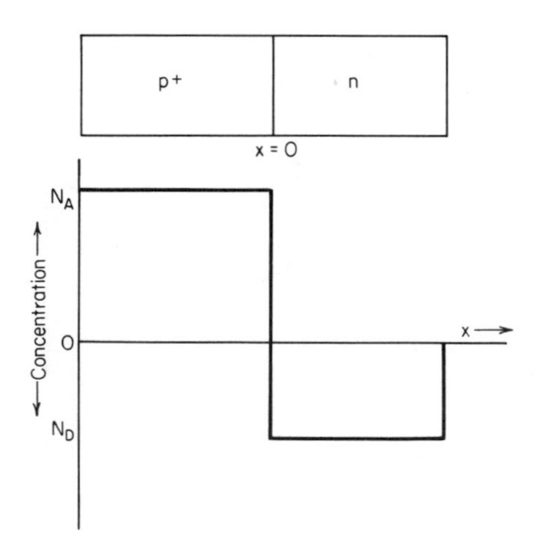

Fig. 1.21. Impurity concentration profile for the step p-n junction.

region. Current equilibrium conditions eventually result when the electric field driving force develops a drift current $Cq\mu\mathcal{E}$ which is equal and opposite to the diffusion current developed by the diffusion gradient *driving force*. The situation can be mathematically described as:

$$j = 0 = -qD \text{ (gradient carrier concentration)} + Cq\mu\mathcal{E} \tag{1-14}$$

or

$$qD \text{ (gradient carrier concentration)} = Cq\mu\mathcal{E} \tag{1-15}$$

where[12] $D = \dfrac{KT\mu}{q}$

To draw the energy-level structure for the p-n junction in equilibrium (zero external bias), first consider the behavior of the Fermi level. The Fermi level can be considered to be the "chemical potential" for electrons or holes. Since the condition for equilibrium in any system is that the chemical potential should be constant throughout the system, the constancy of the Fermi level throughout the semiconductor at equilibrium follows. Another way of stating the foregoing (and of great importance to MOS analyses) is: *A necessary and sufficient condition for the absence of net current flow in a semiconductor is that the Fermi level be constant throughout the system.* With the condition of equilibrium prevailing, the energy-level diagram for the p-n junction can be constructed by bringing Fig. 1.22*a* and *b* together to form Fig. 1.22*c* with the Fermi level throughout being constant.

The fact that E_i is not constant, as represented in Fig. 1.22*c*, indicates the presence of a built-in potential V_B for the junction at equilibrium defined as:

$$V_B \equiv \frac{E_{in} - E_{ip}}{q} \tag{1-16}$$

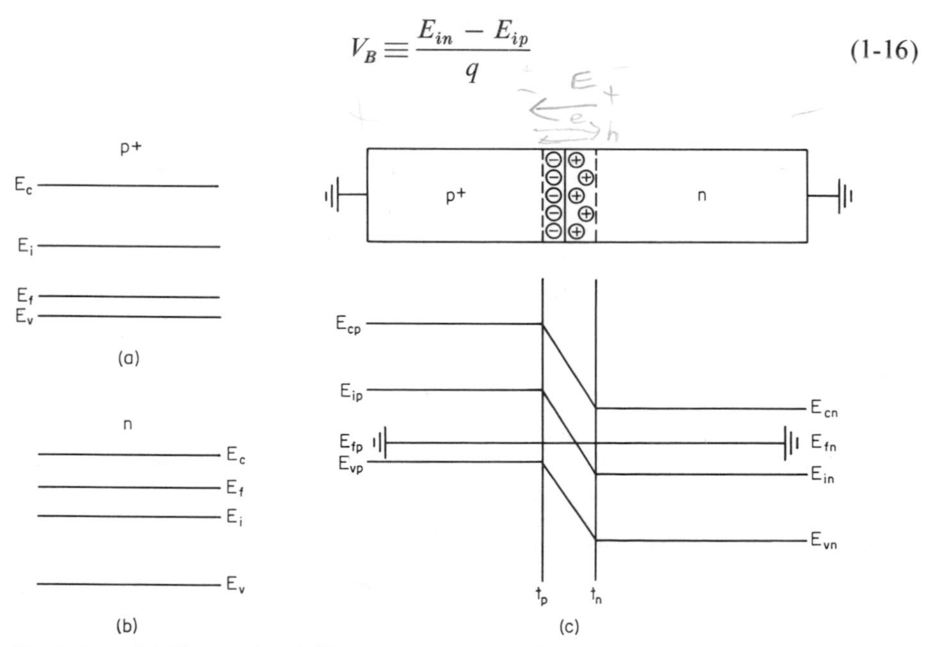

Fig. 1.22. (*a*) Energy-level diagram for n-type silicon; (*b*) energy-level diagram for p-type silicon; (*c*) energy-level diagram for silicon p-n junction with zero applied bias.

To calculate V_B in terms of the doping concentrations on both sides of the junction interface, recall from Eq. (1-2) that the concentration of electrons on the n side is given by:

$$n_n = n_i e^u = n_i e^{(E_{fn}-E_{in})/kT} \tag{1-2}$$

Therefore:

$$\frac{E_{fn} - E_{in}}{kT} = \ln \frac{n_n}{n_i} \tag{1-17}$$

The concentration of holes on the p side is given by:

$$p_p = n_i e^{-u} = n_i e^{-(E_{fp}-E_{ip})/kT} \tag{1-3}$$

Therefore:

$$-\frac{E_{fp} - E_{ip}}{kT} = \ln \frac{p_p}{n_i} \tag{1-18}$$

Adding Eqs. (1-17) and (1-18) yields

$$\frac{E_{fn} - E_{in}}{kT} - \frac{E_{fp} - E_{ip}}{kT} = \ln \frac{n_n}{n_i} + \ln \frac{p_p}{n_i} \tag{1-19}$$

And since $E_{fn} = E_{fp}$ as shown in Fig. 1.22*c* (zero-bias condition), it follows that

$$\frac{E_{in} - E_{ip}}{kT} = -\ln \frac{n_n p_p}{n_i^2} \tag{1-20}$$

and therefore from Eqs. (1-16) and (1-20)

$$V_B = -\frac{kT}{q} \ln \frac{n_n p_p}{n_i^2} \tag{1-21}$$

To calculate a typical value of V_B for a silicon p-n junction, consider

$$n_n = 1.4 \cdot 10^{15}/\text{cm}^3 \qquad p_p = 1.4 \cdot 10^{17}/\text{cm}^3 \qquad T = 300°\text{K}$$

$$V_B = -0.026 \ln \frac{1.4 \cdot 10^{15} \cdot 1.4 \cdot 10^{17}}{(1.4 \cdot 10^{10})^2} \text{ V}$$

$$V_B = -0.7 \text{ V}$$

A built-in voltage value (voltage barrier) of ≈ -0.7 V as calculated above indicates that application of an external voltage of $\approx +0.7$ V forward bias to the junction (p region positive with respect to n region) would reduce the barrier height shown in Fig. 1.22*c*, thereby permitting a copious amount of "forward current" to flow. This of course happens in practice, and the calculated value of V_B (Eq. 1-21) is in accord with the observation that a silicon p-n junction has a forward voltage drop of $\approx +0.7$ V under conditions of heavy current flow in the forward direction.

1.3.2 The p-n Junction with Applied Forward Bias

In Sec. 1.3.1 the concept of a transition region (henceforth referred to as a *depletion-layer* region) was introduced. The depletion-layer region is located at the interface of the p-n junction and for zero-bias conditions is formed in the process of detailed balancing of drift and diffusion carrier current components. The principle of *detailed balancing* states that *under equilibrium conditions, a given process and its reverse occur with equal frequency.* Hence, the drift and diffusion components of the electron and hole currents for the junction must be equal in magnitude but opposite in direction throughout the junction in order that equilibrium be maintained. Thus with zero-bias conditions prevailing, the junction is at equilibrium and the *np product* outside of the depletion region $|t_n - t_p|$ is constant[13] and equal to n_i^2, i.e.:

$$n_{no}p_{no} = n_i^2 \quad \text{on the n side} \quad \begin{cases} n_{no} \text{ are majority carriers} \\ p_{no} \text{ are minority carriers} \end{cases}$$

$$p_{po}n_{po} = n_i^2 \quad \text{on the p side} \quad \begin{cases} p_{po} \text{ are majority carriers} \\ n_{po} \text{ are minority carriers} \end{cases}$$

where the subscript o indicates the equilibrium state. The np product constancy is represented in a plot of the minority- and majority-carrier concentrations on each side of the depletion layer as shown in Fig. 1.23.

When an external forward voltage V_A is applied across the p-n junction, the negative built-in potential barrier is effectively lowered (recall that voltages sum algebraically), and the minority-carrier concentrations at the edges of the transition region $n_p|_{t_p}$ and $p_n|_{t_n}$ are increased over their equilibrium value n_{po} and p_{no}. The increase is described by the law of the junction and is given by

$$p_n|_{t_n} = p_{no}e^{qV_A/kT} \tag{1-22}$$

$$n_p|_{t_p} = n_{po}e^{qV_A/KT} \tag{1-23}$$

(Note, for example, that for $V_A = +0.6$ V, $e^{qV_A/kT} = e^{23} = 9.7 \cdot 10^9$.) The carrier concentrations for the forward-biased p-n junction are shown in Fig. 1.24. Note that the minority-carrier concentrations are reduced to their equilibrium values in a characteristic distance $\mathcal{L}$ by the process of carrier recombination.[14]

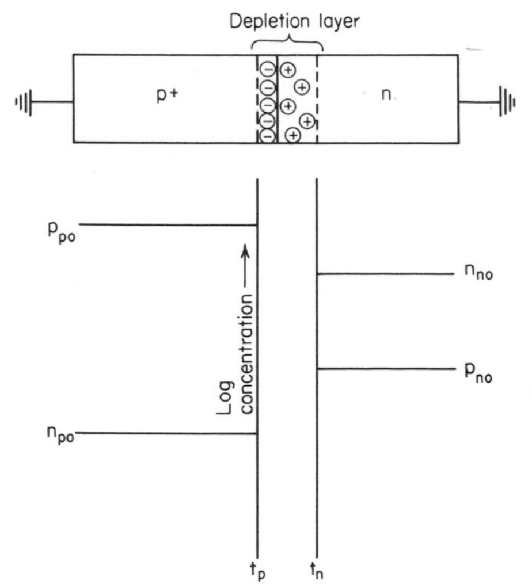

Fig. 1.23. Carrier concentration for the zero-biased p-n junction.

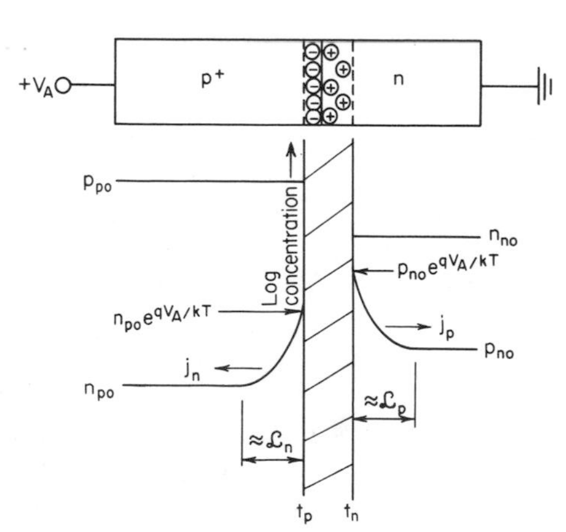

Fig. 1.24. Carrier concentration for the forward-biased p-n junction.

The law of the junction follows readily from the following considerations. From Eq. (1-2) and the assumption that the minority-carrier injection level is small compared to the majority-carrier concentration with n_{no} remaining at its equilibrium value under forward bias, it follows that

$$n_{no} = n_n|_{t_n} = n_i e^u = n_i e^{(E_{fn}-E_{in})/kT} \tag{1-24}$$

Therefore

$$e^{\frac{E_{in}}{kT}} = \frac{n_i e^{E_{fn}/kT}}{n_n|_{t_n}} = \frac{n_i e^{E_{fn}/kT}}{n_{no}} \tag{1-25}$$

From Eq. (1-3)

$$p_n|_{t_n} = n_i e^{-u} = n_i e^{-(E_{fp}-E_{in})/kT} = n_i e^{-E_{fp}/kT} e^{E_{in}/kT} \tag{1-26}$$

Note in Eq. (1-26) that we have arbitrarily extended the Fermi level on the p side of the junction (E_{fp}) across the transition region to the edge of the n region t_n. This extension is referred to as the quasi-Fermi level approximation.[15] Then from Eqs. (1-25) and (1-26)

$$p_n|_{t_n} = \frac{n_i^2}{n_{no}} e^{(E_{fn}-E_{fp})/kT} \tag{1-27}$$

Since V_A is positive and is defined as

$$V_A \equiv \frac{E_{fn} - E_{fp}}{q} \tag{1-28}$$

therefore:

$$p_n|_{t_n} = \frac{n_i^2}{n_{no}} e^{qV_A} = p_{no} e^{qV_A/kT} \tag{1-29}$$

where the relation $n_i^2 = n_{no}p_{no}$ has been utilized.

By similar arguments it can be shown that:

$$n_p|_{t_p} = n_{po} e^{qV_A/kT} \tag{1-30}$$

Minority-carrier injection and resulting build-up of minority-carrier concentrations at the edges of the depletion-layer region as a consequence of applied forward bias create a gradient of carriers that in turn induce minority-carrier diffusion-current flow. If we assume that the build-up of minority carriers has been reduced to its equilibrium value within a diffusion length $\mathcal{L}$ of the depletion-layer edges (the excess minority carriers recombine within that distance and are "consumed"), then a concentration gradient of minority carriers exists as shown in Fig. 1.24, and a hole diffusion current is generated which is described by Fick's Law, i.e.,

$$j_p = -qD_p \frac{\Delta p}{\Delta x} \approx qD_p \frac{p_n|_{t_n} - p_{no}}{\mathcal{L}_p} \tag{1-31}$$

Combining Eqs. (1-29) and (1-31) yields

$$j_p \approx qD_p \frac{p_{no} e^{qV_A/kT} - p_{no}}{\mathcal{L}_p} = \frac{qD_p p_{no}}{\mathcal{L}_p} (e^{qV_A/kT} - 1) \tag{1-32}$$

and similarly an electron diffusion current is given by

$$j_n \approx \frac{qD_n n_{po}}{\mathcal{L}_n}(e^{qV_A/kT} - 1) \tag{1-33}$$

Since j_n and j_p flow in opposite directions, they combine algebraically to yield a total current of

$$I = A(j_p + j_n)$$

$$I = Aq\left(\frac{D_p p_{no}}{\mathcal{L}_p} + \frac{D_n n_{po}}{\mathcal{L}_n}\right)(e^{qV_A/kT} - 1) \tag{1-34}$$

By defining *saturation current* I_o as

$$I_o \equiv Aq\left(\frac{D_p p_{no}}{\mathcal{L}_p} + \frac{D_n n_{po}}{\mathcal{L}_n}\right) = Aqn_i^2\left(\frac{D_p}{\mathcal{L}_p n_n} + \frac{D_n}{\mathcal{L}_n p_p}\right) \tag{1-35}$$

where the expressions:

$$n_{no}p_{no} = n_i^2$$

$$p_{po}n_{po} = n_i^2$$

and

$$n_{no} = n_n$$

$$p_{po} = p_p$$

have been utilized, Eq. (1-34) can therefore be rewritten as

$$I = I_o(e^{qV_A/kT} - 1) \tag{1-36}$$

Equation (1-36) is the Shockley diode equation. It describes current flow generated by the diffusion of injected minority carriers in the forward- or reverse-biased p-n junction.

To complete the discussion of the forward-biased p-n junction, note again that

$$V_A \equiv \frac{E_{fn} - E_{fp}}{q} \quad \text{(a positive quantity for p region positive with respect to n region)} \tag{1-28}$$

Thus the Fermi level is in this case not a constant value across the junction (current is flowing!), and the barrier height has been reduced in comparison to the zero-biased case as shown in Fig. 1.25.

1.3.3 The p-n Junction with Applied Reverse Bias

When V_A is made negative, the p-n junction is reverse biased and the negative built-in potential is compounded by the applied bias. The law of the junction still holds, of course, but now V_A is negative and Eqs. (1-22) and (1-23) become:

$$p_n|_{t_n} = p_{no}e^{-qV_A/kT} \tag{1-37}$$

$$n_p|_{t_p} = n_{po}e^{-qV_A/kT} \tag{1-38}$$

Thus, for example, if a reverse-bias voltage of 1 V is placed across the junction, the equilibrium minority-carrier concentrations at the edges of the depletion region

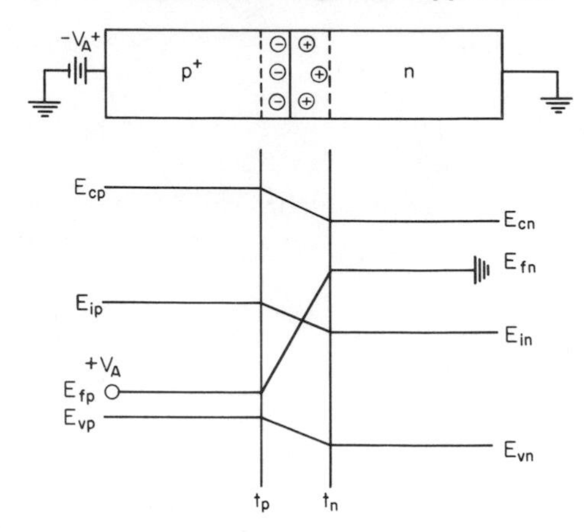

Fig. 1.25. Energy-level diagram for the p-n junction with applied forward bias.

are depressed by a factor of:

$$e^{-1/0.026} \approx e^{-40} \qquad T \approx 300°K$$

A qualitative plot of the minority- and majority-carrier concentrations for the reverse-biased p-n junction is shown in Fig. 1.26. It will be noted in Fig. 1.26 that concentration gradients of minority carriers exist, and thus diffusion-driven minority-carrier current flows. The diffusion current is described by the diode equation (1-36) as developed above. But now for the case of an applied reverse bias of, for example, -1 V, the diode equation can be written as:

$$I = I_o(e^{q(-1)/kT} - 1)$$
$$I \approx I_o(e^{-40} - 1) \qquad T \approx 300°K$$

and $\quad I \approx -I_o \quad$ (negative sign is consistent with reverse current flow)

It now becomes evident that I_o is designated as *saturation leakage current,* since under reverse bias the p-n junction current flow as described by Eq. (1-34) becomes independent of applied reverse bias at relatively small potential values and approaches I_o.

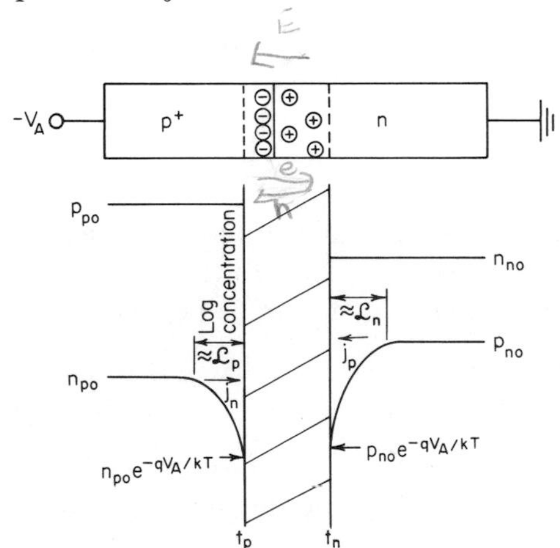

Fig. 1.26. Carrier concentration for the reverse-biased p-n junction.

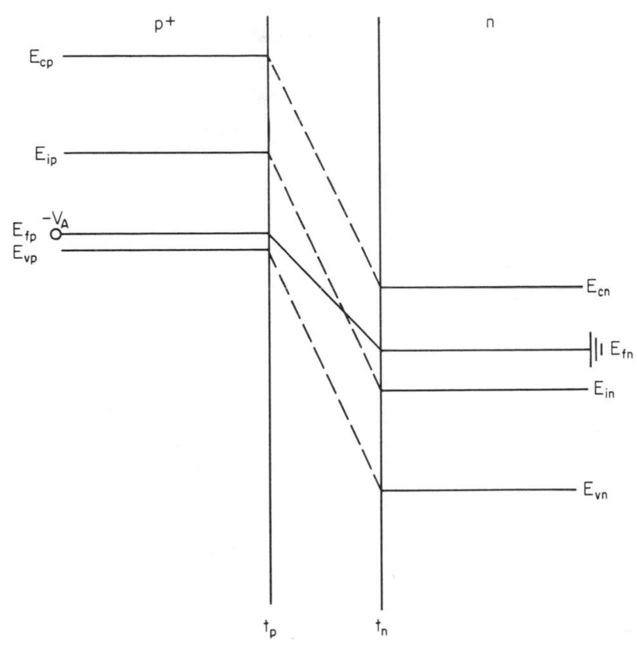

Fig. 1.27. Energy-level diagram for the p-n junction with applied reverse bias.

The energy-level diagram for the reverse-biased p-n junction is shown in Fig. 1.27 and is to be compared to the zero- and forward-biased junction examples of Figs. 1.22*c* and 1.25 respectively.

1.3.4 Junction Leakage Current

(a) Saturation Current I_o. The leakage current for a typical silicon planar p-n junction encountered in MOS circuitry at room temperature is very approximately 10^{-11} Amp per mil^2 of junction area at a reverse bias of -10 V. Let us compare this experimental value to a calculated value of the saturation leakage current I_o. From Eq. (1-35), with the following parameters assumed

$$\begin{cases} A = 1 \text{ mil} \times 1 \text{ mil} = 6.5 \cdot 10^{-6} \text{ cm}^2 \\ q = 1.6 \cdot 10^{-19} \text{ Coul} \\ D_n = D_p = 10 \text{ cm}^2/\text{sec} \\ L_n = L_p = 10^{-3} \text{ cm} \\ n_n = p_p = 10^{15}/\text{cm}^3 \\ n_i = 1.4 \cdot 10^{10}/\text{cm}^3 \end{cases}$$

Then:

$$I_o \approx 5 \cdot 10^{-15} \text{ Amp}$$

The calculated value of I_o is essentially 2,000 times less than the experimental value! In addition, it is experimentally observed that junction leakage current is dependent on the applied voltage in a nonsaturating manner, whereas I_o [Eq. (1-35)] does not theoretically depend on the applied voltage. From these observations, it must be concluded that the saturation current I_o constitutes only a small fraction of the observed leakage current in silicon reverse-biased p-n junctions and that another source of leakage current must be defined.

(b) Generation-recombination Current. The work of Sah, Noyce, and Shockley[16] revealed that reverse-bias junction leakage current can have its origin in the depletion layer of the p-n junction. This particular source of leakage current is called *depletion-layer generation-recombination current* or simply *generation-recombination* (G-R) current. The G-R current has its source in crystal imperfections such as:

1. metallic impurities
2. crystalline defects

The imperfections disturb the otherwise perfect periodicity of the semiconductor lattice, thereby introducing energy levels deep within the forbidden gap (near midband). These deep-lying levels effectively act as stepping stones for the transition of electrons and holes between the valence and conduction bands, by means of which carriers then become available for current conduction. The imperfections populate and depopulate at generation and recombination centers most readily when their energy values lie at midband E_i, as depicted in Fig. 1.28.

Under reverse bias, the depletion-layer region of the junction is virtually swept free of all carriers. With a recombination center at midband, it can be shown that the carrier-generation rate per cubic centimeter of semiconductor is

$$g = \frac{n_i}{2\tau} \tag{1-39}$$

where τ = minority carrier lifetime

The total G-R current originating in the depletion-layer region of width W and area A is, therefore:

$$I_g = \frac{Aqn_i W}{2\tau} \tag{1-40}$$

To determine if the G-R current has a value consistent with that of the experimentally observed leakage current of the planar p-n junction assume: $\tau = 10^{-7}$ sec, $W = 0.15$ mil for $V_A = -10$ V, $T = 300°$K, and $A = 1$ mil^2. Then, substituting in Eq. (1-40) yields $I_g \approx 10^{-11}$ Amp for a 1-mil-square junction. Not only is this theoretical value for I_g in accord with the experimental value for the leakage current of a 1-mil-square junction, but in addition, the depletion-layer width is a function of the applied voltage (see Sec. 1.3.5), thus providing the proper voltage dependence that is ex-

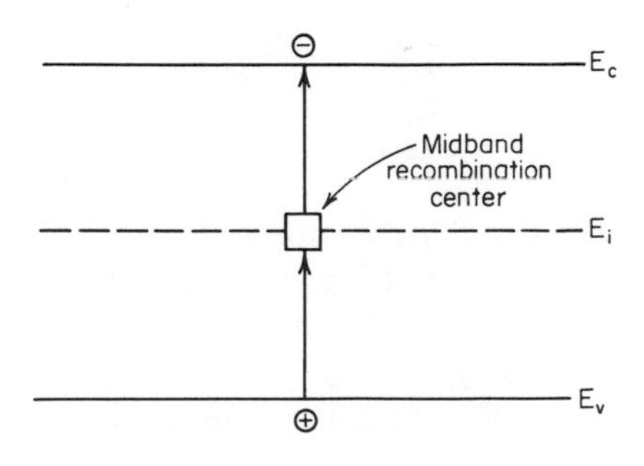

Fig. 1.28. Carrier recombination and generation through a defect center at midband.

hibited by the leakage current. Many detailed studies have now established the validity of the Sah-Noyce-Shockley G-R current as the primary source of leakage current observed in properly constructed silicon planar p-n junctions.

In MOS integrated circuits it is desirable to reduce the G-R leakage current to as low a value as possible. Note in Eq. (1-40) that G-R current is inversely proportional to the minority-carrier lifetime τ. It is therefore beneficial to "preserve" the minority-carrier lifetime at as long a value as possible (in practice, a fraction of a microsecond). Minority-carrier lifetime preservation can be accomplished in the last step of the planar process (prior to contact metallization) by using a gettering method at $\approx$900°C, in which the presence of a boron- or phosphorus-doped surface oxide tends to extract metallic impurities from bulk silicon. The gettering step[17] is usually followed by annealing the silicon wafer from the elevated processing temperature down to room temperature to remove crystal defects, such as dislocations or crystalline slip, which can be introduced during the fabrication of the p-n junction.

(c) Temperature Dependence of Generation-recombination Current. Recalling that the G-R leakage current is given by Eq. (1-40), let us examine the temperature dependence of I_g. The principal temperature-dependent contribution to I_g is through n_i, where n_i is given by

$$n_i = 3.9 \cdot 10^{16} T^{3/2} e^{-1.21/2kT}/\text{cm}^3 \qquad (1\text{-}1)$$

At 300°K, n_i will double for approximately every 10°C increment of temperature increase. I_g will therefore approximately follow the temperature dependence exhibited by n_i. Thus the ratio of leakage current at 125°C to 25°C for a well-behaved planar p-n junction is

$$\frac{I_l(125°\text{C})}{I_l(25°\text{C})} \approx 2^{10} \approx 1{,}000$$

(d) Leakage Current Values for MOS Circuitry. From the above discussion we conclude that generation-recombination of carriers in the depletion layer of the p-n junction serves as the chief contributor to junction leakage current. A room temperature value of $\approx 10^{-11}$ Amp per mil^2 was thus calculated for junction leakage current at a reverse bias of 10 V. In addition, it was stated that G-R current doubles every 10°C increment of temperature increase for silicon. These considerations can be utilized in estimating the leakage current values for MOS/LSI circuitry.

Consider, for example, a large MOS integrated circuit, ratioless type, with 4,000 MOSFETs present. The total junction area of the circuit would be $\approx$10,000 mils^2. A room-temperature leakage current value for this circuit under 10 V reverse bias would be (assuming that source regions are at zero bias)

$$I_l|_{25°\text{C}} \approx \left[\frac{10{,}000 \text{ mils}^2}{2}\right] \cdot (10^{-11} \text{ Amp/mil}^2)$$

$$I_l|_{25°\text{C}} \approx 5 \cdot 10^{-8} \text{ Amp}$$

At 125°C the leakage current is larger than its room-temperature value by a factor of $\approx$1,000, and becomes

$$I_l|_{125°\text{C}} \approx 0.05 \text{ mA}$$

The above comments apply to *well-behaved planar p-n junctions.* Although this is the rule rather than the exception, there are on occasion departures from the leakage-current characteristics described above. The departures have their origin, in many cases, in unremovable metal or oxide precipitates in the p-n junction (mainly at the junction periphery). This phenomenon can enhance by orders of magnitude the normal leakage current values.[18] Ill-behaved leakage currents can disrupt operation of MOS circuitry; for example, the low-frequency operation of an MOS dynamic shift register can be degraded by excess junction leakage current. Inordinately high leakage current values will therefore be considered as a yield-loss mechanism in MOS/LSI circuitry.

1.3.5 Electric Fields in the Reverse-biased, One-sided, Step p-n Junction—The Depletion-layer Width

An external reverse bias applied to the p-n junction will simply add to the built-in potential and will generate a depletion-layer width W as shown in Fig. 1.29. To calculate resulting values for the electric field and for the depletion-layer width, consider the case of the one-sided step junction (infinite doping on the p side). Poisson's equation will permit us to calculate the electric field throughout the depletion layer if certain boundary conditions are established and the principle of charge neutrality is invoked. Thus

$$\nabla^2 V = -\frac{qN_D}{\epsilon_{\text{Si}}\epsilon_o} \tag{1-41}$$

where $\epsilon_o = 8.85 \cdot 10^{-14}$ F/cm
ϵ_{Si} = dielectric constant of silicon

and since the problem is one-dimensional

$$\frac{\partial^2 V}{\partial x^2} = -\frac{qN_D}{\epsilon_{\text{Si}}\epsilon_o} \tag{1-42}$$

By definition

$$\mathcal{E} = -\frac{\partial V}{\partial x} \tag{1-43}$$

Therefore

$$\frac{-\partial \mathcal{E}}{\partial x} = \frac{-qN_D}{\epsilon_{\text{Si}}\epsilon_o} \tag{1-44}$$

Integrating:

$$\int_0^{\mathcal{E}} d\mathcal{E} = \int_0^x \frac{qN_D}{\epsilon_{\text{Si}}\epsilon_o} dx \tag{1-45}$$

and

$$\mathcal{E}_{\max} = \frac{qN_D x}{\epsilon_{\text{Si}}\epsilon_o}\Bigg|_0^W \tag{1-46}$$

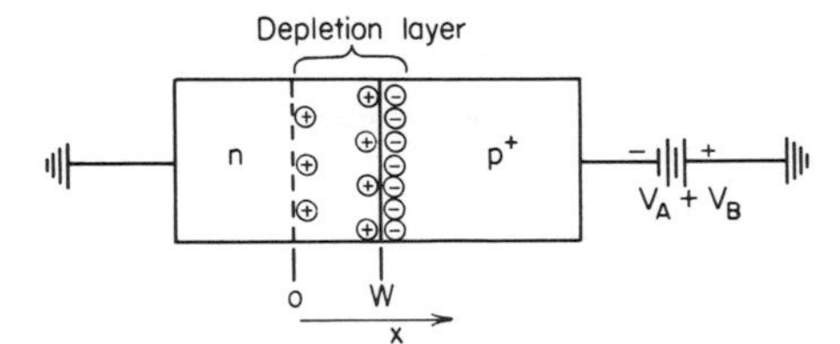

Fig. 1.29. The reverse-biased p-n junction.

$$\mathcal{E}_{max} = \frac{qN_D W}{\epsilon_{Si}\epsilon_o} \tag{1-47}$$

A graph of $\mathcal{E}$ versus x through the junction is shown in Fig. 1.30. To define $\mathcal{E}_{max}$, an expression for the depletion-layer width W must be determined for the junction under reverse bias. From Eqs. (1-43) and (1-45),

$$\mathcal{E} = -\frac{\partial V}{\partial x} = \frac{qN_D x}{\epsilon_{Si}\epsilon_o} \tag{1-48}$$

Integrating

$$-\int_0^V dV = \int_0^W \frac{qN_D x dx}{\epsilon_{Si}\epsilon_o} \tag{1-49}$$

yields

$$-V = \frac{qN_D W^2}{2\epsilon_{Si}\epsilon_o} \tag{1-50}$$

and since V_A and V_B are negative

$$W = \sqrt{\frac{2\epsilon_{Si}\epsilon_o|V_A + V_B|}{qN_D}} \tag{1-51}$$

where $\epsilon_{Si} = 12$
$\epsilon_o = 8.85 \cdot 10^{-14}$ Farad/cm

The maximum field becomes, from Eqs. (1-47) and (1-51)

$$\mathcal{E}_{max} = \sqrt{\frac{2qN_D|V_A + V_B|}{\epsilon_{Si}\epsilon_o}} \tag{1-52}$$

To calculate typical values for $\mathcal{E}_{max}$ and depletion-layer width for typical MOS circuitry, consider the junctions to be one-sided, with a step gradient, and that substrate resistivity is 4 Ohm-cm n type, and

$$N_D \approx 10^{15}/\text{cm}^3$$
$$V_A = -10 \text{ V}$$
$$V_B = -0.5 \text{ V}$$

Then from Eq. (1-51)

$$W = \sqrt{\frac{2\epsilon_{Si}\epsilon_o|V_A + V_B|}{qN_D}}$$
$$= \sqrt{\frac{(2)(12)(8.85 \cdot 10^{-14})|10.5|}{(1.6 \cdot 10^{-19})(10^{15})}} = 0.37 \cdot 10^{-3} \text{ cm}$$
$$= 0.15 \text{ mil}$$

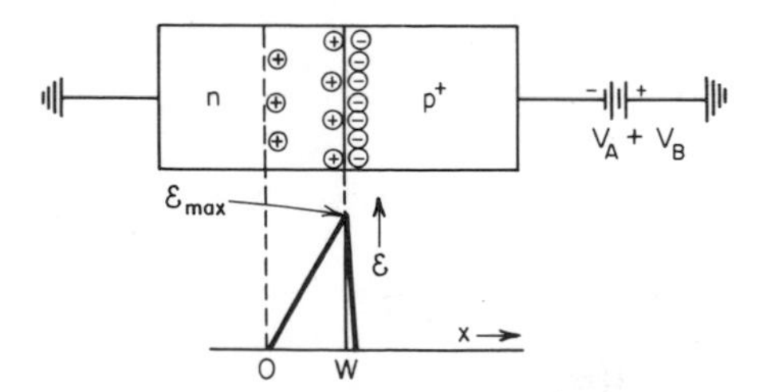

Fig. 1.30. Electric-field distribution in the one-sided step p-n junction.

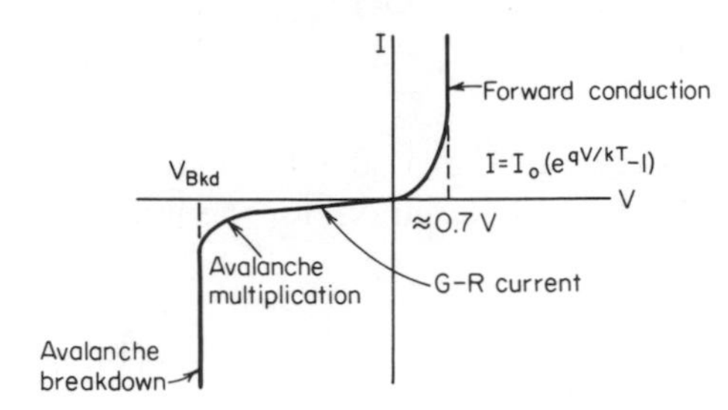

Fig. 1.31. General current-voltage characteristics of the silicon diode.

Note the compatibility of the narrow depletion-layer width and the concept of monolithic microelectronic circuitry! From Eq. (1-52)

$$\mathcal{E}_{max} = 5.6 \cdot 10^4 \text{ V/cm for this example}$$

1.3.6 Avalanche Breakdown of the One-sided Step Junction

(a) Introduction. The presence of the junction electric field accelerates carriers to velocities such that they are capable of carrier-pair generation upon impact with silicon atoms in the depletion layer. The carriers so created are in turn accelerated and produce additional pairs. This multiplicative process results in an avalanche effect (similar to the Townsend avalanche evidenced in the electrical breakdown of a gas). Avalanching in the p-n junction results in carrier multiplication M, which can be shown[19] to be represented empirically as

$$M = \frac{1}{1 - \left(\dfrac{V_A}{V_{Bkd}}\right)^n} \tag{1-53}$$

where $n \approx 4$ for most of the practical examples in silicon

V_{Bkd} = reverse-voltage value where current tends toward infinity and avalanche breakdown occurs

If the applied voltage approaches a critical value V_{Bkd}, then M and the reverse current approach ∞, resulting in the situation shown in Fig. 1.31. Avalanche breakdown is, in general, nondestructive to the junction, provided the acceptable power dissipation of the junction is not exceeded.

Avalanche breakdown occurs when $\mathcal{E}_{max} \approx 2 \cdot 10^5$ V/cm. An experimentally determined plot of junction breakdown voltage versus impurity concentration is shown in Fig. 1.32.[20]

To test the concepts and equations developed in this section, let us calculate $\mathcal{E}_{max}$

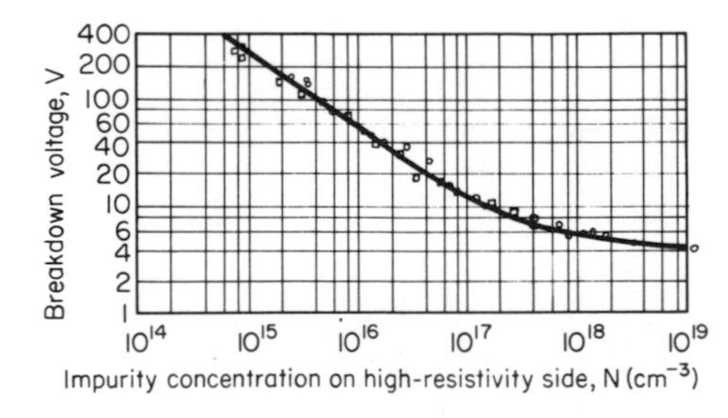

Fig. 1.32. The breakdown voltage of a silicon step junction (n^+p or p^+n) as a function of impurity concentration on the lightly doped side.[20]

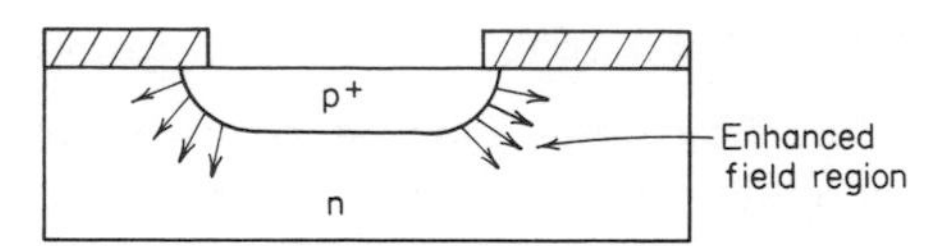

Fig. 1.33. The planar junction with enhanced field regions.

at avalanche from the data of Fig. 1.32 and then compare the calculated value to an empirical value of $\mathcal{E}_{max} \approx 2 \cdot 10^5$ V/cm at breakdown. Again, consider the case $N_D = 10^{15}/\text{cm}^3$. Then from Fig. 1.32,

$$V_{Bkd} = 300 \text{ V}$$

From Eq. (1-52)

$$\mathcal{E}_{max} = \sqrt{\frac{2qN_D|(V_A + V_B)|}{\epsilon_{Si}\epsilon_o}}$$

$$= \sqrt{\frac{(2)(1.6 \cdot 10^{-19})(10^{15})(300)}{(12)(8.85 \cdot 10^{-14})}}$$

$$= 3 \cdot 10^5 \text{ V/cm}$$

Note the calculated $\mathcal{E}_{max}$ is consistent with the empirical critical field value of $\approx 2 \cdot 10^5$ V/cm characteristic of avalanche breakdown.

(b) The Effect of Junction Curvature. If a p-n junction departs from true planarity as it does when formed by the planar process (Fig. 1.33), the junction fields will be enhanced in regions of junction curvature, and avalanche breakdown voltage values will be reduced from the values shown in Fig. 1.32. The effect of junction curvature on the breakdown voltage of a planar one-sided step junction is shown in Fig. 1.34.[21] The planar junctions in a conventional MOS circuit are diffused to a depth of approximately 0.08 mil (= 2 microns). From Fig. 1.34 it is apparent that *substrate silicon having a background doping level of $10^{15}/cm^3$ and planar diffusion depth of 2 microns will experience avalanche breakdown at ≈ 70 V.* The junction under these conditions is referred to as "diffusion limited" with respect to avalanche breakdown.

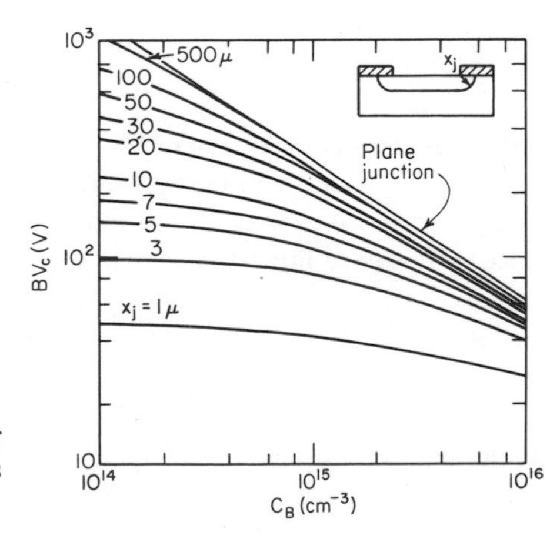

Fig. 1.34. Breakdown voltage of planar one-sided step junctions possessing various radii of curvature.[21]

1.3.7 The Depletion-layer Capacitance

(a) Capacitance of the Reverse-biased, One-sided Step p-n Junction. When an increment of reverse voltage is applied to a junction, the depletion layer expands, and charge from uncovered impurity atoms is "stored" on both sides of the depletion layer. If the applied voltage is returned to its original value, carriers will flow in a direction to neutralize the previous increment of charge. (In effect, the depletion-layer width has been expanded and then contracted to its original dimension as a result of the applied voltage.) The response of the p-n junction to incremental reverse bias thus results in generation of an effective capacitance referred to as depletion-layer capacitance (or as transition capacitance).

To derive an expression for depletion-layer capacitance, recall that the fundamental definition for capacitance per unit area in terms of incremental charge per unit area dQ induced by incremental change in applied voltage dV is:

$$\frac{C}{A} = \frac{dQ}{dV} \tag{1-54}$$

And since, for the one-sided step junction

$$Q = qN_D W; \quad \frac{dQ}{dW} = qN_D \tag{1-55}$$

and from Eq. (1-51)

$$V = \frac{qN_D W^2}{2\epsilon_{\text{Si}}\epsilon_o}; \quad \frac{dV}{dW} = \frac{qN_D W}{\epsilon_{\text{Si}}\epsilon_o} \tag{1-56}$$

Therefore, from Eqs. (1-54), (1-55), and (1-56)

$$\frac{C}{A} = \frac{dQ}{dV} = \frac{qN_D}{qN_D W/\epsilon_{\text{Si}}\epsilon_o} = \frac{\epsilon_{\text{Si}}\epsilon_o}{W}$$

and hence

$$C = \frac{A\epsilon_{\text{Si}}\epsilon_o}{W} \tag{1-57}$$

Since

$$W = \sqrt{\frac{2\epsilon_{\text{Si}}\epsilon_o|V_A + V_B|}{qN_D}} \tag{1-51}$$

therefore

$$C = A\sqrt{\frac{\epsilon_{\text{Si}}\epsilon_o qN_D}{2|V_A + V_B|}} \tag{1-58}$$

(b) Junction Capacitance of MOS Circuitry. To calculate the approximate junction capacitance value for planar p^+-n junctions formed in MOS/LSI with p-type dopant diffused into n-type 4-Ohm-cm material ($N_D \approx 10^{15}/\text{cm}^3$), consider junction capacitance per square mil for the case of 10 V reverse bias, then from Eq. (1-58)

$$C = 10^{-6}\,(2.54)^2\sqrt{\frac{(12)(8.85\cdot 10^{-14})(1.6\cdot 10^{-19})(10^{15})}{(2)(10.5)}}$$

$$C \approx 0.02 \text{ pF/mil}^2 \text{ (at 10 V reverse bias)}$$

We will frequently use this capacitance value in MOS circuit design.

1.4 THE SILICON SURFACE

1.4.1 Introduction

Consider a p-channel MOSFET structure as in Fig. 1.1, where the work function between metal gate and silicon $\Phi_{MS} = 0$, charge in the gate oxide $Q_{0x} = 0$, and charge at the silicon–silicon dioxide interface $Q_{SS} = 0$. The energy-level structure for the device in the n region under the gate oxide, with source, substrate, gate, and drain grounded, is shown in Fig. 1.35. The horizontal structure of the energy levels is referred to as the flat-band condition. It is to be noted that the energy-level diagram is now two-dimensional, with distance into silicon represented in the horizontal direction, and electron energy represented along the vertical direction. The *on* state for the MOSFET is achieved by applying a gate voltage V_G of proper polarity between the metal gate (often referred to as the "field plate") and the silicon substrate. An applied negative-gate voltage V_G thus sets up an electric field which in turn bends the energy bands in the bulk silicon to the extent that E_i crosses E_f and the surface region becomes p type (inverted) as shown in Fig. 1.36. The p channel (inversion layer) electrically connects the p-type source and drain regions to one another. The inversion layer is considered to be effectively present when the silicon at the Si-SiO_2 interface is as heavily concentrated p type as is the bulk n type. Mathematically this condition is stated as

$$\phi_s = 2\phi_f \qquad \text{(cf. Fig. 1.36)} \tag{1-59}$$

where the surface potential $\phi_s = (E_{i\,\text{bulk}} - E_{i\,\text{surface}})/q$ and the Fermi potential $\phi_f = (E_{i\,\text{bulk}} - E_{f\,\text{bulk}})/q$. Equation (1-59) will thus serve as a definition for onset of inversion for n- or p-channel enhancement-mode devices.

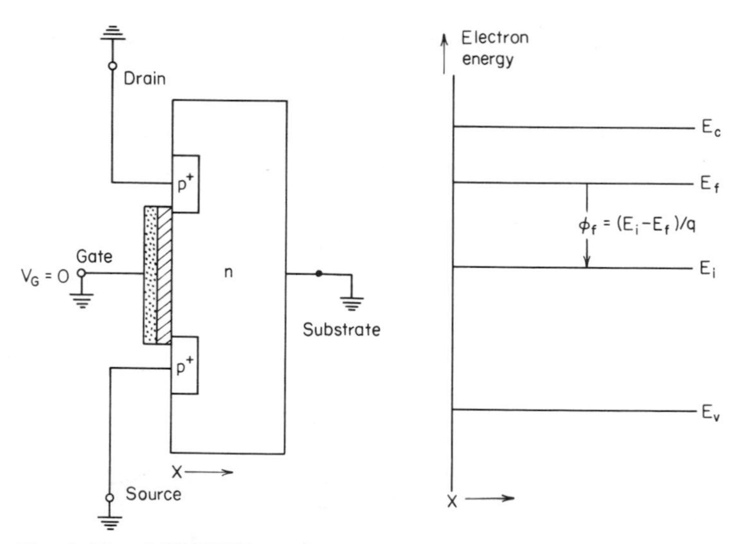

Fig. 1.35. MOSFET configuration with gate, source, drain, and substrate at ground potential with accompanying energy-level diagram for device terminals grounded and $Q_{SS} = 0$, $Q_{0x} = 0$, and $\Phi_{MS} = 0$.

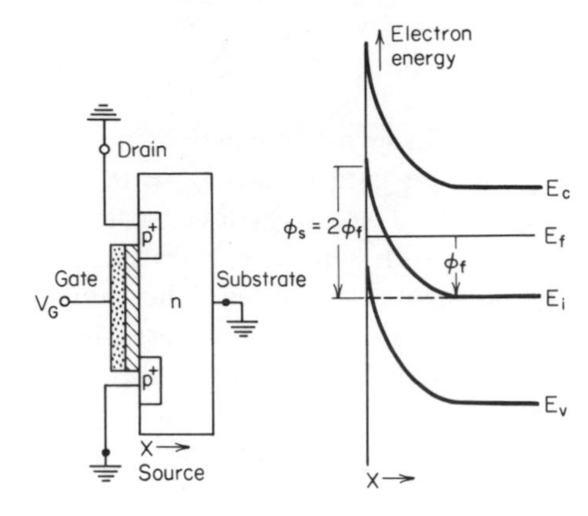

Fig. 1.36. MOSFET configuration with V_G negative and MOSFET energy-level diagram with V_G negative, $Q_{SS} = 0$, $Q_{ox} = 0$, and $\Phi_{MS} = 0$. The channel is inverted.

Note in Fig. 1.36 that zero current flows in the *x* direction if the SiO_2 dielectric is considered to be an ideal electrical insulator. It thus follows for the MOS structure shown in Fig. 1.36 that the Fermi level is constant in the *x* direction. (Recall that a necessary and sufficient condition for the absence of current flow in a semiconductor is that the Fermi level be constant throughout the system.)

Band-bending to accumulate, deplete, or set the surface of n-type silicon at intrinsicity is depicted in Fig. 1.37*a*, *b*, and *c*, respectively. The above considerations serve as a basis for establishing the following definitions for the MOS device:

1. The silicon surface is defined as that *region* of semiconductor material encompassing the outermost layer of structural silicon atoms inward to that position within the bulk interior where the bands become flat. For example—conduction takes place in the p-channel MOSFET between the source and drain regions on the silicon surface. The silicon surface in this example is composed of a p-type inversion layer of *finite thickness*.
2. The silicon–silicon dioxide interface is defined as that infinitely thin region between an ordered structure of silicon and a pure amorphous silicon dioxide.

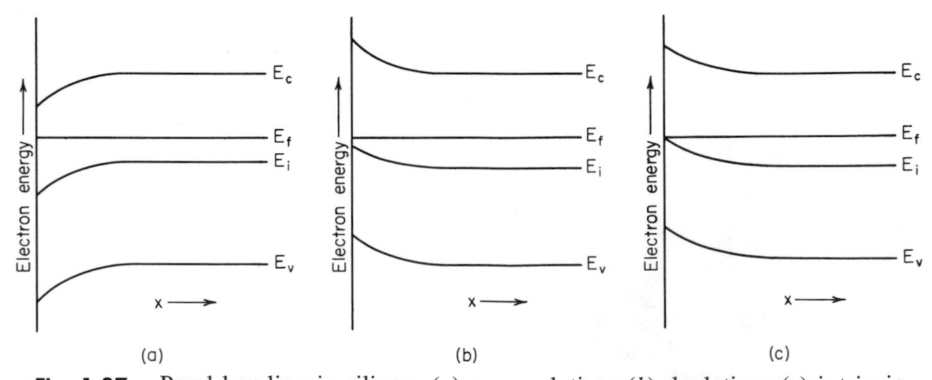

Fig. 1.37. Band-bending in silicon: (*a*) accumulation; (*b*) depletion; (*c*) intrinsic.

1.4.2 Band-bending with Applied Gate Voltage for the Case $\Phi_{MS} = 0$, $Q_{SS} = 0$

To gain quantitative insight as to band-bending in silicon when a voltage is applied to the gate of a p-channel MOS structure (n-type substrate), consider how the applied gate voltage is distributed across the gate-oxide and silicon-surface region of the structure shown in Fig. 1.38. Applied gate voltage V_G will be divided across the gate-oxide and silicon-surface region as:

$$V_G = V_{\text{Ox}} + \phi_s \tag{1-60}$$

where V_{Ox} = voltage across the gate oxide

ϕ_s = voltage across the silicon-surface region (equivalent to the surface potential)

The continuity of electric displacement vector **D** across the Si-SiO_2 interface requires that at the interface

$$\mathcal{E}_{\text{Ox}}\epsilon_{\text{Ox}} = \mathcal{E}_{\text{Si}}\epsilon_{\text{Si}} \tag{1-61}$$

The electric field directed *into* the silicon surface at the interface is, from the theorem of Gauss

$$\mathcal{E}_{\text{Si}} = -\frac{Q_s}{\epsilon_{\text{Si}}\epsilon_o} \tag{1-62}$$

where $\epsilon_{\text{Si}} = 12$

$\epsilon_o = 8.85 \cdot 10^{-14}$ F/cm

Q_s = charge density (Coul/cm^2) in silicon resulting from band-bending at the silicon surface. Band-bending is induced by potential drop ϕ_s across the silicon-surface region.

The electric field is constant in the gate oxide and is by definition:

$$\mathcal{E}_{\text{Ox}} \equiv \frac{V_{\text{Ox}}}{t_{\text{Ox}}} \tag{1-63}$$

where t_{Ox} is the gate-oxide thickness.

Then combining Eqs. (1-61), (1-62), and (1-63)

$$V_{\text{Ox}} = -\frac{Q_s t_{\text{Ox}}}{\epsilon_{\text{Ox}}\epsilon_o} \tag{1-64}$$

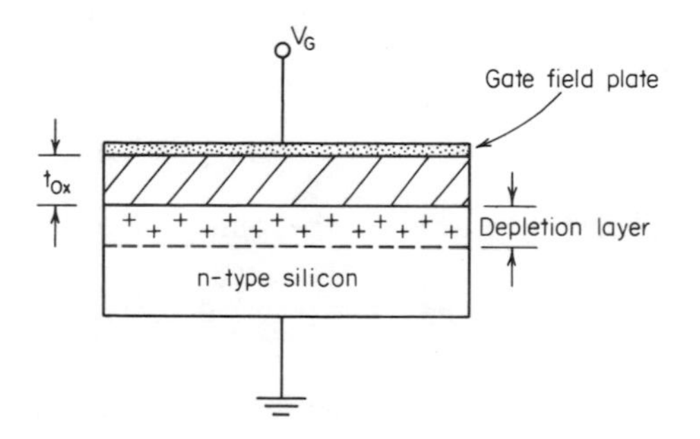

Fig. 1.38. The MOS structure.

and therefore
$$V_{Ox} = -\frac{Q_s}{C_o} \tag{1-65}$$

where the gate capacitance per unit area C_o is defined as

$$C_o \equiv \frac{\epsilon_{Ox}\epsilon_o}{t_{Ox}} \tag{1-66}$$

Eq. (1-60) therefore becomes

$$V_G = \frac{-Q_s}{C_o} + \phi_s \tag{1-67}$$

The total charge induced in the silicon Q_s is equal, by definition, to the charge induced in the surface depletion layer generated by ionized donor atoms Q_B, plus the charge contributed by holes in the inversion layer Q_P, i.e.,

$$Q_s \equiv Q_B + Q_P \tag{1-68}$$

The voltage ϕ_s developed across the silicon-surface region will in turn induce a surface depletion layer whose depth, in analogy to Eq. (1-51), is given by:

$$x_d = \sqrt{\frac{2\epsilon_{Si}\epsilon_o|\phi_s|}{qN_D}} \tag{1-69}$$

If we make the assumption that for $|\phi_s| < 2|\phi_f|$, all charge induced by the applied gate voltage accrues to the surface depletion layer with none accruing to the inversion layer; whereas, for $|\phi_s| > 2\,|\phi_f|$, all further charge induced by the applied gate voltage accrues to the surface inversion layer with no additional amount accruing to the depletion layer then

$$x_{d\,\max} = \sqrt{\frac{2\epsilon_{Si}\epsilon_o|2\phi_f|}{qN_D}} \tag{1-70}$$

[Compare Eq. (1-70) to Eq. (1-51).]

Now to determine the gate voltage required for the onset of inversion, the conditions become:

$\phi_s = 2\phi_f$
$Q_s = Q_B$ (all charge resides in the depletion layer up to the point of onset of inversion)

From Eq. (1-70)

$$Q_B = qN_D x_{d\,\max} = \sqrt{2\epsilon_{Si}\epsilon_o qN_D|2\phi_f|} \tag{1-71}$$

Therefore Eq. (1-67) becomes for onset of inversion

$$V_G = \frac{-Q_B}{C_o} + 2\phi_f = \frac{-\sqrt{2\epsilon_{Si}\epsilon_o qN_D|2\phi_f|}}{C_o} + 2\phi_f \tag{1-72}$$

Further increase of the absolute value of applied gate voltage results in enhancement of the inversion-layer charge density, Eq. (1-68), described by

$$V_G = \frac{-Q_B}{C_o} - \frac{Q_P}{C_o} + 2\phi_f = \frac{-\sqrt{2\epsilon_{Si}\epsilon_o q N_D |2\phi_f|}}{C_o} - \frac{Q_P}{C_o} + 2\phi_f \qquad (1\text{-}73)$$

As an illustration of the above concepts, consider the following example where the gate voltage required to bend the bands through $2\phi_f$ and produce onset of inversion is calculated:

n-type substrate with $N_D = 10^{15}/\text{cm}^3$
$2\phi_f = -0.58$ V (from Fig. 1.12) $t_{ox} = 1000$ Å
$C_o = 35 \cdot 10^{-9}$ F/cm²
From Eq. (1-70), $x_{d\,max} = 0.88 \cdot 10^{-4}$ cm $= 0.035$ mil

At onset of inversion for this example, Eq. (1-72) becomes

$$V_G = -\frac{1}{35 \cdot 10^{-9}} \sqrt{(2)(12)(8.85 \cdot 10^{-14})(1.6 \cdot 10^{-19})(10^{15})(0.58)} - 0.58 \text{ V}$$
$$= -0.98 \text{ V}$$

and hence a gate voltage of -0.98 V would be required to just bend the bands to the $\phi_s = 2\phi_f$ position. Further negative increase of V_G to a value, for example, of -5 V would result in formation of inversion-layer charge density Q_P given by Eq. (1-73) of

$$-5 = -\frac{Q_P}{C_o} - 0.98 \text{ V}$$

$$Q_P = C_o\,(4.02 \text{ V}) = (35 \cdot 10^{-9})(4.02) = 1.4 \cdot 10^{-7} \text{ Coul/cm}^2$$

$$\frac{Q_P}{q} = 0.88 \cdot 10^{12} \text{ positive charges/cm}^2$$

These mobile Q_P charges would be available for contribution to current flow in the MOSFET channel on application of potential difference between source and drain. It is of interest to convert the calculated value of these $0.88 \cdot 10^{12}$ positive charges per cm² to an "effective doping level" in the inversion layer. Thus if we assume the inversion layer to be ≈50 Å thick and uniformly doped (for accurate calculation of inversion-layer thickness and distribution of carrier concentration therein the reader is referred to the classic paper by Ihantola and Moll[22]) the effective doping level is

$$\frac{Q_P}{q x_d} = \frac{0.88 \cdot 10^{12}/\text{cm}^2}{50 \cdot 10^{-8}/\text{cm}} \approx 2 \cdot 10^{18}/\text{cm}^3$$

Thus for this example we could, in first approximation, consider the inversion layer to be a sheet of p-type material 50 Å thick with an effective doping level of $\approx 2 \cdot 10^{18}/\text{cm}^3$, which would electrically connect the p-type source and drain regions.

1.4.3 Band-bending with Applied Gate Voltage for the Case of Finite Φ_{MS} with $Q_{SS} = 0$

As an initial example of the effect on band-bending in the MOS structure when the work function between the semiconductor body and gate field plate has a finite

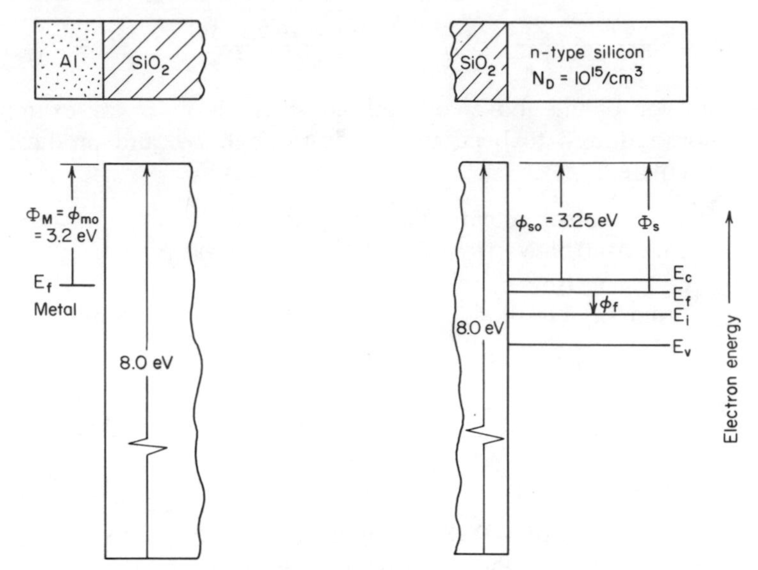

Fig. 1.39. The aluminum-SiO_2—SiO_2-Si system.

value, consider the aluminum gate, SiO_2 dielectric, n-type silicon semiconductor (with $N_D = 10^{15}/\text{cm}^3$ and $\phi_f = -0.29$ V) system as the two separate entities shown in Fig. 1.39.

Note the following important definitions pertaining to Fig. 1.39: $\Phi_M = \phi_{mo} =$ (metal-SiO_2) barrier height (in Volts) equal to the difference between the conduction-band edge of SiO_2 and the Fermi level of the metal gate region. For the aluminum-SiO_2 system, $\phi_{mo} = 3.2$ V,[23] and ϕ_{so} = potential energy difference (in Volts) between the conduction-band edge of SiO_2 and the conduction-band edge of silicon. $\phi_{so} =$ 3.25 V for the Si-SiO_2 system.

Φ_S = potential difference between the conduction-band edge of SiO_2 and the Fermi level in silicon.

$$\Phi_S = \phi_{so} + \left(\frac{E_g}{2q} + \phi_f\right) \tag{1-74}$$

Then on bringing the two halves of Fig. 1.39 together:

$$\begin{aligned} \Phi_{MS} &= \Phi_M - \Phi_S \\ &= \phi_{mo} - \left(\phi_{so} + \frac{E_g}{2q} + \phi_f\right) \\ &= 3.2 - (3.25 + 0.55 - 0.29)\text{ V} \\ &\approx -0.3\text{ V} \end{aligned} \tag{1-75}$$

The flat-band condition in this example is thus maintained with an applied gate voltage $V_G = -0.3$ V, as shown in Fig. 1.40.

As a second example, consider an MOS structure consisting of an aluminum gate,

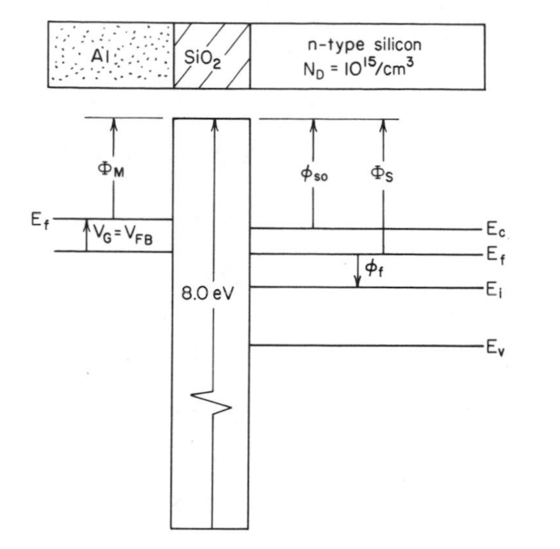

Fig. 1.40. Flat-band condition for the aluminum-SiO_2-silicon system ($N_D = 10^{15}/cm^3$).

SiO_2 dielectric, p-type silicon semiconductor (with $N_A = 10^{15}/cm^3$ and $\phi_f = +0.29$ V). Then from Eq. (1-75)

$$\begin{aligned}\Phi_{MS} &= 3.2 - (3.25 + 0.55 + 0.29) \text{ V}\\ &= -0.9 \text{ V}\end{aligned}$$

For this example the flat-band condition is maintained with $V_G = -0.9$ V, as shown in Fig. 1.41.

As a third example, consider an MOS structure consisting of a p-type silicon gate (with $N_A = 10^{15}/cm^3$ and $\phi_f = +0.29$ V), SiO_2 dielectric, silicon semiconductor (with $N_D = 10^{15}/cm^3$ and $\phi_f = -0.29$ V). Note in this example that the gate is to be

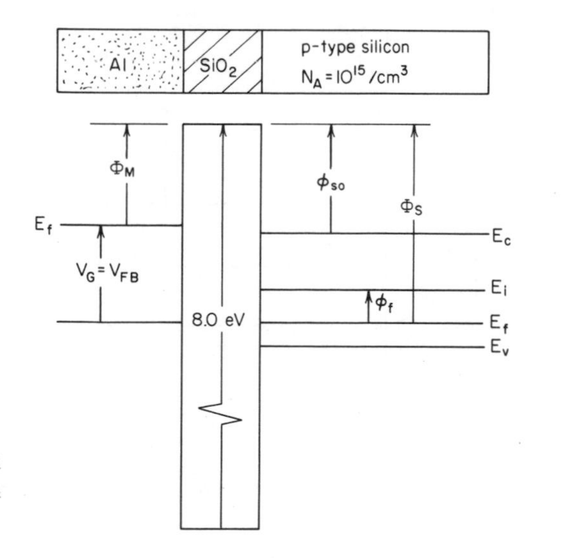

Fig. 1.41. Flat-band condition for the aluminum-SiO_2-silicon system ($N_A = 10^{15}/cm^3$).

fabricated from p-type silicon. Thus

$$\Phi_M = \phi_{so} + \left(\frac{E_g}{2q} + \phi_f\right)$$
$$= 3.25 + (0.55 + 0.29)\ \text{V} = 4.1\ \text{V}$$
$$\Phi_S = \phi_{so} + \left(\frac{E_g}{2q} + \phi_f\right)$$
$$= 3.25 + (0.55 - 0.29)\ \text{V} = 3.5\ \text{V}$$

So
$$\Phi_{MS} = \Phi_M - \Phi_S = 4.1 - 3.5\ \text{V}$$
$$= +0.6\ \text{V}$$

For this example, the flat-band condition is maintained with a *positive* V_G value of 0.6 V as shown in Fig. 1.42. The silicon-gate structure will be shown (in Chap. 2) to be extremely useful in bringing about low threshold voltage values for the p-channel enhancement-mode MOSFET.

1.4.4 Band-bending with Applied Gate Voltage for the Case of Finite Φ_{MS} and Q_{SS}

(a) Introduction. Consideration must now be given to the effect on band-bending when a finite surface-charge density Q_{SS} exists at the boundary of the Si-SiO_2 interface. Q_{SS} is found to be positive for both n- and p-type varieties of thermally oxidized silicon. Q_{SS} is dependent on crystalline orientation, and typical values are given in Table 1.1. In general, $Q_{SS}\,(111) > Q_{SS}\,(110) > Q_{SS}\,(100)$. It is to be noted that Q_{SS} is also process dependent. The empirical dependence of Q_{SS} on process

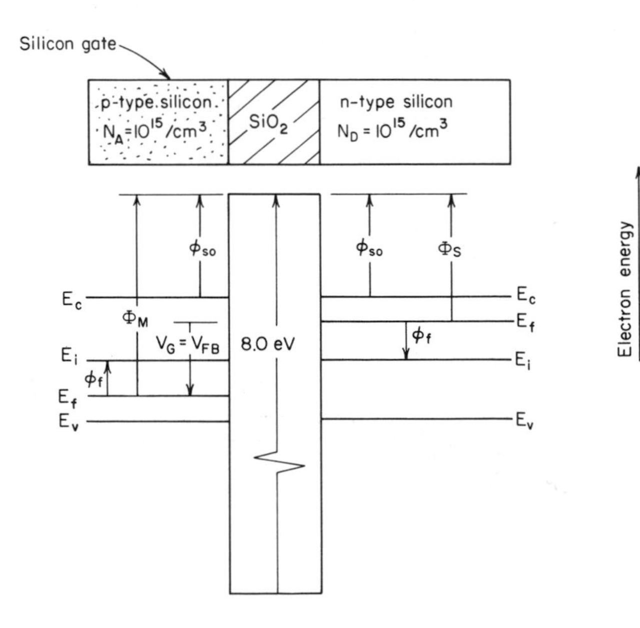

Fig. 1.42. Flat-band condition for the silicon gate ($N_A = 10^{15}/cm^3$)-SiO_2-silicon system ($N_D = 10^{15}/cm^3$).

Table 1.1. Typical Q_{SS} Values as a Function of Crystal Orientation

Crystal orientation	Q_{SS}, Coulombs	Q_{SS}/q
(111)	$+8.0 \cdot 10^{-8}/cm^2$	$5 \cdot 10^{11}/cm^2$
(110)	$+3.2 \cdot 10^{-8}/cm^2$	$2 \cdot 10^{11}/cm^2$
(100)	$+1.4 \cdot 10^{-8}/cm^2$	$9 \cdot 10^{10}/cm^2$

parameters is summarized by Deal et al.[24] Considerable work remains to be done, however, in securing a complete understanding of the nature and source of Q_{SS}.

(b) The Effect of Q_{SS} on the Flat-band Voltage. Because of the presence of the positive fixed charge Q_{SS} at the Si-SiO_2 interface, it is necessary to employ a negative gate voltage to bring about the flat-band condition (no charge induced in the silicon substrate). The gate voltage required to bring the bands to their flat-band condition if the Φ_{MS} term equals zero is given by

$$V_G = V_{FB} = -\frac{Q_{SS}}{C_o} \tag{1-76}$$

The flat-band voltage contribution generated by Q_{SS} in an MOS structure where, for example, $Q_{SS}/q = 5 \cdot 10^{11}/cm^2$, $t_{ox} = 10^{-5}$ cm, $C_o = 35 \cdot 10^{-9}$ F, and $\Phi_{MS} =$ 0 V is, from Eq. (1-76)

$$V_{FB} = -\frac{(1.6 \cdot 10^{-19})(5 \cdot 10^{11})}{35 \cdot 10^{-9}} = -2.3 \text{ V}$$

For the case where Q_{SS} and Φ_{MS} are finite, for example: $Q_{SS}/q = 5 \cdot 10^{11}/cm^2$, and the MOS structure is composed of aluminum gate, SiO_2 thickness of 10^{-5} cm, and n-type silicon substrate with $N_D = 10^{15}/cm^3$ (i.e., $\Phi_{MS} = -0.3$ V), the gate voltage required to bring about the flat-band condition would be given by

$$V_G = V_{FB} = \Phi_{MS} - \frac{Q_{SS}}{C_o} = -0.3 - 2.3 \text{ V} = -2.6 \text{ V} \tag{1-77}$$

(c) Summary. Finally it is important to note that a finite value of flat-band voltage serves to modify Eq. (1-67) to the form

$$V_G = V_{FB} + \phi_s - \frac{Q_s}{C_o} \tag{1-78}$$

Threshold voltage of the p-channel MOSFET can be calculated by first determining the gate voltage required to counterbalance the effects of Φ_{MS} and Q_{SS}, thereby establishing the flat-band condition. The gate voltage required to establish the flat-band condition is then added to the gate voltage required to bend the bands through a potential of $\phi_s = 2\phi_f$. Under these conditions a surface inversion layer is said to be established and the MOSFET is at the threshold of its *on* state. The threshold voltage V_T is thus obtained by combining Eqs. (1-77) and (1-78).

$$V_T = \Phi_{MS} - \frac{Q_{SS}}{C_o} + 2\phi_f - \frac{Q_B}{C_o} \tag{1-79}$$

Voltage required to establish flat-band condition (first two terms); Voltage required to just bend the bands through a potential of $2\phi_f$ (last two terms)

As an example of a threshold voltage calculation, consider an MOS structure with aluminum gate, SiO_2 thickness of 10^{-5} cm ($C_o = 35 \cdot 10^{-9}$ F/cm²), n-type silicon with $N_D = 10^{15}/\text{cm}^3$, ($\phi_f = -0.29$ V, $\Phi_{MS} = -0.3$ V), and $Q_{SS} = 5 \cdot 10^{11}/\text{cm}^3$. The threshold voltage is then, from Eq. (1-79)

$$V_T = -0.3 - \frac{(1.6 \cdot 10^{-19})(5 \cdot 10^{11})}{35 \cdot 10^{-9}} + 2(-0.29) - \frac{1}{35 \cdot 10^{-9}} \sqrt{(2)(12)(8.85 \cdot 10^{-14})(1.6 \cdot 10^{-19})(10^{15})(2)|-0.29|} \text{ V}$$

$$= -0.3 - 2.3 - 0.58 - 0.4 \text{ V}$$
$$= -3.6 \text{ V}$$

1.5 THE MOS CAPACITOR

1.5.1 Theory*

It is worthwhile to develop an expression for the capacitance-voltage (C-V) characteristics of the MOS capacitor since:

1. The C-V characteristics of the MOS structure are frequently used both in MOS/LSI process development and for in-process monitoring during product manufacture.
2. An understanding of MOSFET device action will be strengthened through analysis of the MOS capacitor.

The gate region of the MOSFET constitutes an MOS capacitor (Fig. 1.38). One plate of the capacitor consists of the gate field plate which is separated by a dielectric insulating layer from the alternate plate, which consists of a silicon-surface region. The silicon-surface region can, under the appropriate gate bias, assume a state of accumulation, depletion, or depletion-inversion.

The total capacitance exhibited by the MOS structure of Fig. 1.38 is

$$C_T = \frac{dQ_{\text{gate}}}{dV_G} \tag{1-80}$$

Note that $dQ_{\text{gate}} = -dQ_s$ (i.e., the charge on the gate side of the MOS capacitor is equal to the negative of the charge on the silicon side of the capacitor). Therefore

$$C_T = \frac{-dQ_s}{dV_G} \tag{1-81}$$

*This derivation follows, in part, that presented by A. S. Grove, Physics and Technology of Semiconductor Devices, pp. 271–274, John Wiley and Sons, Inc., New York, 1967.

Then recalling
$$V_G = \frac{-Q_s}{C_o} + \phi_s \tag{1-67}$$

it follows that:
$$dV_G = \frac{-dQ_s}{C_o} + d\phi_s \tag{1-82}$$

Combining Eqs. (1-81) and (1-82)

$$C_T = \frac{-dQ_s}{-dQ_s/C_o + d\phi_s} = \frac{1}{1/C_o - d\phi_s/dQ_s} \tag{1-83}$$

The capacitance of the silicon side of the gate region is defined as

$$C_s \equiv -\frac{dQ_s}{d\phi_s} \equiv \frac{\epsilon_{Si}\epsilon_o}{x_d} \tag{1-84}$$

where x_d = surface depletion layer depth.

Therefore from Eqs. (1-83) and (1-84)

$$C_T = \frac{1}{1/C_o + 1/C_s} \tag{1-85}$$

and hence:
$$\frac{1}{C_T} = \frac{1}{C_o} + \frac{1}{C_s} \tag{1-86}$$

Equation (1-86) states that the MOS capacitor is in essence composed of two capacitances (C_o, the capacitance of the gate oxide, and C_s, the capacitance of the silicon depletion layer) connected in series.

We will find it useful to employ the expression for C_T/C_o in experimental analysis of the MOS capacitor; therefore, from Eq. (1-86)

$$\frac{C_T}{C_o} = \frac{1}{1 + C_o/C_s} \tag{1-87}$$

To determine an expression for C_o/C_s and hence C_T/C_o as the silicon surface is depleted, combine the following equations and eliminate x_d

$$V_G = \phi_s + \frac{-Q_s}{C_o} \tag{1-67}$$

$$-\phi_s = \frac{qN_D x_d^2}{2\epsilon_{Si}\epsilon_o} \tag{1-69}$$

$$Q_s = qN_D x_d \tag{1-71}$$

($Q_s = Q_B$, and we will develop an expression for the capacitance for the depletion region condition, from accumulation to onset of inversion)

$$C_s = \frac{\epsilon_{Si}\epsilon_o}{x_d} \tag{1-84}$$

Then
$$\frac{C_o^2}{C_s^2} + \frac{2C_o}{C_s} + \frac{2V_G C_o^2}{qN_D\epsilon_{Si}\epsilon_o} = 0 \tag{1-88}$$

Solving Eq. (1-88) for C_o/C_s and taking the meaningful root, we obtain

$$\frac{C_o}{C_s} = -1 + \sqrt{1 - \frac{2\epsilon_{\text{Ox}}{}^2\epsilon_o V_G}{qN_D\epsilon_{\text{Si}}t_{\text{Ox}}{}^2}} \qquad (1\text{-}89)$$

Now from Eq. (1-87) and (1-89)

$$\frac{C_T}{C_o} = \frac{1}{\sqrt{1 - 2\epsilon_{\text{Ox}}{}^2\epsilon_o V_G/qN_D\epsilon_{\text{Si}}t_{\text{Ox}}{}^2}} \qquad (1\text{-}90)$$

It should be noted that when using Eq. (1-90), V_G values to be substituted are negative for the n substrate situation summarized in Fig. 1.43 for which Eq. (1-90) was developed. Therefore on application of negative gate voltage, the value of C_T/C_o will decrease approximately as the square root of the applied gate voltage as long as operation is confined to the depleted surface condition; i.e., to the onset of inversion ($\phi_s = 2\phi_f$).

We have previously assumed that at onset of inversion no further charge accrues to depletion. The question then arises both as to the behavior of the MOS capacitance in the inversion region as well as the source of holes which will constitute the surface inversion layer. The source of holes (minority carriers in the n-type substrate) for the structure of Fig. 1.38 originates from the generation-recombination process in the surface depletion layer. The generation mechanism cannot supply holes to the inversion layer instantaneously, and thus if the frequency used for the small-signal C-V measurements is too high (greater than a few hundred Hertz), C_T/C_o can remain at a constant minimum value given by

$$\frac{C_T}{C_o} = \frac{1}{\sqrt{1 - 2\epsilon_{\text{Ox}}{}^2\epsilon_o V_T/qN_D\epsilon_{\text{Si}}t_{\text{Ox}}{}^2}} \qquad (V_T \text{ is negative for n-type substrate}) \qquad (1\text{-}91)$$

The theoretical results are shown in Fig. 1.43 and are compared with a pictorial representation of experimental results. Experimental agreement is in full accord with a more exact theory than we have presented here.[25]

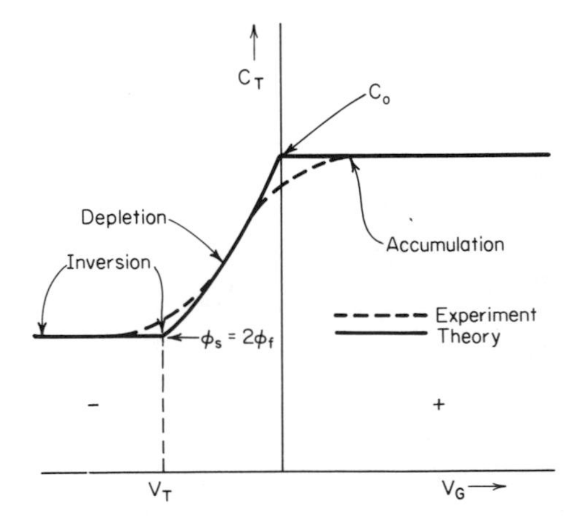

Fig. 1.43. Experimental versus theoretical (depletion-layer approximation) high-frequency C-V characteristics (n-type substrate).

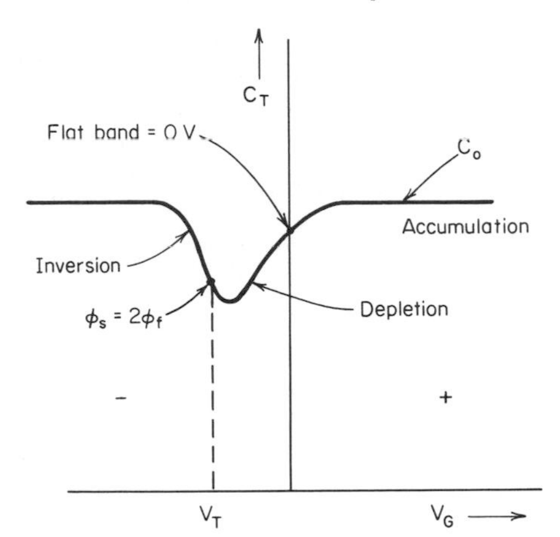

Fig. 1.44. Pictorial representation of low-frequency C-V characteristics (n-type substrate).

If MOS C-V measurements are made at sufficiently low frequencies (<100 Hz), then minority carriers are supplied by the generation-recombination process in sufficient quantity to the inversion layer so that they can adequately "follow" the small-signal ac voltage used in the experimental determination of the capacitance. The capacitance value at strong inversion will eventually approach that of the oxide capacitance ($C_o = \epsilon_{ox}\epsilon_o/t_{ox}$) as shown in Fig. 1.44. If C-V measurements are performed on the channel region of a MOSFET structure where source and drain regions are present, then of course minority carriers can be supplied readily to the channel region from the source and drain regions. The characteristics of Fig. 1.44 will thus be obtained for the MOSFET device at both low *and* high frequencies, since a copious supply of minority carriers is readily available.

1.5.2 Utility of the C-V Method

The theoretical development of the MOS capacitor in Sec. 1.5.1 was based on the condition

$$V_{FB} = 0 \qquad \text{(i.e., } Q_{SS} = 0,\ \Phi_{MS} = 0\text{)}$$

where V_{FB} is given by

$$V_{FB} = \Phi_{MS} - \frac{Q_{SS}}{C_o} \tag{1-77}$$

Interface surface-charge density Q_{SS} and the metal-semiconductor work-function difference Φ_{MS} will both contribute to a translation of the entire C-V trace along the gate voltage axis through a distance V_{FB} as shown in Fig. 1.45. The translation of the C-V trace through V_{FB} is of practical importance and is best illustrated by considering a specific example.

example

An MOS capacitor has an n-type substrate with $N_D = 10^{15}/\text{cm}^3$, $t_{ox} = 1000$ Å, and $\Phi_{MS} = -0.3$ V. The measured flat-band voltage shift of the C-V trace is found to be -2.5 V. From this information determine the value of Q_{SS} for the sample.

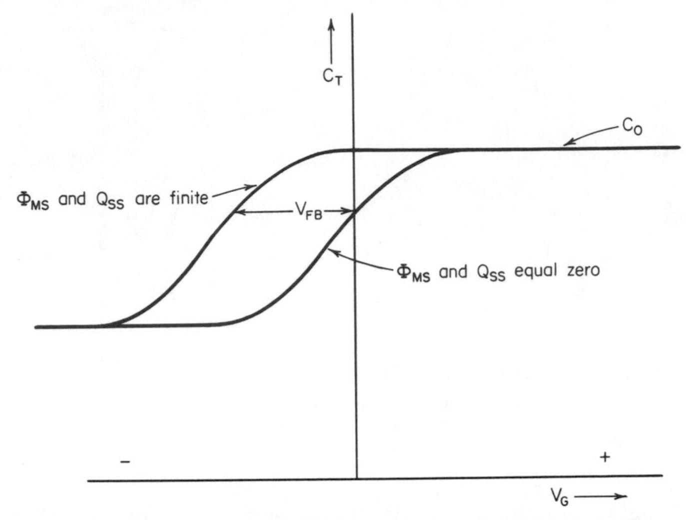

Fig. 1.45. Translation of C-V characteristic through V_{FB}.

solution

From Eq. (1-77)

$$V_{FB} = \Phi_{MS} - \frac{Q_{SS}}{C_o}$$

$$-2.5\text{ V} = -0.3 - \frac{Q_{SS}}{35 \cdot 10^{-9}}\text{ V}$$

Therefore $$\frac{Q_{SS}}{q} = 4.8 \cdot 10^{11}\text{ positive charges/cm}^2$$

The C-V method has been found to be extremely useful for monitoring silicon surface-state conditions during MOS/LSI processing. In fact, MOS/LSI chips often have an MOS capacitor included thereon for process monitoring. Undoubtedly the C-V method will also continue to serve as one of the main analytical tools in the research and development of MOS and MIS (metal-insulator-semiconductor) structures and devices.

1.6 THE MOSFET DEVICE EQUATIONS

1.6.1 The Current-voltage Characteristics (Enhancement Mode)

(a) The Linear Region (Triode Region). To develop the fundamental current-voltage characteristics for the MOSFET, let us initially consider a p-channel, enhancement-mode device. By definition, *the enhancement-mode device is normally off until sufficient gate voltage is applied to induce an inversion layer between source and drain.* The enhancement-mode device is the desired configuration for a logic element since it is normally *off*. To achieve the *on* condition for the device, an appropriate voltage must be applied to the gate field plate, and the gate field plate is required to overlap the source and drain regions to ensure formation of a continuous inversion layer between source and drain. A number of technological peculiarities such as the sign and size of Q_{SS}, ϕ_f, and Φ_{MS} have resulted (at least in the initial phase of MOS/LSI development) in the channel of the silicon enhancement-mode MOSFET being p-type.

For the derivation of the MOSFET current-voltage characteristics we shall assume that an inversion layer is present in the channel, the source is connected to the substrate, and the device is operated in its linear region with current flowing between source and drain. Then from Fig. 1.46 the voltage across an incremental element Δy of the channel is given by

$$\Delta V = I_D \Delta R \tag{1-92}$$

where $\Delta R = \Delta \dfrac{\rho y}{A} = \dfrac{\rho \Delta y}{A}$

$A = xW$

So

$$\Delta V = I_D \rho \frac{\Delta y}{A} \tag{1-93}$$

where $\rho = \dfrac{1}{P(y) q \mu}$

$P(y)$ = hole concentration in the channel as a function of the distance y. Then from Eqs. (1-92) and (1-93)

$$\Delta V = \frac{I_D \Delta y}{P(y) q \mu W x} \tag{1-94}$$

Since $P(y)$ = concentration of mobile carriers per cm^3 in the channel at point y, it follows that

$$q x P(y) = Q_p(y) \tag{1-95}$$

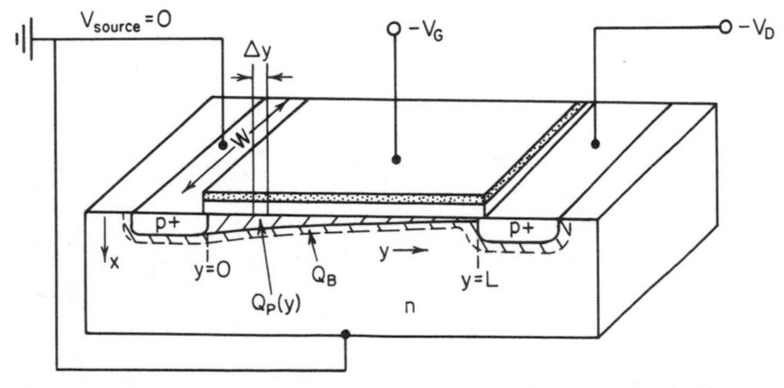

Fig. 1.46. Cross-sectional view and geometrical factors employed in determining the I-V characteristics of the MOSFET.

where $Q_p(y)$ = mobile surface charge density per cm^2 in the channel at point y. Therefore from Eqs. (1-94) and (1-95)

$$\Delta V = \frac{I_D \Delta y}{Q_p(y)\mu W} \tag{1-96}$$

and therefore

$$I_D \Delta y = Q_p(y)\mu W \Delta V \tag{1-97}$$

From Eq. (1-68) $Q_s \equiv Q_B + Q_P$, and utilizing the result of Eq. (1-97) we obtain

$$I_D \Delta y = (Q_s - Q_B)\mu W \Delta V \tag{1-98}$$

where Q_s and Q_B are now a function of position y.

From Eq. (1-78), $Q_s(y) = -C_o[V_G - V_{FB} - \phi_s(y)]$, and combining this result with Eq. (1-98) yields

$$I_D \Delta y = -C_o[V_G - V_{FB} - \phi_s(y)]\mu W \Delta V - Q_B \mu W \Delta V \tag{1-99}$$

At the sacrifice of some accuracy, but at the gain of considerable simplicity and clarity of the physical situation, we will *assume that Q_B is a constant* (independent of position y in the channel) and is given by the previously developed

$$Q_B = \sqrt{2\epsilon_{Si}\epsilon_o q N_D |2\phi_f|} \tag{1-71}$$

Since it has been assumed for this derivation that an inversion layer is present with current I_D flowing in the channel, it follows that

$$\phi_s(y) = 2\phi_f - V(y) \tag{1-100}$$

where ϕ_f = Fermi potential of the substrate and $V(y)$ = the reverse bias between the incremental section of the inversion layer at the point y and the substrate (caused by the potential drop generated by the positive current flow down the channel). Then Eq. (1-99) becomes

$$I_D \int_0^L dy = -\mu W C_o \int_0^{-V_D} \left[V_G - V_{FB} - 2\phi_f + \frac{Q_B}{C_o} + V(y) \right] dV \tag{1-101}$$

if the threshold voltage is defined as

$$V_T \equiv V_{FB} + 2\phi_f - \frac{Q_B}{C_o} \qquad \text{cf. Eq. (1-79)}$$

then

$$I_D \int_0^L dy = -\mu W C_o \int_0^{-V_D} [V_G - V_T + V(y)]\, dV \tag{1-102}$$

and

$$I_D L = \mu W C_o \left[(V_G - V_T)V_D - \frac{V_D^2}{2} \right] \tag{1-103}$$

And since

$$C_o = \frac{\epsilon_{Ox}\epsilon_o}{t_{Ox}}$$

$$I_D = \frac{\mu \epsilon_{Ox}\epsilon_o W}{t_{Ox} L} \left[(V_G - V_T)V_D - \frac{V_D^2}{2} \right] \tag{1-104}$$

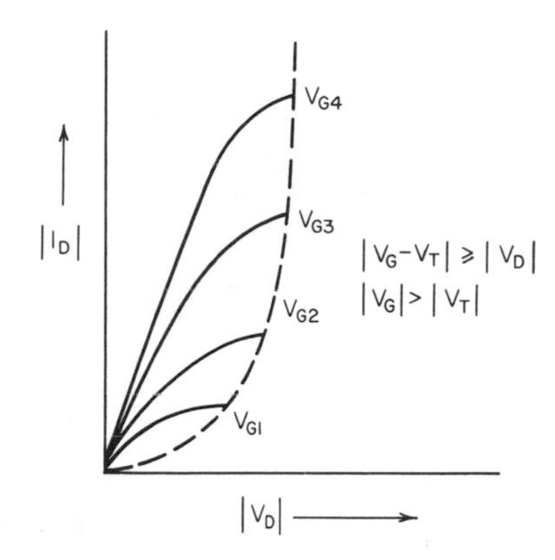

Fig. 1.47. I_D *vs.* V_D for the region $|V_G - V_T| \geqslant |V_D|$.

A pictorial representation of I_D *vs.* V_D to the point where $|V_D| = |V_G - V_T|$ yields Fig. 1.47.

The I-V characteristics of Fig. 1.47 have a linear structure in the initial portion of their formation, which prompts the description of this domain of opcration as the *linear region.* Furthermore, the characteristics are reminiscent of those of a vacuum-tube triode which suggests the nomenclature *triode region.*

(b) The Saturation Region. Plotting of Eq. (1-104) as shown in Fig. 1.47 to the point where

$$|V_D| = |V_G - V_T| \tag{1-105}$$

represents a convenient cut-off point as will be seen from the following derivation. Consider that:

1. Potential drop along the channel is generated by current flow in the channel (IR drop).
2. Potentials are always combined algebraically.

Then note (Fig. 1.48) that adjacent to the drain there exists no inversion layer under the condition that Eq. (1-105) holds, since the potential available is only V_T (essentially all induced charge is in the depletion layer).

When the condition described by Eq. (1-105) prevails, the device is said to be operating in the *saturation region* with the channel "pinched off." At point A of

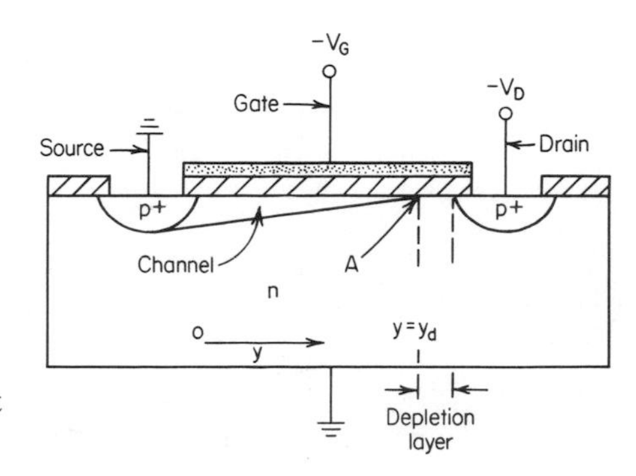

Fig. 1.48. The MOSFET channel at pinch-off.

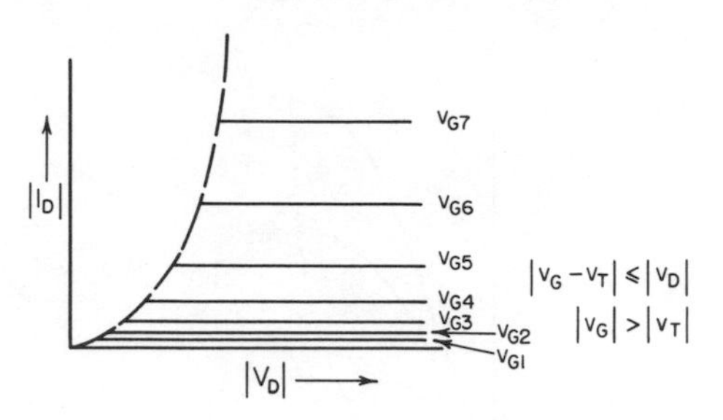

Fig. 1.49. MOSFET saturation characteristics.

Fig. 1.48, Q_P approaches zero (it becomes very small, but is finite), thus resulting in a high-resistance region close to the drain. As a first approximation, we shall assume that at and beyond pinch-off the differential drain resistance $(\Delta V_D/\Delta I_D)$ becomes infinite while the drain current remains constant at a value equal to that achieved at pinch-off.

An expression for the pinch-off voltage can be derived formally by finding the value of V_D at which I_D given by Eq. (1-104) reaches a maximum. The pinch-off voltage referred to as $V_{D(\text{sat})}$ is thus obtained by maximizing I_D of Eq. (1-104). Maximizing yields:

$$\left.\frac{dI_D}{dV_D}\right|_{V_D=V_{D(\text{sat})}} = 0 = \frac{\mu\epsilon_{\text{Ox}}\epsilon_o W}{t_{\text{Ox}}L}[(V_G - V_T) - V_{D(\text{sat})}] \tag{1-106}$$

and hence,

$$|V_{D(\text{sat})}| = |V_G - V_T| \tag{1-107}$$

It is reassuring to note that Eq. (1-107) is identical to our intuitive argument given for the pinch-off condition, Eq. (1-105).

The saturation current-voltage characteristics are obtained by substituting $|V_D| = |V_G - V_T|$ in Eq. (1-104), thereby obtaining

$$I_D = \frac{\mu\epsilon_{\text{Ox}}\epsilon_o W}{t_{\text{Ox}}L} \cdot \frac{(V_G - V_T)^2}{2} \tag{1-108}$$

These results are pictorially represented in Fig. 1.49.

Note in Fig. 1.49 and Eq. (1-108) that the saturation current obeys a "square law" relationship, i.e., $I_D \approx (V_G - V_T)^2$. Furthermore, if the enhancement-mode MOSFET has gate connected to drain as shown in Fig. 1.50, $|V_G| = |V_D|$. This

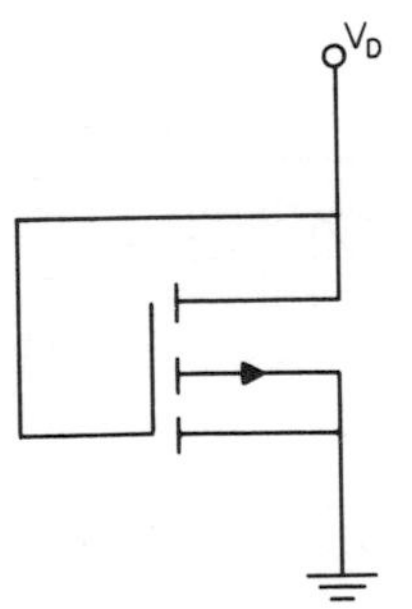

Fig. 1.50. MOSFET with gate connected to drain.

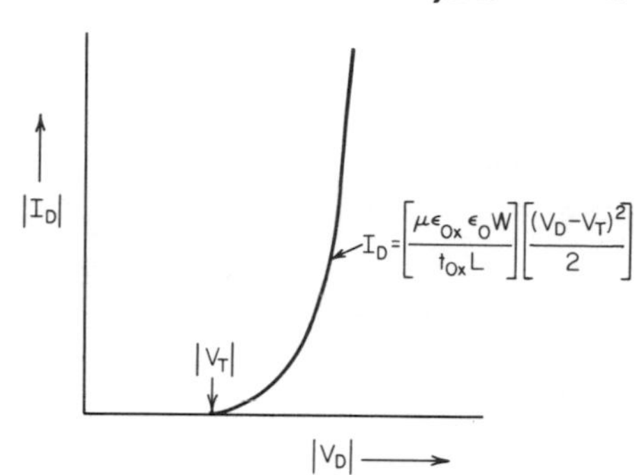

Fig. 1.51. Transfer characteristics for the p-channel enhancement-mode MOSFET with gate connected to drain.

implies that $|V_G - V_T| < |V_D|$ and the device will always be in the saturation region, and the transfer characteristic will therefore result as shown in Fig. 1.51.

Any additional voltage that is applied between drain and source beyond the pinch-off condition will appear across the drain-substrate depletion layer. As the drain-to-source voltage increases past pinch-off, the depletion layer edge y_D shown in Fig. 1.48 will move closer to the source. If the resulting fractional change in the length of the channel is small, the current will be essentially constant for drain-to-source voltage above pinch-off. If, however, the channel length is appreciably modified or "modulated" by the depletion-layer width, the channel will be shortened, resulting in an increase in the effective value of W/L with a consequential increase in I_D [cf. Eq. (1-108)]. The increase in I_D caused by channel-length modulation results in a finite output impedance characteristic for the MOSFET as shown in Fig. 1.52. The existence of finite output conductance should be expected in integrated MOSFETs where channel length is small ($\approx$0.2 mils), and hence subject to modulation by the drain-substrate reverse bias.[26]

(c) The Avalanche Region. Since the drain-substrate junction is normally reverse biased, it is to be expected that the avalanche breakdown region will eventually be reached as reverse bias is increased. Note that the presence of the gate field plate will alter the normally occurring electric-field distribution in the depletion region of the drain-substrate junction.[27] The situation is summarized in Fig. 1.53. The effect of placing a positive voltage on the gate of a p-channel MOSFET will be to increase the field in the drain-substrate region, thereby decreasing the fundamental avalanche breakdown capability of the drain-substrate junction. A negative voltage applied to the gate with respect to the substrate will decrease the field in

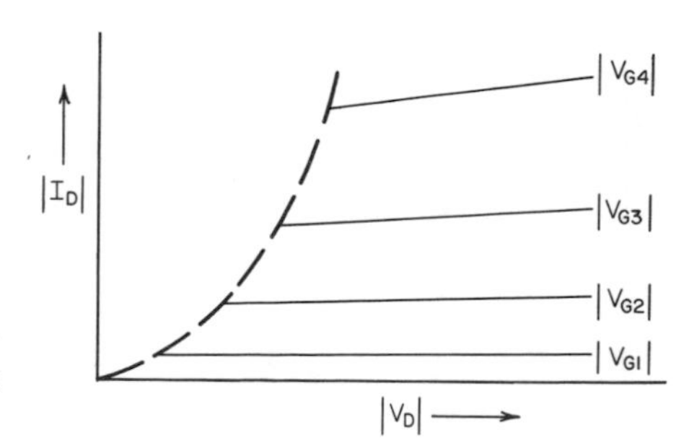

Fig. 1.52. Channel length modulation resulting in finite output impedance in the saturation region.

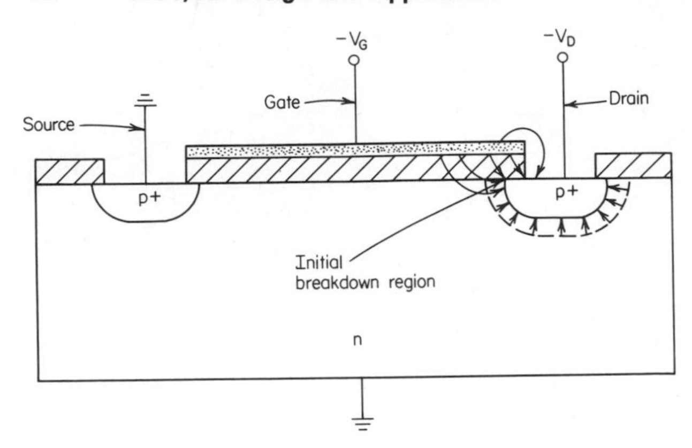

Fig. 1.53. Modification of electric-field distribution in the drain-substrate junction by gate field plate.

the drain-substrate region and thereby tend to restore the avalanche breakdown capability of the junction. Since negative voltage is applied to the gate to activate the p-channel enhancement-mode device, the avalanche breakdown voltage will increase in absolute value as gate voltage increases in absolute value (Fig. 1.54). The discussion of the avalanche region of MOSFET electrical characteristics has been presented for the sake of completeness. In integrated-circuit operation, the devices are normally not subjected to avalanche.

1.6.2 The Current-voltage Characteristics (Depletion Mode)

The depletion-mode MOSFET is in the conducting state (*on* state) with zero voltage applied to the gate with respect to source and substrate. The depletion-mode device can be turned on harder by applying the appropriate gate voltage, or can be turned off by applying the threshold voltage value (Fig. 1.55). Note that the term *threshold voltage,* as it pertains to depletion-mode devices, defines a condition at which conduction is terminated; whereas, with respect to enhancement-mode devices, it defines a condition at which conduction is initiated. Table 1.2 summarizes the voltage polarities encountered in p- and n-channel enhancement- or depletion-

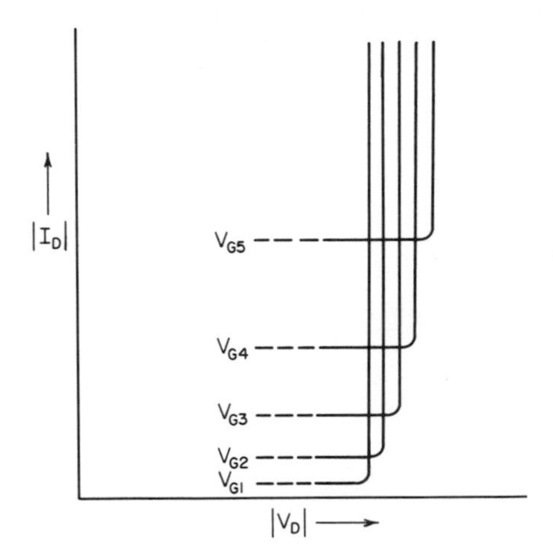

Fig. 1.54. The avalanche region of the p-channel enhancement-mode device.

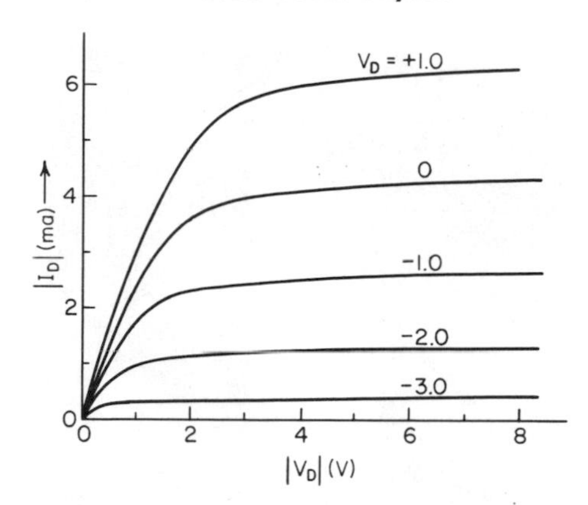

Fig. 1.55. Typical I-V characteristics of the n-channel depletion-mode MOSFET.

mode devices. The set of device equations developed in Sec. 1.6.1 applies equally well to either depletion- or enhancement-mode devices of n- or p-channel structures for voltage polarities given in Table 1.2.

1.7 SURFACE CARRIER MOBILITY

1.7.1 p-Channel Devices

It was shown in Sec. 1.6.1 that in the triode region

$$I_D = \frac{\mu\epsilon_{\mathrm{Ox}}\epsilon_o W}{t_{\mathrm{Ox}}L}\left[(V_G - V_T)V_D - \frac{V_D{}^2}{2}\right] \tag{1-104}$$

and in the saturation region

$$I_D = \frac{\mu\epsilon_{\mathrm{Ox}}\epsilon_o W}{t_{\mathrm{Ox}}L}\frac{(V_G - V_T)^2}{2} \tag{1-108}$$

From these results we note that I_D can be readily calculated if the mobility value μ is known. All remaining terms in Eqs. (1-104) and (1-108) are either known geometrical factors established in device processing, or the V_T value which can be experimentally determined by appropriate extrapolation of I_D as a function of V_G

Table 1.2. Voltage Polarities for p- and n-Channel Enhancement- or Depletion-mode Devices in the *on* State Condition

	V_G	V_T	V_D
p-Channel:			
Enhancement	−	−	−
Depletion	±	+	−
n-Channel:			
Enhancement	+	+	+
Depletion	±	−	+

Fig. 1.56. p-channel enhancement-mode MOSFET in Hall geometry for measuring surface carrier mobility.[28]

to $I_D = 0$, or are known voltage factors such as V_G or V_D. The experimental determination of surface carrier mobility will thus permit calculation of I_D values in circuit design through the use of the device equations.

The surface carrier mobility values should not be expected to be similar to those observed in the bulk, since the transport of carriers in MOSFET inversion layers involves movement through an infinitesimally thin conducting sheet in contrast to the unconstrained motion in the three-dimensional bulk case.

To determine hole mobility in p-channel inversion layers experimentally, a p-channel enhancement-mode device was fabricated in Hall geometry as shown in Fig. 1.56.[28] The results of the investigation are summarized in Fig. 1.57, where it is observed that:

1. Mobility values are dependent on the silicon-surface orientation.
2. Mobility values are azimuthally dependent on the (110) plane.

One should ask if the above results are consistent with Neumann's principle which defines symmetry properties of physical quantities in single-crystalline material. To answer the question, we must realize that transport of carriers in an inversion layer may be thought of as a quantized motion of the carriers in a two-dimensional electron gas. The electric field resulting from the applied gate voltage forms an infinitely thin ($\approx$50 Å) inversion layer and is thus said to remove the degeneracy of the cubic-crystal structure with respect to the transport of carriers. Carriers are thereby confined to two-dimensional motion and can exhibit a plane-dependent mobility value which is characteristic of the physical properties of the plane. Furthermore, by applying Neumann's principle to the symmetry properties of the given planes, it is found (in agreement with experiment) that azimuthally independent mobility should be observed on the (111) and (100) planes. Conversely, the (110) plane can

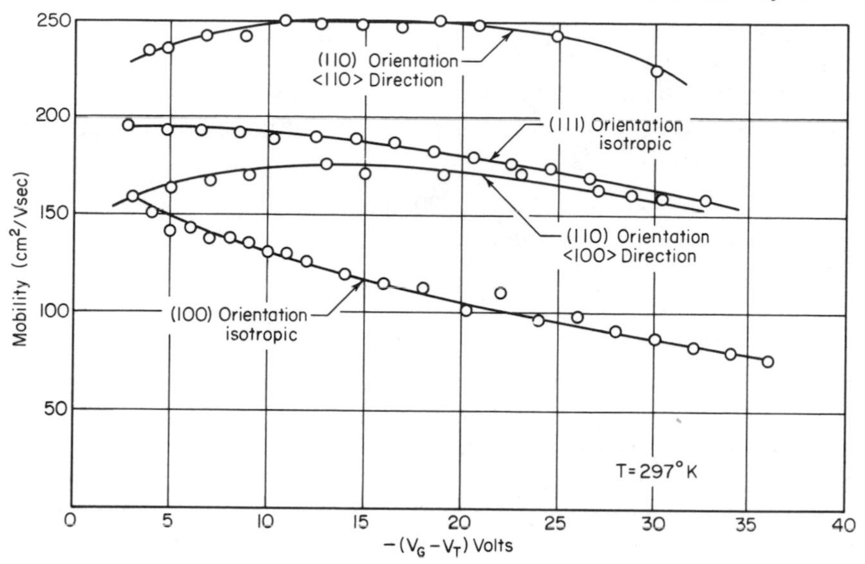

Fig. 1.57. Experimental values of inversion-layer hole mobility versus effective gate voltage ($V_G - V_T$) for $T = 297°K$.[28]

possess azimuthally dependent mobility values. The latter effect is experimentally observed on the (110) plane, where mobility values (Fig. 1.57) are higher for current flow in the ⟨110⟩ direction than in the ⟨100⟩ direction.

It is also to be noted that surface hole mobility is not constant with variation in MOS gate voltage. The dependence on gate voltage is strongly accentuated by cooling the sample, Fig. 1.58. The mobility is observed to increase with decrease in temperature and is found to vary as[29]

$$\mu \propto T^{-1} \tag{1-109}$$

in the vicinity of room temperature. Below 60°K, surface hole mobility shows little dependence on temperature.

The largest surface hole mobility observed at room temperature is only ≈50 percent of the bulk silicon value for the same substrate resistivity. This is unfortu-

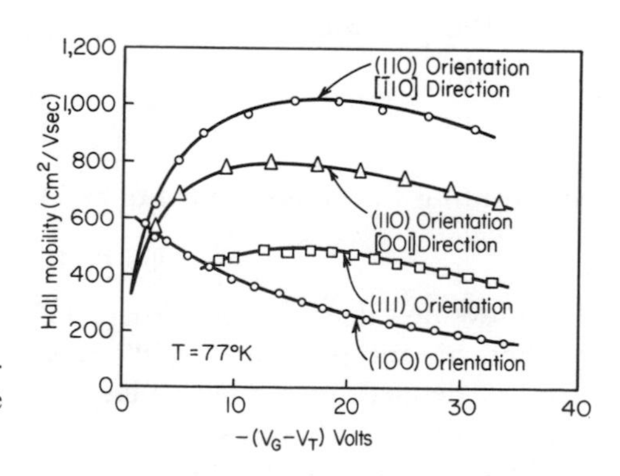

Fig. 1.58. Experimental values of inversion-layer hole mobility versus effective gate voltage ($V_G - V_T$) for $T = 77°K$.[28]

nate, since a large mobility value is required if high cut-off frequency is desired for the MOSFET.

Essentially all p-channel MOS circuitry is presently fabricated on the (100) or (111) planes of silicon. For design purposes we will therefore conclude from the data of Fig. 1.57 that

$$\mu\,(111) \approx 190\ \text{cm}^2/\text{V-sec}$$
$$\mu\,(100) \approx 130\ \text{cm}^2/\text{V-sec}$$

at $T = 300°\text{K}$

1.7.2 n-Channel Devices

Investigation of surface electron mobility in n-type inversion layers indicates that the maximum mobility values are obtained on the (100) orientation.[30] Although the results to date are not as detailed as those given for holes in p-type inversion layers, it can be stated that surface electron mobility values are approximately three times larger than surface hole mobility values. For design purposes we will consider the surface mobility value of electrons to be: $\mu_n = 600\ \text{cm}^2$ per V-sec, a value which is approximately one-half of the electron bulk mobility value.

1.8 THE MOSFET AS AN ACTIVE DEVICE

1.8.1 The Linear Region (Triode Region)

(a) Introduction. The current-voltage expression Eq. (1-104) derived for MOSFET operation in the triode region can be written as

$$I_D = \beta\left[(V_G - V_T)V_D - \frac{{V_D}^2}{2}\right] \tag{1-110}$$

where

$$\beta = \frac{\mu\epsilon_{\text{Ox}}\epsilon_o W}{t_{\text{Ox}} L} \tag{1-111}$$

In addition to Eq. (1-110), two other characteristics are found useful in describing device operation in the triode region. They are: (1) the source-drain conductance g_{sd}, and (2) the transconductance g_m. Following is a description of these two parameters:

(b) The Source-drain Conductance g_{sd} is defined as

$$g_{sd} \equiv \left.\frac{\partial I_D}{\partial V_D}\right|_{V_G=\text{constant}} \tag{1-112}$$

Substituting Eq. (1-110) into Eq. (1-112) yields

$$g_{sd} = \beta|V_G - V_T - V_D| \tag{1-113}$$

The absolute value is taken in Eq. (1-113) since the sign of g_{sd} must always be positive for both p- and n-channel MOSFETs.

g_{sd} is defined for a fixed value of gate voltage and is analytically the slope of the I-V characteristics as shown in Fig. 1.59. The conductance g_{sd} is the reciprocal of the channel resistance (*on* resistance) of the device.

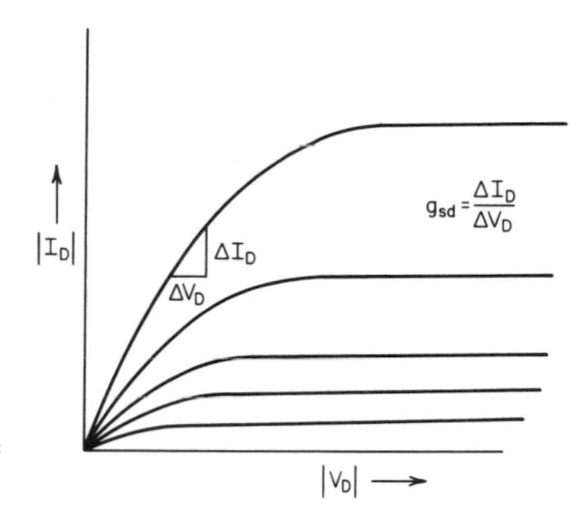

Fig. 1.59. Graphic interpretation of g_{sd} in the triode region.

To calculate g_{sd} for a typical p-channel enhancement-mode MOSFET, (111) orientation, operating in the triode region, consider the following example:

$$\begin{aligned} W &= 1 \text{ mil} \\ L &= 0.2 \text{ mil} \\ \epsilon_{\text{Ox}} &= 4 \\ \epsilon_o &= 8.85 \cdot 10^{-14} \text{ F/cm} \\ t_{\text{Ox}} &= 1000 \text{ Å} \\ \mu &= 190 \text{ cm}^2/\text{V-sec} \\ V_G &= -10 \text{ V} \\ V_T &= -4 \text{ V} \\ V_D &= -1 \text{ V} \end{aligned}$$

then $g_{sd} = \dfrac{(190)(4)(8.85 \cdot 10^{-14})(1)}{(10^{-5})(0.2)} \left| (-10 + 4 + 1) \right|$

$g_{sd} \approx 168 \cdot 10^{-6}$ mhos

and the channel resistance would be

$$r_d = \frac{1}{g_{sd}} = \frac{1}{168 \cdot 10^{-6} \text{ mhos}} = 5.9 \cdot 10^3 \text{ ohms}$$

Equation (1-113) describes the triode region g_{sd} with $\approx \pm 30$ percent accuracy if the appropriate mobility values (Fig. 1.57) are utilized. Frequent use will be made of the conductance parameter g_{sd} in the design of MOS integrated circuits.

(c) The Transconductance g_m is defined as

$$g_m \equiv \frac{\partial I_D}{\partial V_G} \bigg|_{V_D = \text{constant}} \tag{1-114}$$

Substituting Eq. (1-110) into Eq. (1-114) yields

$$g_m = \beta |V_D| \tag{1-115}$$

The absolute value is taken in Eq. (1-115), since the sign of g_m must always be positive for both p-channel and n-channel MOSFET.

The value g_m is defined for a fixed value of drain voltage and is a measure of the spacing between members of the set of I-V traces at the value of V_D. The concept of transconductance g_m is used in defining the amplification of the active device and accompanying circuitry.

To calculate g_m for a typical MOSFET, (111) orientation, operating in the triode region, consider the following example:

$$W = 1 \text{ mil}$$
$$L = 0.2 \text{ mil}$$
$$\epsilon_{\text{Ox}} = 4$$
$$\epsilon_o = 8.85\ 10^{-14} \text{ F/cm}$$
$$t_{\text{Ox}} = 1000 \text{ Å}$$
$$\mu = 190 \text{ cm}^2\text{/V-sec}$$
$$V_G = -10 \text{ V}$$
$$V_T = -4 \text{ V}$$
$$V_D = -1 \text{ V}$$

then $$g_m = \frac{(190)(4)(8.85 \cdot 10^{-14})(1)}{10^{-5}(0.2)} \cdot |-1|$$
$$= 34 \cdot 10^{-6} \text{ mhos}$$

Equation (1-115) describes the triode region g_m rather inaccurately since:

1. Mobility has been assumed to be independent of the gate voltage.
2. The term Q_B has been assumed to be constant.

A discussion of the approach to a more accurate equation for g_m will be presented in Sec. 1.9.2. The corrections which can be made to Eq. (1-115) are somewhat academic, however, since the concept of g_m in the triode region finds infrequent use in the design of MOS/LSI.

1.8.2 The Saturation Region

The current-voltage expression Eq. (1-108) derived for MOSFET operation in the saturation region can be written as

$$I_D = \frac{\beta(V_G - V_T)^2}{2} \tag{1-116}$$

where β has been defined by Eq. (1-111). In addition to Eq. (1-116), transconductance g_m is found to be useful in describing device operation in the saturation region. Conductance g_{sd} is zero by definition in the saturation region, since Eq. (1-116) is independent of V_D. The MOSFET does, however, exhibit a finite value of g_{sd} in the saturation region[26] as briefly described in Sec. 1.6.1*b*. Equation (1-116) does not take into account the physical mechanisms that generate g_{sd} values in the saturation region.

If Eq. (1-116) is utilized in conjunction with the definition for g_m, the following expression for g_m in the saturation region is obtained:

$$g_m = \beta|V_G - V_T| \tag{1-117}$$

To calculate g_m for a typical p-channel enhancement-mode MOSFET, (111) orientation, operating in the saturation region, consider the following example:

$$\begin{aligned} W &= 1 \text{ mil} \\ L &= 0.2 \text{ mil} \\ \epsilon_{\text{Ox}} &= 4 \\ \epsilon_o &= 8.85 \cdot 10^{-14} \text{ F/cm} \\ t_{\text{Ox}} &= 1000 \text{ Å} \\ \mu &= 190 \text{ cm}^2/\text{V-sec} \\ V_G &= -10 \text{ V} \\ V_T &= -4 \text{ V} \\ V_D &= -20 \text{ V} \end{aligned}$$

then $$g_m = \frac{(190)(4)(8.85 \cdot 10^{-14})(1)}{(10^{-5})(0.2)}|-10 + 4|$$
$$= 200 \cdot 10^{-6} \text{ mhos}$$

Equation (1-117) describes the saturation region with an accuracy sufficient for many of the engineering calculations required in MOS/LSI circuit design. Refinement of the expression for g_m in the saturation region will be presented in Sec. 1.9.3.

1.8.3 Frequency Response

It can be shown that the maximum frequency of operation of a MOSFET is given by:[31]

$$f_o = \frac{g_m}{2\pi C_G} \tag{1-118}$$

where g_m is the transconductance of the device and C_G is the total gate capacitance. The g_m-to-C_G ratio is also frequently referred to as the *figure of merit*. An explicit relationship for the figure of merit of a MOSFET operating in the saturation region can be obtained as follows:

$$f_o = \frac{g_m}{2\pi C_G} = \frac{(\mu\epsilon_{\text{Ox}}\epsilon_o W/t_{\text{Ox}} L)|V_G - V_T|}{(2\pi\epsilon_{\text{Ox}}\epsilon_o/t_{\text{Ox}})W \cdot L} \tag{1-119}$$

$$f_o = \frac{\mu|V_G - V_T|}{2\pi L^2} \tag{1-120}$$

Note that for a high figure of merit it is desirable to have as high a mobility as possible and a channel length as narrow as possible.

It is of interest to calculate the theoretical cut-off frequency of a typical p-channel enhancement-mode MOSFET, (111) orientation, employed in MOS/LSI circuitry. To perform the calculation, assume:

$$\begin{aligned} V_G &= -10 \text{ V} \\ V_T &= -4 \text{ V} \\ L &= 0.2 \text{ mil} \\ \mu &= 190 \text{ cm}^2/\text{V-sec} \end{aligned}$$

Then substituting in Eq. (1-120) yields

$$f_o = \frac{(190)(6)}{(2)(\pi)(0.2 \cdot 10^{-3})^2(2.54)^2} \approx 0.7 \cdot 10^9 \text{ Hz}$$

From the above calculation, it is noted that the intrinsic cut-off frequency of the p-channel MOSFET can be in the GHz region. Currently, however, one rarely observes digital MOS circuits operating much above 10 MHz, which is a factor of $\approx$100 below the inherent capability of the device. It will show in future discussions (Chap. 4) that the present speed limitation of MOS digital circuitry is primarily caused by stray capacitance and the difficulty which the MOSFET experiences (because of its relatively low transconductance) in charging and discharging this capacitance.

1.9 SECOND-ORDER EFFECTS AND THE MOSFET CHARACTERISTICS

1.9.1 Back-gate Bias

The channel conductance of the MOSFET can be modulated by application of voltage V_{BG} to the substrate in a reverse-bias sense (Fig. 1.60). The effect is known as *back-gate bias* or *substrate bias,* and plays an important role in the operation of MOS/LSI circuitry (cf. Chaps. 4 and 9). The back-gate bias serves to uncover (i.e., ionize) additional dopant impurities in the depletion layer beneath the conducting channel. Electric field lines from the uncovered positively ionized n-type impurity centers in the substrate (p-channel device) terminate on the negatively charged field plate. If it is desired to keep the p-type inversion layer carrier density constant, and hence conduction constant, additional negative gate voltage must be applied in order to terminate the field lines generated by the additional depletion-layer charges. Back-gate bias thus affects the threshold voltage of Eq. (1-79) through the term Q_B. The Q_B expression given by Eq. (1-71) becomes, for n- or p-channel devices:

$$Q_B = \sqrt{2\epsilon_{\text{Si}}\epsilon_o q N_{\text{substrate}}(|2\phi_f| + V_{BG})} \tag{1-121}$$

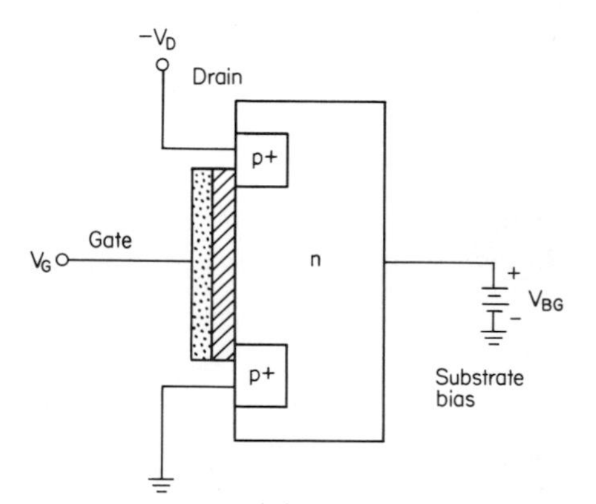

Fig. 1.60. Back-gate bias configuration.

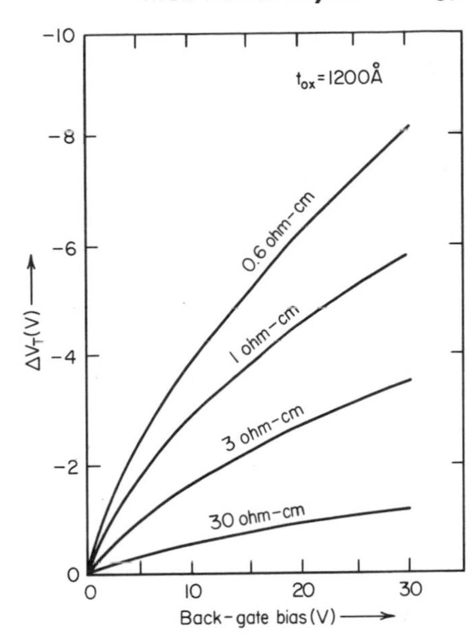

Fig. 1.61. Quantitative effect of back-gate bias on the threshold voltage of the p-channel enhancement-mode device.

The expression for threshold voltage is thus modified to

$$V_T = \Phi_{MS} \frac{-Q_{SS}}{C_o} + 2\phi_f - \frac{\sqrt{2\epsilon_{Si}\epsilon_o q N_{\text{substrate}}(|2\phi_f| + V_{BG})}}{C_o} \tag{1-122}$$

As back-gate bias increases in absolute value, the threshold voltage for the enhancement-mode device increases in absolute value, whereas the depletion-mode threshold decreases in absolute value. The quantitative effect of back-gate bias on the threshold voltage of p-channel enhancement-mode devices is summarized in Fig. 1.61.

1.9.2 Nonconstant Mobility

It was shown in Sec. 1.7 that surface carrier mobility is dependent on effective gate voltage ($V_G - V_T$). In developing the device equations in Secs. 1.6 and 1.8, voltage-dependent mobility values should be included. For example, empirical expressions can be written for mobility as a function of effective gate voltage from Fig. 1.57. The empirical expressions for mobility can then be substituted into Eq. (1-101), the integration performed, and resulting theoretical expression for the I-V characteristics obtained. In addition, the partial derivative of I_D with respect to V_G must be taken to analytically determine g_m. This in turn will involve the derivative of mobility with respect to gate voltage, since mobility is gate-voltage dependent. These calculations[32] when combined with additional corrections for nonconstant Q_B (to be discussed in the next section), result in a set of device equations which are in agreement with experimental dc characteristics to within $\approx \pm 10$ percent. If device modeling to within that accuracy is required, then nonconstant mobility must be taken into account. Fortunately most MOS/LSI designs can be treated analytically with device equations wherein constant mobility values typical of the given crystalline orientation are employed.

1.9.3 Consequences of the Assumption Q_B = Constant

In the development of Eqs. (1-104) and (1-108), it was assumed that the Q_B term was constant [Eq. (1-71)] and independent of position y along the direction of current flow in the channel (cf. Fig. 1.46). Actually, Q_B is a function of position y in the channel, since a voltage drop along the channel generated by current flow modifies the depletion-layer depth induced by the gate field-plate potential. The assumption of constant Q_B can lead to error in the analysis of MOS/LSI devices, particularly those operated in the saturation region. A detailed treatment of the effect has been performed[33,34] with the result that Eq. (1-104) for the triode region becomes:

$$I_D = \beta\left[(V_G - V_T)V_D - \frac{1}{2}V_D{}^2 - \frac{2}{3}\phi V_D{}^{3/2}\right] \tag{1-123}$$

where

$$\phi = \frac{(2q\epsilon_{\text{Si}}\epsilon_o N)^{1/2} t_{\text{Ox}}}{\epsilon_{\text{Ox}}\epsilon_o} \tag{1-124}$$

with N = donor (or acceptor) dopant density of substrate material per cm^3. Thus in the triode region for constant mobility

$$g_{sd} = \beta|V_G - V_T - V_D - \phi V_D{}^{1/2}| \tag{1-125}$$

$$g_m = \beta|V_D| \tag{1-126}$$

and in the saturation region for constant mobility

$$I_D = \frac{\beta(V_G - V_T')^2}{2} \tag{1-127}$$

where

$$V_T' = V_T + \frac{2}{3}\phi(V_G - V_T)^{1/2} \tag{1-128}$$

and

$$g_m = \beta\left|\left[V_G - V_T - \frac{2}{3}\phi(V_G - V_T)^{1/2}\right]\cdot\left[1 - \frac{\phi}{3(V_G - V_T)^{1/2}}\right]\right| \tag{1-129}$$

Examples of the practical use of these equations have been given by Greene and Soldano.[34] In summary, the effect has its greatest importance in the calculation of device characteristics in the saturation region.

1.9.4 Temperature Effects

It was noted in Sec. 1.7 that surface carrier mobility is temperature dependent [Eq. (1-109)]. Furthermore, threshold voltage is dependent on temperature through terms such as ϕ_f.[29] Since device characteristics intimately depend on mobility and threshold voltage values, we would expect that device characteristics would in turn be temperature dependent. Analysis and discussion of MOSFET I-V characteristics as a function of temperature are presented in Chapter 3.

REFERENCES

1. G. Koerber, "Properties of Solids," p. 141, Prentice-Hall, Inc., Englewood Cliffs, N.J., 1962.
2. A. B. Phillips, "Transistor Engineering," Chap. 3, McGraw-Hill Book Company, New York, 1962.

3. It is assumed that the reader has a working knowledge of the Fermi function and the Fermi level concepts. A brief review of the subject can be found in: S. N. Levine, "Principles of Solid-state Microelectronics," Chap. 2, Holt, Rinehart and Winston, Inc., New York, 1963.
4. J. Lindmayer and C. Y. Wrigley, "Fundamentals of Semiconductor Devices," p. 709, D. Van Nostrand Company, Inc., Princeton, N.J., 1965.
5. R. A. Smith, "Semiconductors," Chap. 4, McGraw-Hill Book Company, New York, 1959.
6. A. B. Phillips, "Transistor Engineering," p. 93, McGraw-Hill Book Company, New York, 1962.
7. E. M. Conwell, Properties of Silicon and Germanium, *Proc. IRE,* **46:** 1281(1958).
8. A. B. Phillips, "Transistor Engineering," p. 69, McGraw-Hill Book Company, New York, 1962.
9. W. W. Gartner, "Transistors: Principles Design and Applications," p. 46, D. Van Nostrand Company, Inc., Princeton, N.J., 1960.
10. F. J. Morin and J. P. Maita, Electrical Properties of Silicon Containing Arsenic and Boron, *Phys. Rev.,* **96:** 28(1954).
11. J. C. Irvin, Resistivity of Bulk Silicon and of Diffused Layers in Silicon, *Bell System Tech. J.,* **41:** 387(1962).
12. A. Nussbaum, "Semiconductor Device Physics," p. 86, Prentice-Hall, Inc., Englewood Cliffs, N.J., 1962.
13. A. B. Phillips, "Transistor Engineering," Chap. 3, McGraw-Hill Book Company, New York, 1962.
14. *Ibid.,* Chap. 4.
15. R. W. Warner, Jr. and J. N. Fordenwalt, "Integrated Circuits Design Principles and Fabrication," Chap. 2, McGraw-Hill Book Company, New York, 1965.
16. C. T. Sah, R. N. Noyce, and W. Shockley, Carrier Generation and Recombination in P-N Junctions and P-N Junction Characteristics, *Proc. IRE,* **45:** 1228(1957).
17. W. W. Ing, R. E. Morrison, L. L. Alt, and R. W. Aldrich, Gettering of Metallic Impurities from Planar Silicon Diodes, *J. Electro Chem. Soc.,* **110:** 533(1963).
18. A. Goetzberger and W. Shockley, Metal Precipitates in Silicon P-N Junctions, *J. Appl. Phys.,* **31:** 1821(1960).
19. S. L. Miller, "Effects of Avalanche Multiplication in Silicon Transistors," presented at AIEE-IRE Semiconductor Research Conference, Purdue University, June 1956.
20. After Warner and Fordenwalt. The squares are the data of Wilson and Pearson as reported by K. G. McKay, Avalanche Breakdown in Silicon, *Phys. Rev.,* **94:** 877(1954). The circles are the data of S. L. Miller, Ionization Rates for Holes and Electrons in Silicon, *Phys. Rev.,* **105:** 1246(1957).
21. O. Leistiko and A. S. Grove, Breakdown Voltage of Planar Silicon Junctions, *Solid State Electronics,* **9:** 847(1966).
22. H. K. J. Ihantola and J. L. Moll, Design Theory of a Surface Field-effect Transistor, *Solid State Electronics,* **7:** 423(1964).
23. B. E. Deal, E. H. Snow, and C. A. Mead, Barrier Energies in Metal-Silicon Dioxide-Silicon Structures, *J. Phys. Chem. Solids,* **27:** 1873–1879, December 1966.
24. B. E. Deal, M. Sklar, A. S. Grove, and E. H. Snow, Characteristics of Surface-state Charge (Q_{SS}) of Thermally Oxidized Silicon, *J. Electro Chem. Soc.,* **114:** 266(1967).
25. A. S. Grove, B. E. Deal, E. H. Snow, and C. T. Sah, Investigation of Thermally Oxidized Silicon Surfaces Using Metal-Oxide-Semiconductor Structures, *Solid State Electronics,* **8:** 145(1965).
26. D. Frohman-Bentchkowsky and A. S. Grove, Conductance of MOS Transistors in Saturation, *IEEE Trans. Electron Devices,* **ED-16:** 100(1969).

27. A. S. Grove, O. Leistiko, and W. W. Hooper, Effect of Surface Fields on the Breakdown Voltage of Planar Silicon P-N Junctions, *IEEE Trans. Electron Devices,* **ED-14:** 157, March 1967.
28. D. Colman, R. T. Bate, and J. P. Mize, Mobility Anisotropy and Piezoresistance in Silicon p-Type Inversion Layers, *J. Appl. Phys.,* **39:** 1923(1968).
29. L. Vadasz and A. S. Grove, Temperature Dependence of MOS Transistor Characteristics below Saturation, *IEEE Trans. Electron Devices,* **ED-13:** 863(1966).
30. E. Arnold and G. Abowitz, *Appl. Phys. Letters,* **9:** 344(1966).
31. C. T. Sah, Characteristics of the Metal-Oxide-Semiconductor Transistors, *IEEE Trans. Electron Devices,* **ED-11:** 324(1964).
32. J. P. Mize and D. Colman, "Fundamental Revisions to the MOSFET Device Equations," International Electron Devices Meeting, Washington, D.C., 1966.
33. H. K. J. Ihantola, Design Theory of a Surface Field-effect Transistor, *Tech. Rept.,* **1661-1,** Stanford Electronics Labs, Stanford, Calif., 1961.
34. R. Greene and T. Soldano, Increasing the Accuracy of MOS Calculations, *Proc. IEEE,* **53:** 1241(1965).

2

The MOS Technology Arsenal

2.1 INTRODUCTION

The purpose of this chapter is to describe and summarize the array of elements and processing methods which have in essence become an arsenal of MOS technology. Clarification of terms and concepts that define this assemblage of MOS technology, and which at times undoubtedly confound the system designer, will also be presented. Since the arsenal of MOS technology continues to expand with no end presently visible, the chapter also explores the motivating factors which are maintaining the drive toward further additions to the arsenal.

MOS integrated circuits were initiated in the mid-1960s with the simplistic technology of p-channel enhancement-mode device fabrication on (111) oriented silicon. The rapid proliferations of the technology which followed have at times been no less startling to the MOS technologist than they have been to the system designer. In retrospect, we can understand the ensuing course of technological events by realizing that the MOSFET is a surface-oriented device and by its very nature differs markedly from the bipolar transistor. A misleading factor which tended to deemphasize the profound differences between the two devices was that the basic silicon planar technology which had been developed for the discrete and integrated circuit bipolar transistor was adopted "in total" for the initial fabrication of MOS circuits.

It was not long, however, before the surface orientation features of the MOSFET stimulated an avalanche of technological development. For example, the important MOSFET parameters of threshold voltage and mobility were discovered to be dependent on the orientation of the silicon on which the devices were fabricated. In addition it was found that gate insulating dielectric materials other than silicon dioxide could be used to advantage in fabricating MOS circuitry. Furthermore, the gate field plate of the device when fabricated with conducting materials other than aluminum resulted in improved device performance for certain applications. (Recall that silicon dioxide and aluminum metalization are the major constituents of silicon planar processing for bipolar devices.) It was also discovered that methods for forming p-n junctions in silicon aside from that of diffusion proved effective in producing MOSFETs with unique characteristics. Essentially, the only facet of bipolar planar technology that survived was the semiconductor material on which the MOS circuits were constructed, i.e., *silicon!*

As it turned out, however, no single combination of the above technical trends has produced the optimal form of MOS/LSI. Therefore, when the combinational possibilities of silicon-surface orientations, gate dielectrics, gate field-plate materials, and impurity doping techniques are considered in overview, it is understandable how we have come to experience the establishment of an MOS technology arsenal.

The thrusts of MOS technology development have been directed at performance improvement of the MOSFET as an integrated-circuit element. All aspects of performance (electrical, economical, reliability, etc.) are encompassed in this context. To appreciate the directions which MOS technology innovations take, we must first understand the basic features of the MOSFET as a digital integrated-circuit element; these will be considered in the following section.

2.2 FEATURES OF THE MOSFET AS A DIGITAL INTEGRATED-CIRCUIT ELEMENT

Several of the more important properties of the MOSFET that make it attractive as an integrated-circuit element are summarized below.

1. Self isolation. Electrical isolation occurs naturally between MOS devices in integrated-circuit form since all p-n junctions are operated under reverse bias or zero bias in existing MOS circuit designs. Space-consuming isolation diffusions are therefore not required. *On* state conduction takes place only when the gate region is activated with the proper voltage (enhancement-mode device).
2. Normally *off* condition. The normally *off* condition of the enhancement-mode device is fully compatible with the functional properties of the device as a logic element. Logic circuitry becomes intolerably complex when implemented with normally *on* devices. (This is one reason why the JFET, a normally *on* device, is seldom if ever used in logic circuits.)
3. High input impedance. The dc input impedance of the typical LSI MOSFET is $>10^{14}$ Ohms. Essentially no input current is required to maintain the device in the *on* or *off* state. The only current required then is the displacement current, which serves to charge the gate capacitance and associated parasitics such as *gate-lead* capacitance in the process of turn *on* or turn *off*. The device therefore has a very high capability of dc fan-out to other MOS devices.
4. Inherent memory storage. Charge stored on the gate capacitor can be used to hold the enhancement-mode device in a conducting state. This "memory feature" is used to advantage in MOS/LSI circuit design where charge-storage intervals of milliseconds to years are utilized in accord with circuit design.
5. Bilateral symmetry. The source and drain regions of the device are in principle electrically interchangeable. The device can be used to charge or discharge (with equal effectiveness) the gate region of the following stage. This feature is of great utility in the design and operation of memory circuitry.
6. Active or passive operation. The MOSFET is an active three-terminal device exhibiting g_m. It can also function as a passive resistor with sheet resistance values of $\approx 10{,}000$ Ohms per square. Since the squares can be a fraction

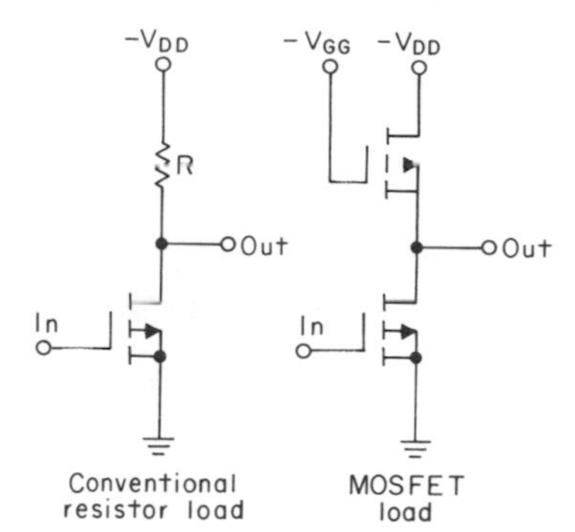

Fig. 2.1. The MOSFET as an active and passive element.

of a mil on a side, the resistance function can be highly compacted on an MOS/LSI chip. Load resistors for active components are thus rendered in the form of inactive MOS devices (Fig. 2.1).

7. Inherently high fabrication yield. MOS processing of standard p-channel enhancement-mode circuitry utilizes a minimum of one diffusion and four photomasking steps. The accompanying relatively low number of process steps as compared to bipolar integrated-circuit requirements, coupled with inherently small device size, results in high manufacturing yields for MOS/LSI circuitry of 100 to 200 equivalent-gate complexity (Fig. 2.2).
8. High functional complexity. As a consequence of a number of ameliorating factors such as yield, device size, and number of circuit elements required to implement a function, the number of effective gates per package pin is

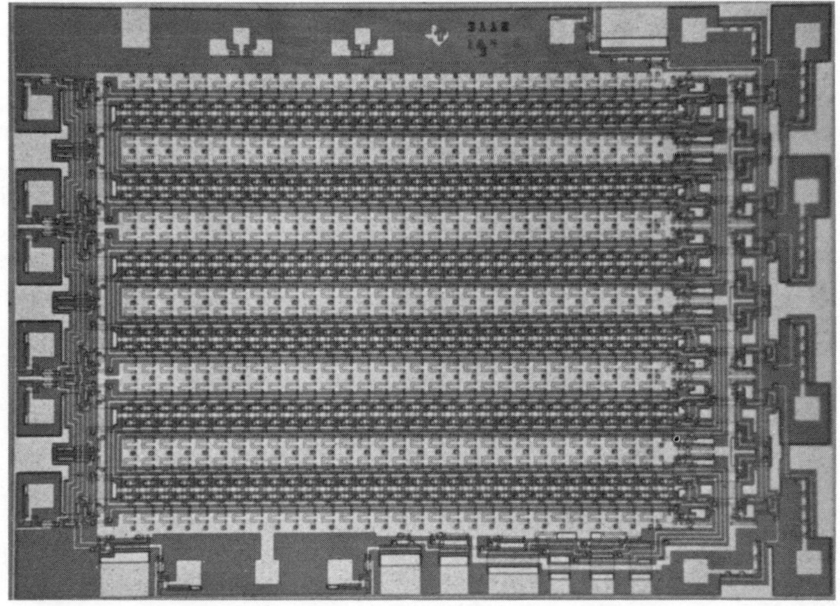

Fig. 2.2. MOS/LSI hex 32-bit accumulator.

large. MOS/LSI is thus said to have high functional complexity and this is favorably reflected in certain aspects of system reliability and economics discussed in Chaps. 3 and 10, respectively.

2.3 MOTIVATING FACTORS FOR FURTHER DEVELOPMENT OF MOS INTEGRATED-CIRCUIT TECHNOLOGY

The semiconductor industry had, as of 1970, fairly well explored and exploited the attractive features of the MOSFET as an integrated-circuit element. The circuit of Fig. 2.2 typified MOS/LSI at that time. These accomplishments, however, did not slow the advance of MOS circuit design and technology; they only seemed to stimulate additions to the arsenal! In view of this course of events it is worthwhile to define the motivating factors which continue to spur the drive toward further refinements of the technology. The factors can be briefly summarized as follows:

1. Ability to interface with TTL and DTL circuitry. A given system may require several MOS-bipolar interfaces. The cost effectiveness of the system design is reduced if interfacing components such as level shifters, auxiliary power supplies, and circuit buffering are required. Trends toward lowering p-channel MOS threshold values to the −1.5- to −2.0-V region will simplify the interfacing problem.
2. Speed/power product improvement. Although the power consumption of MOS circuits is lower than bipolar circuits, the faster switching speed of bipolar devices results in their overall speed/power figure of merit being at least an order of magnitude more favorable than that of MOS. The desire to improve this figure of merit has provided a stimulus to depart from the hitherto conventional circuit implementation with p-channel enhancement-mode devices.
3. Circuit innovation. The possibilities for circuit innovation seem to continually expand with the ensuing advances of technology and techniques. We refer here to developments which range from circuit design with complementary MOS to the growing variety of MOS/LSI memory configurations.
4. Economy. Continued efforts toward MOS/LSI process perfection result in improved manufacturing yields. This is in turn reflected in an improved economic situation for the system designer. In addition, the ever-present goal to shrink existing circuit size per logic function to achieve high functional density either by technology or circuit design innovations continues to spur further developments which result in both economic and reliability improvement of systems.

2.4 THE (111) STANDARD PROCESS (p-CHANNEL ENHANCEMENT MODE)

The (111) standard process is by definition the method utilized for fabricating p-channel enhancement-mode MOS/LSI on (111)-oriented silicon. The process sequence is shown in Fig. 2.3. This *thick-oxide* process was used in the introduction of MOS circuits in 1964 and it accounted for approximately 50 percent of all MOS/LSI manufactured in 1971. Silicon oriented on the (111) plane was chosen for initial MOS development since bipolar transistors and integrated circuits are fabricated on that orientation, and hence that particular material is readily available.

The process realizes many of the attractive features of the MOSFET as an integrated-circuit element described in Sec. 2.2. A notable by-product of the process is that two layers of interconnections (tunnels and crossovers) are available in the form of p-diffused regions and aluminum metalization, respectively, as shown in Fig. 2.3, step 10. The pertinent threshold values for the process can be readily calculated from the concepts developed in Chap. 1. Since numerous references to Eq. (1-79) will be made in this chapter, it is rewritten here for convenience.

$$V_T = \Phi_{MS} - \frac{Q_{SS}}{C_o} + 2\phi_f - \frac{Q_B}{C_o} \tag{1-79}$$

Then from Chap. 1, if

$$\begin{aligned} T &= 300°\text{K} \\ \Phi_{MS} &= -0.3 \text{ V} \\ t_{\text{Ox}} &= 1200 \text{ Å} \\ C_o &= 2.9 \cdot 10^{-8} \text{ F/cm}^2 \\ N_D &= 10^{15}/\text{cm}^3 \\ \phi_f &= -0.29 \text{ V} \\ Q_{SS}/q &= 5 \cdot 10^{11}/\text{cm}^2 \end{aligned}$$

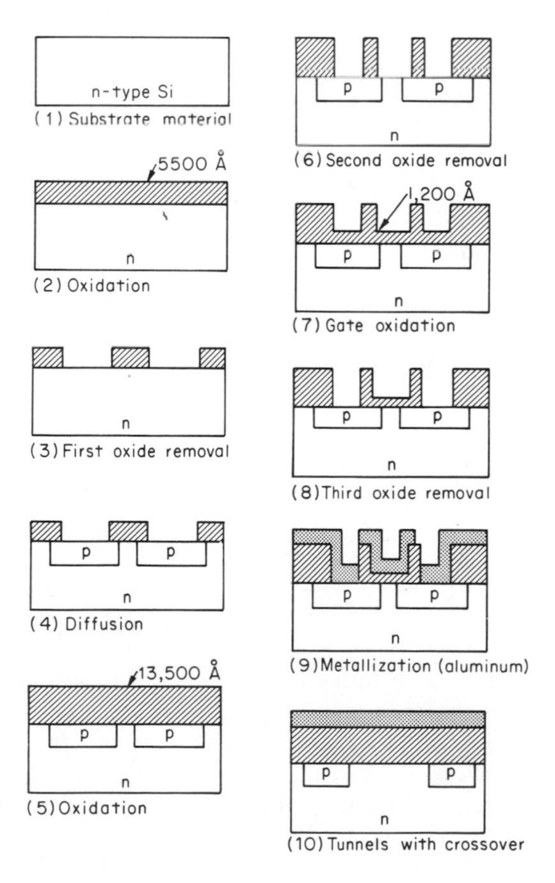

Fig. 2.3. Standard process sequence.

and

$$Q_B = \sqrt{2\epsilon_{Si}\epsilon_o q N_D 2|\phi_f|} = 14 \cdot 10^{-9}\ \text{Coul/cm}^2 \qquad (1\text{-}71)$$

the threshold voltage is, from Eq. (1-79),

$$\begin{aligned} V_T &= -0.30 - \frac{(5 \cdot 10^{11})(1.6 \cdot 10^{-19})}{2.9 \cdot 10^{-8}} - 0.58 - \frac{14 \cdot 10^{-9}}{2.9 \cdot 10^{-8}}\ \text{V} \qquad (2\text{-}1) \\ &= -0.30 - 2.75 - 0.58 - 0.48\ \text{V} \\ &= -4.1\ \text{V} \end{aligned}$$

To calculate the *field threshold* V_{TF}, the threshold voltage which results for aluminum interconnection leads over 13,500 Å thick oxide, Eq. (1-79) will again be employed. Note, however, that C_o now becomes $2.6 \cdot 10^{-9}$ F per cm^2 in accord with the 13,500-Å thick-oxide value. Thus:

$$\begin{aligned} V_{TF} &= -0.30 - \frac{(5 \cdot 10^{11})(1.6 \cdot 10^{-19})}{2.6 \cdot 10^{-9}} - 0.58 - \frac{14 \cdot 10^{-9}}{2.6 \cdot 10^{-9}}\ \text{V} \\ &= -0.30 - 30.8 - 0.58 - 5.7\ \text{V} \\ &\approx -37\ \text{V} \end{aligned}$$

It is to be pointed out that power supplies chosen for operation of the resulting MOS circuitry must be smaller in absolute value than the absolute value of V_{TF}. If this condition is not satisfied, the aluminum crossover leads will be reflected as inversion layers on the silicon surface, and the required isolation between integrated MOSFETs will be destroyed.

The above calculations involved very specific device fabrication parameters. In actual processing, these values vary; e.g., the tolerance on Q_{SS} can be as high as $\approx \pm 40$ percent. Therefore we will characterize the standard (111) process as

$$V_T = -3.0 \text{ to } -5.0\ \text{V} \qquad (2\text{-}2)$$

$$V_{TF} = -30 \text{ to } -50\ \text{V} \qquad (2\text{-}3)$$

Power-supply requirements for p-channel enhancement mode circuitry (Chap. 4) are

$$V_{DD} \approx 3\ V_T \qquad (2\text{-}4)$$

$$V_{GG} \approx 6\ V_T \qquad (2\text{-}5)$$

Then from Eqs. (2-1), (2-4), and (2-5),

$$\left.\begin{aligned} V_{DD} &= -12\ \text{V} \\ V_{GG} &= -24\ \text{V} \end{aligned}\right\} \quad \text{Power-supply requirements for (111) standard process MOS/LSI}$$

The most important motivation for modification of the (111) standard process is that the required power supplies are not TTL-compatible (Chap. 6). Hence, interfacing with TTL will require buffering. It is therefore necessary to develop a process which has inherently lower V_T values. To achieve this goal, efforts were directed toward fabricating MOS integrated circuits on the (100) orientation.

2.5 THE (100) PROCESS (p-CHANNEL ENHANCEMENT MODE)

The (100) process follows the same sequence shown in Fig. 2.3. The major advantage of the (100) orientation is that Q_{SS}/q values are the smallest in that particular plane. If we choose a Q_{SS}/q value of $9 \cdot 10^{10}/\text{cm}^2$, which is fairly typical for the (100) plane, then from Eq. (1-79) and the results of Sec. 2.4,

$$\begin{aligned} V_T &= -0.30 - \frac{(9 \cdot 10^{10})(1.6 \cdot 10^{-19})}{2.9 \cdot 10^{-8}} - 0.58 - 0.48 \text{ V} \\ &= -0.30 - 0.50 - 0.58 - 0.48 \text{ V} \\ &= -1.9 \text{ V} \end{aligned} \tag{2-6}$$

In anticipation of difficulties with the lower bound of V_{TF}, the thick-oxide value shown in Fig. 2.3 will be increased to 15,000 Å with accompanying C_o value of $2.3 \cdot 10^{-9}$ F/cm². Then from Eq. (1-79) and Sec. 2.4,

$$\begin{aligned} V_{TF} &= -0.30 - \frac{(9 \cdot 10^{10})(1.6 \cdot 10^{-19})}{2.3 \cdot 10^{-9}} - 0.58 - \frac{14 \cdot 10^{-9}}{2.3 \cdot 10^{-9}} \text{ V} \\ &= -0.30 - 6.3 - 0.58 - 6.1 \text{ V} \\ &\approx -13 \text{ V} \end{aligned}$$

Again, in actual fabrication of devices, the tolerances and resulting process parameters vary and the (100) process is characterized as

$$V_T = -1.5 \text{ to } -2.2 \text{ V} \tag{2-7}$$

$$V_{TF} = -10 \text{ to } -18 \text{ V} \tag{2-8}$$

and from Eqs. (2-4), (2-5), and (2-6),

$$\left.\begin{aligned} V_{DD} &= -6 \text{ V} \\ V_{GG} &= -12 \text{ V} \end{aligned}\right\} \quad \begin{aligned}&\text{Power-supply requirements} \\ &\text{for (100) standard process MOS/LSI}\end{aligned}$$

Advantages of the (100) process are therefore:

1. The low threshold voltage value makes the resulting circuitry compatible with TTL supply requirements.
2. Lower power dissipation is required in comparison to the (111) standard process because supply voltages are lower.

Disadvantages of the (100) process are:

1. The field threshold voltage V_{TF} is uncomfortably low. This introduces a constraint on the noise margin of the power supplies.
2. The (100) plane suffers lower hole mobility μ than that of the (111) plane; cf. Fig. 1.57. Lowered mobility results in lowered speed, since

$$T \approx RC \approx \frac{C}{g_m} = \frac{C}{(\mu\epsilon_{\text{ox}}\epsilon_o W/t_{\text{ox}}L)|V_G - V_T|}$$

Therefore a larger layout (increase in W/L values) is required (more silicon real estate is consumed) in raising circuit speed to that of the (111) case. (This is an ≈20 percent effect.)

2.6 THE NITRIDE PROCESS, (111) ORIENTATION (p-CHANNEL ENHANCEMENT MODE)

To simultaneously achieve TTL compatibility and exploit the high mobility values obtained on the (111) orientation, silicon nitride (Si_3N_4) with $\epsilon = 7.5$ is employed in conjunction with SiO_2 ($\epsilon = 4.0$) for gate dielectric formation. Thus if an initial SiO_2 layer $\approx$200 Å thick is placed on the silicon to preserve the Q_{SS} value of the Si-SiO_2 interface, and $\approx$800 Å of Si_3N_4 is placed on top of the SiO_2, an effective gate dielectric constant of 6.8 results for the configuration (Fig. 2.4) and $C_o = 6.0 \cdot 10^{-8}$ F/cm². From Eq. (1-79) and the results of Sec. 2.4,

$$V_T = -0.30 - \frac{(5 \cdot 10^{11})(1.6 \cdot 10^{-19})}{6.0 \cdot 10^{-8}} - 0.58 - \frac{14 \cdot 10^{-9}}{6.0 \cdot 10^{-8})} \text{ V}$$

$$= -0.30 - 1.33 - 0.58 - 0.23 \text{ V}$$

$$= -2.4 \text{ V} \tag{2-9}$$

V_{TF} is calculated assuming that 15,000 Å of SiO_2 constitutes the thick dielectric (the 800 Å of Si_3N_4 perturbs this only slightly), and thus

$$V_{TF} = -0.30 - \frac{(5 \cdot 10^{11})(1.6 \cdot 10^{-19})}{2.3 \cdot 10^{-9}} - 0.58 - \frac{14 \cdot 10^{-9}}{2.3 \cdot 10^{-9}} \text{ V}$$

$$= -0.30 - 35 - 0.58 - 6.1 \text{ V}$$

$$\approx -42 \text{ V}$$

Allowing for process variations, the *nitride process* on (111) silicon is characterized as

$$V_T = -1.9 \text{ to } -2.9 \text{ V} \tag{2-10}$$

$$V_{TF} = -30 \text{ to } -50 \text{ V} \tag{2-11}$$

and from Eqs. (2-4), (2-5), and (2-9),

$$\left.\begin{aligned} V_{DD} &= -7 \text{ V} \\ V_{GG} &= -15 \text{ V} \end{aligned}\right\} \quad \text{Power-supply requirements for the (111) nitride process MOS/LSI}$$

Advantages of the nitride process on (111) silicon are:

1. Direct interfacing to TTL is made possible.
2. V_{TF} is high.

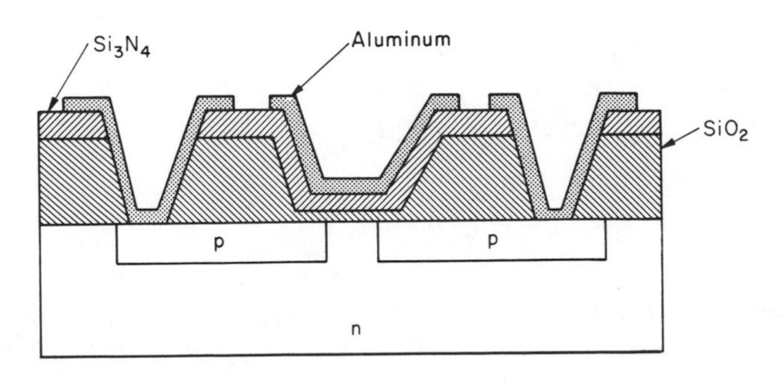

Fig. 2.4. Nitride-SiO_2 gate structure.

3. g_m per unit area is higher than (111) or (100) standard processes since the effective dielectric constant has been increased to $\approx$6.8. This in turn permits a reduction in size of circuit elements leading to higher packing densities and improved economy.
4. The high mobility of the (111) plane is enjoyed.
5. The junction and surface seal provided by the silicon nitride enhances circuit reliability.[1]

Disadvantage:

1. Processing steps are more involved in comparison to the standard process.

2.7 THE SILICON-GATE PROCESS, (111) ORIENTATION (p-CHANNEL ENHANCEMENT MODE)

A particularly important result is obtained if the conventional aluminum gate of the MOSFET is replaced with a silicon gate.[2] The doping of the silicon gate with boron gives the material fairly good conducting properties (sheet resistance <100 Ohms per square) and also provides a Φ_{MS} term which contributes in a favorable direction to lower threshold voltage values. Recall from the third example given in Sec. 1.4.3 of Chap. 1 that for a p-doped silicon gate ($N_A = 10^{15}/\text{cm}^3$), $\Phi_{MS} = 0.60$ V. Thus if a 1000-Å silicon dioxide gate dielectric ($C_o = 3.5 \cdot 10^{-8}$ F/cm^2) is employed with a silicon gate on (111) oriented n-type silicon ($N_D = 10^{15}/\text{cm}^3$), the threshold voltage will be, from Eq. (1-79) and the results of Sec. 2.4,

$$
\begin{aligned}
V_T &= +0.60 - \frac{(5 \cdot 10^{11})(1.6 \cdot 10^{-19})}{3.5 \cdot 10^{-8}} - 0.58 - \frac{14 \cdot 10^{-9}}{3.5 \cdot 10^{-8}} \text{ V} \\
&= +0.60 - 2.28 - 0.58 - 0.40 \text{ V} \\
&\approx -2.5 \text{ V} \qquad \text{(2-12)}
\end{aligned}
$$

Since the Φ_{MS} term is relatively innocuous in the calculation of V_{TF}, the resulting V_{TF} value will be essentially independent of whether silicon or aluminum interconnections are employed over thick oxide. Therefore, from Sec. 2.4,

$$V_{TF} \approx -37 \text{ V}$$

The silicon-gate process on (111)-oriented silicon will be characterized as

$$V_T = -1.5 \text{ to } -3.5 \text{ V} \qquad \text{(2-13)}$$

$$V_{TF} = -30 \text{ to } -50 \text{ V} \qquad \text{(2-14)}$$

From Eqs. (2-4), (2-5), and (2-12),

$$\left.\begin{aligned} V_{DD} &= -7.5 \text{ V} \\ V_{GG} &= -15 \text{ V} \end{aligned}\right\} \quad \text{Power-supply requirements for silicon-gate process, (111) MOS/LSI}$$

Significant advantages aside from low threshold voltage values result from the silicon-gate process. One of the most important of these is that the process is self-aligning, i.e., the gate field plate does not have to be aligned in a photomasking step to the source and drain regions. This feature inherently results from the

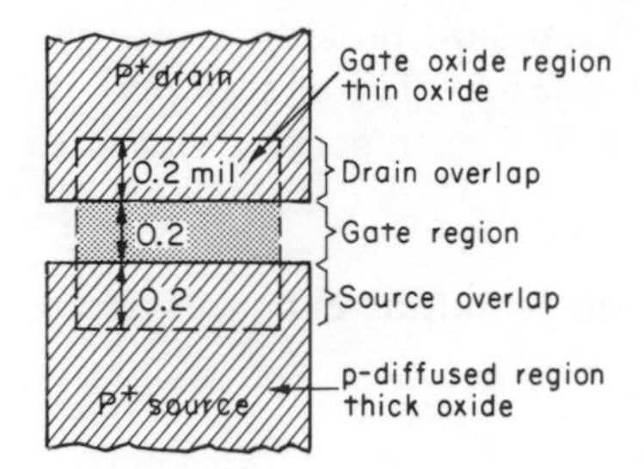

Fig. 2.5. Source, gate, and drain regions defined.

refractory properties of silicon. Thus the silicon gate can be used as an integral part of the mask that determines the position of the diffused source and drain regions.

The aluminum gate of the previously described processes does not have the refractory properties of silicon and will vaporize if exposed to temperatures characteristic of source and drain formation by diffusion ($T > 1000°C$). Since the aluminum gate region is formed after the source and drain regions are positioned by diffusion, a photomasking step (with accompanying alignment tolerance) is of necessity involved in positioning the aluminum gate with respect to the source and drain. An overlap of 0.2 mil is required to ensure, in the alignment process, that the metal gate covers the source-to-drain spacing (Fig. 2.5). In contrast, the silicon gate constitutes the upper surface of the mask which defines the source-drain spacing (i.e., channel length). The silicon gate is positioned *prior* to source-drain diffusion. A resulting gate-to-source or gate-to-drain overlap of only 0.07 mil results from the lateral diffusion of source and drain regions.

A comparison drawing of the aluminum-gate and silicon-gate structures is shown in Fig. 2.6.[3] Note that aluminum interconnections to source and drain regions are employed in both processes. The silicon-gate method can thereby provide three levels of interconnection, with the constraint that a p-diffused tunnel cannot run under a silicon gate lead since the silicon gate lead is an integral part of the p-diffusion mask.

In summary, the attendant advantages of the silicon gate process are:

1. The resulting low threshold voltage permits direct interface to TTL.
2. V_{TF} is high.
3. The high mobility of the (111) plane is enjoyed.

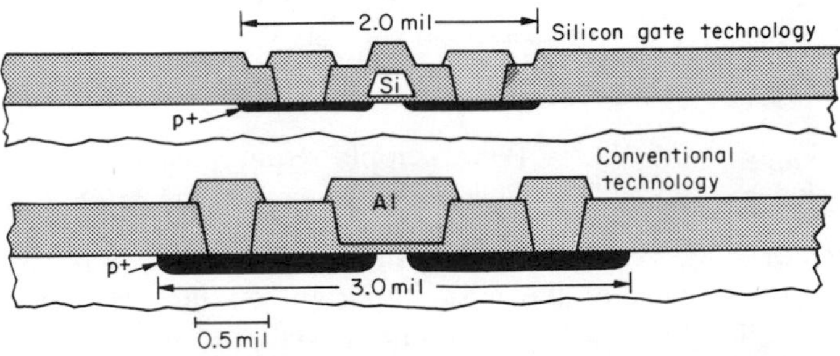

Fig. 2.6. Silicon gate versus aluminum gate structure.[3]

4. The critical alignment of the gate field plate in a photomasking step is eliminated. This implies improved economy in processing.
5. Gate field plate overlap of source and drain regions is reduced by $\approx$50 percent in comparison to previously described processes. Overlap is $\approx$0.07 mil, and this provides a circuit speed improvement (Chap. 4) of as much as 50 percent in comparison to the (111) standard process.
6. Three levels of interconnection result, thereby reducing chip size by as much as 50 percent in comparison to the (111) standard process.

The disadvantage is that processing is more involved than that of the standard (111) process. (But the process lies directly in the mainstream of well-developed silicon technology!)

2.8 THE METHOD OF ION IMPLANTATION

Ion implantation is a technique for doping a silicon wafer with n- or p-type impurities. It is inherently a low-temperature process and offers excellent control of impurities over the concentration range of 10^{15} to 10^{20} per cm^3. The method as it pertains to MOS/LSI involves surprisingly small perturbations to the standard process of Fig. 2.3. It is presently used to achieve two physical effects in MOS/LSI which differ considerably from one another. They are: (1) self-aligned gate, and (2) channel doping. The ion implantation techniques used to achieve these two effects will be discussed separately.

2.8.1 Self-aligned Gate with Ion Implantation

In the process of ion implantation, the kinetic energy of the ions is accurately controlled by an ion accelerator and hence the range of the ions in a given material can be precisely adjusted. Thus if boron ions are accelerated to an energy of approximately 80 kilo electron Volts (KeV), the aluminum gate or thick oxide shown in Fig. 2.7 will serve to stop the incident ion beam. If only a thin oxide (1200 Å) is positioned over the remaining portion of the silicon surface as shown in Fig. 2.7, the accelerated boron ions will penetrate this oxide, be deposited in the silicon surface

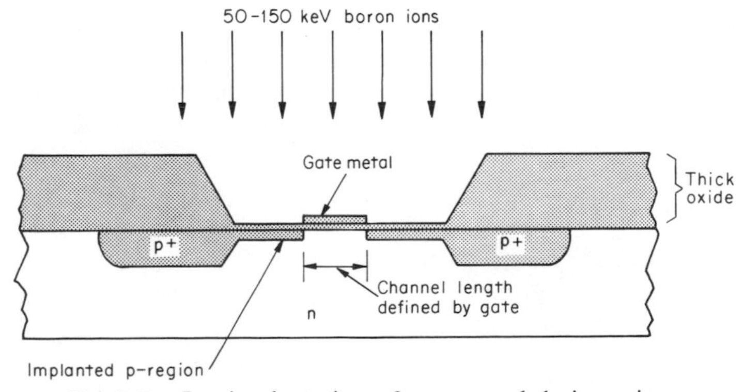

Fig. 2.7. Ion implantation of source and drain regions.

under the oxide, and thereby extend the source and drain regions to the gate periphery. The gate field plate has served to mask the ions from the gate region and the 1200-Å oxide will provide passivation for the newly formed p-n junction region. The boron ion bombardment of $\approx 10^{14}/\text{cm}^2$ is followed by a 400°–500°C anneal for approximately 30 minutes to remove radiation damage to the oxide and semiconductor by the ion bombardment.

Boron diffusion at these time and temperature values is negligible and hence the overlap of the source-drain regions by the metal field plate remains essentially zero. (Overlap is sufficient, however, to ensure gate activation of the entire channel region between source and drain.) Essentially perfect alignment of the source and drain regions to the gate and its field plate result. The gate-to-drain feedback capacitance is lowered by a factor of ≈ 40 in comparison to the standard process (Sec. 2.4), and hence the Miller capacitance (Chap. 9) is decreased and the switching speed (Chap. 4) is increased.

Advantages gained by this method are:

1. The critical gate field-plate step has been replaced by a near-perfect self-aligned gate technique.
2. Essentially zero overlap capacitance results. This in turn enhances circuit speed by a factor of 30 to 40 percent over a non-self-aligned gate process.

Disadvantages are:

1. The technique requires a sophisticated and expensive ion accelerator.
2. The method has not as yet been made adaptable to multiwafer batch processing. Therefore the accompanying economic advantages of batch processing have not been enjoyed.

2.8.2 Channel Doping by Ion Implantation

Ion implantation can be used to advantage in doping the MOSFET channel with boron atoms. The technique again employs only minor modifications of the standard process (Fig. 2.3). With boron-ion energy, in this case ≈ 50 KeV, the range of the ions will again be such that they will penetrate the 1200-Å gate oxide (Fig. 2.8) and stop in the silicon surface region. Because of ion straggling in-the-range, an ≈ 0.2-micron-thick layer of boron ions can be obtained at the silicon surface. Annealing at several hundred degrees centigrade will repair radiation damage introduced in the channel and gate oxide regions by the bombardment.

The utility of the method is demonstrated in the following calculation. Suppose

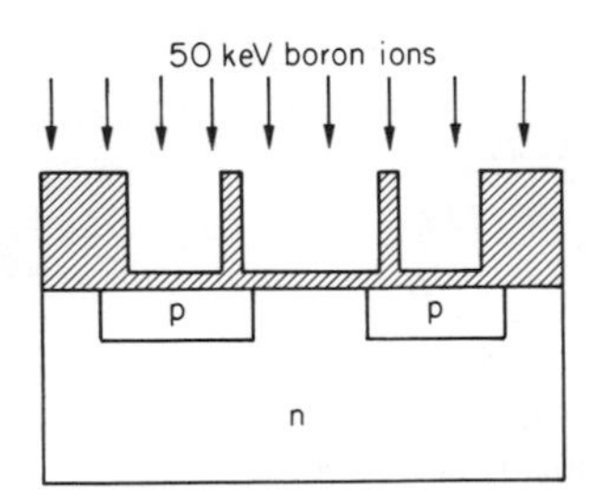

Fig. 2.8. Channel doping by ion implantation.

that $5 \cdot 10^{16}$ boron atoms per cm^3 are implanted in the gate region of a p-channel MOSFET (Fig. 2.8) which is fabricated on (111)-oriented n-type material with $N_D = 10^{15}/\text{cm}^3$. The resulting compensation of the channel region, assuming each implanted atom is electrically active, will be $N_A = 4.9 \cdot 10^{16}/\text{cm}^3$ and the channel will be p-type. With the channel assumed to be uniformly implanted to a depth $\approx$0.2 micron, only $\approx 10^{12}$ boron atoms per cm^2 are involved. Thus if the accelerator has a current density of $10^{-2}\ \mu\text{A/cm}^2$ of singly charged boron ions, a bombardment time of approximately 16 sec will be required for the implant since:

$$\text{Number of implanted atoms/cm}^2 = \left(\frac{\text{beam current density}}{q}\right)\text{time}$$

$$10^{12} = \frac{10^{-8}}{1.6 \cdot 10^{-19}}\,t$$

$$t = 16 \text{ sec}$$

To calculate the resulting threshold value for a p-doped channel of concentration $4.9 \cdot 10^{16}/\text{cm}^3$ and $T = 300°\text{K}$, note that

$$\phi_f = +0.39 \text{ V}$$
$$\Phi_{MS} = -0.99 \text{ V}$$
$$Q_B = -11.5 \cdot 10^{-8} \text{ Coul/cm}^2$$

and for (111)-oriented silicon with $Q_{SS} = 5 \cdot 10^{11}/\text{cm}^2$ we obtain from Eqs. (1-79) and (2-4)

$$\begin{aligned} V_T &= -0.99 - \frac{(5 \cdot 10^{11})(1.6 \cdot 10^{-19})}{2.9 \cdot 10^{-8}} + 2(0.39) + \frac{11.5 \cdot 10^{-8}}{2.9 \cdot 10^{-8}} \text{ V} \\ &= -0.99 - 2.75 + 0.78 + 3.97 \text{ V} \\ &= +1.0 \text{ V} \end{aligned} \tag{2-15}$$

The threshold voltage of this p-channel device is positive, thereby resulting in depletion-mode operation.

It is evident from the above calculation that by adjusting the beam current value or bombardment time, threshold voltage values ranging from low threshold enhancement to arbitrary depletion operation can be readily obtained. Furthermore, with the addition of only one photomasking step for selective gate shielding with photoresist, enhancement- and depletion-mode p-channel MOSFETs can be formed on the same chip! The enhancement- and depletion-load inverter configurations with their resulting load lines are shown in Fig. 2.9. Resulting advantages of the depletion-load inverter are significant and will receive detailed attention in Chap. 4.

Advantages offered by channel doping with ion implantation are:

1. Threshold voltage can be adjusted to accommodate direct interfacing to TTL circuitry.
2. A high ratio of field threshold (V_{TF}) to device threshold (V_T) is obtained.
3. Depletion-mode-load/enhancement-mode-driver configurations are obtainable on a chip. Advantages derived from this circuit combination are:

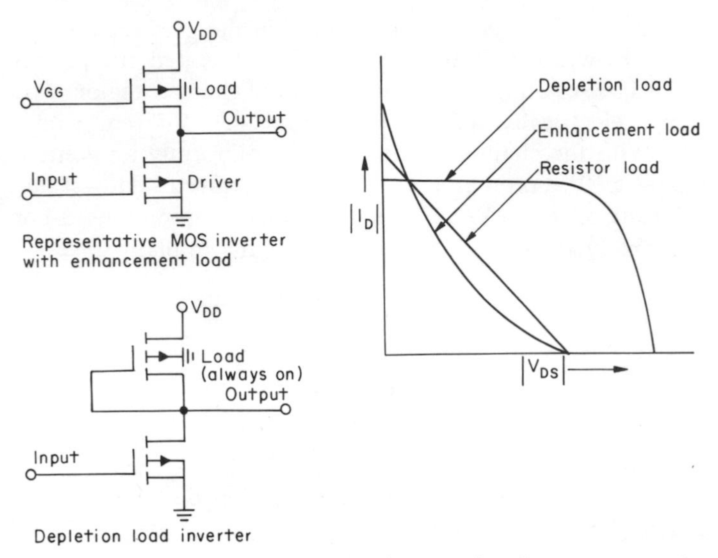

Fig. 2.9. Comparison of depletion- and enhancement-mode load devices and load lines.

a. Circuit operation from a *single* 5-V supply can be realized.
b. The near-ideal load line of the depletion-mode load device results in reduction of the size of the load device in comparison to that of the enhancement-load circuit, thereby providing increased circuit density.
c. Since the gate-to-source voltage of the depletion-mode load remains fixed during the switching transient, the switching speed and speed/power product are thereby improved over the enhancement-mode load circuit.

Disadvantages of the method are again primarily economical, in that an expensive ion accelerator is required and the technique is difficult to adapt to multiwafer batch processing.

2.9 MOS-BIPOLAR ON THE CHIP

By the addition of a phosphorus diffusion and accompanying photomask step inserted in the appropriate sequence of any of the p-channel processes discussed to this point, NPN bipolar transistors can be formed on the MOS/LSI chip. The n^+ emitter of the bipolar transistor is thereby readily positioned in selected p-diffused regions. The remaining p-diffused regions of course serve as source-drain structures for the MOS circuitry. The resulting vertical NPN configuration is shown in Fig. 2.10.

It will be demonstrated in Chap. 9 that the transconductance g_m per unit area of the bipolar transistor is at least two orders of magnitude greater than that of the MOSFET. The integrated bipolar device can thereby effectively supplement an MOS/LSI chip in sinking or driving current from or to external circuitry with

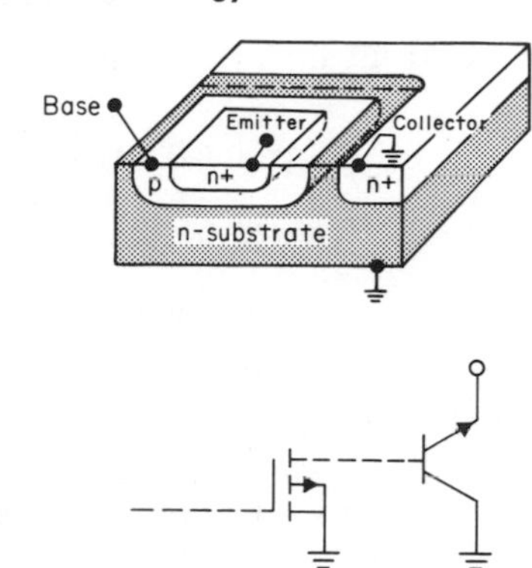

Fig. 2.10. Emitter-follower bipolar transistor on MOS chip.

which the MOS/LSI chip must interface. Note from Fig. 2.10 that device polarity and required electrical isolation constrain the integrated bipolar transistor to emitter-follower operation. This mode of bipolar operation proves, however, to be quite effective in interfacing with external circuitry. The capability of the bipolar device is therefore fully realized and conservation of MOS/LSI chip area is achieved. High circuit density is thus further enhanced with this technique.

2.10 n-CHANNEL PROCESS

One of the enticing features of n-channel MOS circuitry is that electron mobility on (100)-oriented n-type inversion layers is almost three times as large as is hole mobility on (111)-oriented p-type inversion layers. This results in faster circuits for the same chip layout in comparison to standard p-channel circuitry. The n-channel MOSFET can be constructed as shown in Fig. 2.3 by exchanging p regions for n regions and *vice versa*. The resulting structure (Fig. 2.11) would be a depletion-mode, rather than an enhancement-mode, device as evidenced both in experiment and by the following calculation. Given:

$$T = 300°\text{K}$$
$$\text{Substrate doping } N_A = 10^{15}/\text{cm}^3$$
$$\phi_f = +0.29 \text{ V}$$
$$\Phi_{MS} = -0.90 \text{ V (second example of Sec. 1.4.3 of Chap. 1)}$$
$$Q_{SS}/q = 9 \cdot 10^{10}/\text{cm}^2 \text{ for (100) orientation}$$
$$t_{\text{ox}} = 1200 \text{ Å}$$
$$C_o = 2.9 \cdot 10^{-8} \text{ F/cm}^2$$
$$Q_B = -14 \cdot 10^{-9} \text{ Coul/cm}^2$$

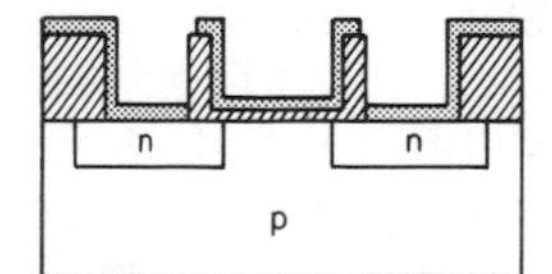

Fig. 2.11. Structure of the n-channel MOSFET.

Then from Eqs. (1-79) and (2-4)

$$\begin{aligned} V_T &= -0.90 - \frac{(9 \cdot 10^{10})(1.6 \cdot 10^{-19})}{2.9 \cdot 10^{-8}} + 0.58 + \frac{14 \cdot 10^{-9}}{2.9 \cdot 10^{-8}} \text{ V} \\ &= -0.90 - 0.50 + 0.58 + 0.48 \text{ V} \\ &= -0.34 \text{ V} \end{aligned} \tag{2-16}$$

The negative value of threshold voltage is indicative of depletion-mode operation for n-channel devices.

As stated earlier in this chapter, one of the basic requirements for a logic device is that it be normally *off*, which implies enhancement-mode operation for the MOSFET. The fact that the conventionally fabricated n-channel device is normally *on* points up one of the basic problems that has faced n-channel MOS/LSI. Aggravation of the problem is caused by the sign and magnitude of Q_{SS} being such that a surface inversion layer is generated under the oxide, resulting in all devices being shorted together. This problem can be corrected by appropriate diffusion of p-type "channel stoppers," but this, of course, contributes to process complexity.

Several procedures can be used to convert the normally depletion-mode devices to enhancement-mode operation. The more important methods are summarized as follows:

1. Substrate bias. Eq. (1-122) indicates that back-gate biasing the p substrate by a few volts will result in threshold shift to the enhancement-mode regime. But the additional power supply and associated circuitry required for substrate biasing are generally unappealing to the system designer.
2. Substrate doping. The surface of the p-type substrate can be doped more heavily p-type with the result that the ϕ_f and Q_B/C_o terms of Eq. (1-79) dominate and V_T becomes positive. The additional doping, however, degrades the electron mobility in the inversion layer to an extent that generally precludes this approach.
3. Silicon gate. If a p-type silicon gate is used to replace the conventional aluminum gate on the n-channel device (Fig. 2.11), the Φ_{MS} value becomes from Eq. (1-75)

$$\Phi_{MS} = 4.1 - (3.25 + 0.55 + 0.29) \approx 0 \text{ V}$$

Then from Eqs. (1-79), and (2-16) and Sec. 2.4

$$\begin{aligned} V_T &= 0 - 0.50 + 0.58 + 0.48 \text{ V} \\ &= +0.56 \text{ V} \end{aligned} \tag{2-17}$$

and the n-channel device operates in the enhancement mode! The silicon-gate technique thus appears attractive for n-channel MOS/LSI. Note that the high electron mobility has been maintained, the resulting threshold voltage is low thereby indicating TTL compatibility, and the advantageous features of the self-aligned gate will be enjoyed. It would appear that this method with accompanying innovations (one, perhaps, being ion implantation for "channel stopping") will receive close scrutiny during the next few years.

2.11 COMPLEMENTARY MOS

The complementary MOS (C/MOS) process combines p- and n-channel enhancement-mode transistors on the same chip. A cross-sectional view of the structure is depicted in Fig. 2.12*a*. The unit cell circuit formed with a p-channel device as the load is shown in Fig. 2.12*b*. It is to be noted that only one power supply is required and the circuit is operated so that one transistor is always on. Significant power is dissipated only during the change of state. The arrangement provides high-speed operation and low standby power dissipation in the nanowatt range. The circuitry exhibits speeds comparable to many bipolar devices (about 10 to 20 MHz) and offers the ultimate in speed/power product for MOS/LSI. C/MOS thus serves as an ideal solution for specific applications that cannot be handled by the more conventional MOS technologies.

C/MOS fabrication poses a problem in that the addition of n-channel transistors and the need to isolate them electrically from p-channel transistors requires two extra photomasking steps. The entire array of n- and p-channel technologies discussed in this chapter can be utilized for threshold adjustment of both active devices used in C/MOS. For example, C/MOS technology appears fully compatible with ion-implantation and silicon-gate technology.

C/MOS circuitry offers:

1. The best speed/power product available for existing MOS/LSI.
2. Very low static power dissipation, typically a few microwatts for 100 gates.
3. Full TTL compatibility.

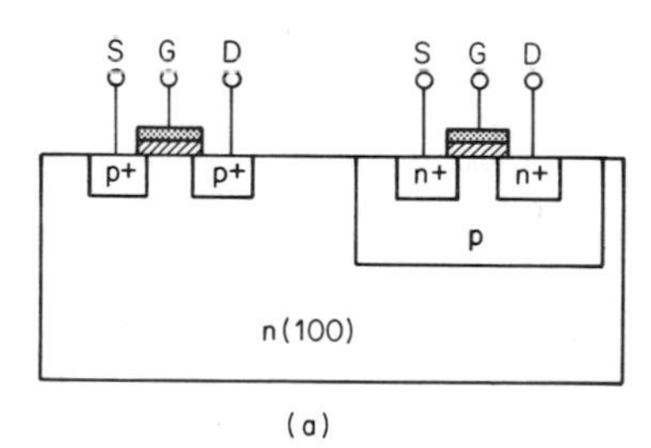

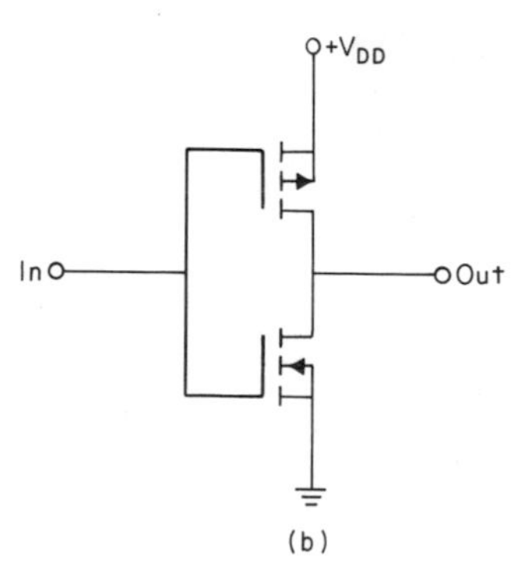

Fig. 2.12. (*a*) CMOS structure; (*b*) CMOS cell.

4. Single power-supply operation. $|V_{DD}| >$ higher of the two absolute-valued threshold voltages, $|V_T|$.
5. High noise immunity ($\approx$40 percent of supply voltage).

Disadvantages of the method are:

1. Two additional diffusions are required in comparison to the standard process. Process sophistication for adjustment of threshold voltages also contributes to circuit costs.
2. Circuit density is decreased because of the isolation or channel separation required to preserve n- and p-channel devices as separate entities.

2.12 SILICON ON INSULATOR

Single-crystalline silicon (required for all devices discussed in this book) can be grown epitaxially on single-crystalline materials such as spinel or sapphire. These latter materials are insulators and their thermal expansion coefficients and crystalline properties match silicon to the extent that satisfactory epitaxial silicon growth can take place. With resulting silicon layers of a few microns in thickness coupled with appropriate photomasking steps for diffusion and etching of silicon islands, the structure pictured in Fig. 2.13 can be obtained.

This technology is referred to as *silicon on insulator* and can be used for single-channel MOS/LSI or complementary MOS/LSI as shown in Fig. 2.13. It is to be noted that source and drain capacitance values of the elements in Fig. 2.13 are greatly reduced in comparison to those of Fig. 2.12*a*, since p-n junction interfaces have been eliminated in the horizontal plane of Fig. 2.13. In addition, stray capacitance associated with metal interconnections is negligible for silicon on insulator circuitry since the interconnections overlap the several-mil-thick insulator (spinel or sapphire). Every order of magnitude reduction of parasitic capacitance is essentially accompanied by an order of magnitude improvement in speed/power product. C/MOS digital circuits fabricated on insulator have thereby exhibited satisfactory operation to 75 MHz.

The reduced parasitic capacitance realized when using silicon on insulator offers an order of magnitude increase in circuit speed. The technique when used for C/MOS circuitry provides the ultimate speed/power product for MOS/LSI. But this promising performance is bought at the cost of attendant process complexity. In addition, the materials cost is appreciable. The result to date is that applications of this sophisticated technology have not been cost effective . . . but motivations to improve this situation are strong, indeed.

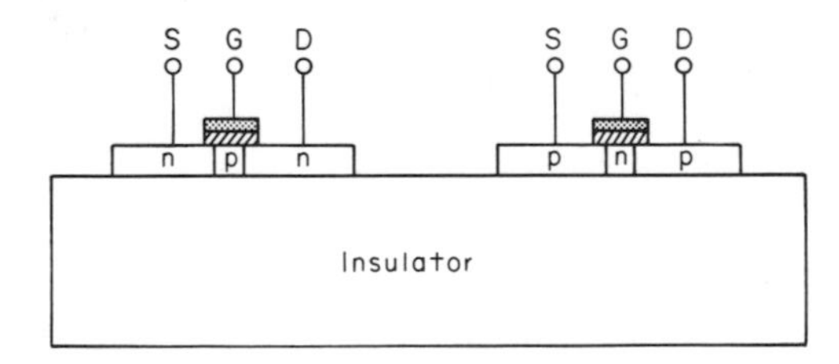

Fig. 2.13. Silicon on insulator (SOI) configuration.

2.13 ADDITIONAL TRENDS IN MOS TECHNOLOGY

2.13.1 The MOSFET as a Memory Storage Element

Conventional MOS read-only memory circuits (ROMs) as discussed in Chap. 7 are programed permanently at the gate-oxide photomasking step of the fabrication process. If a program change is required, the gate-oxide mask must be redrawn and a new photomask generated. Batch processing of silicon wafers is then resumed at the gate-oxide step of the process. This approach to programing the ROM is both expensive and time consuming.

Flexibility in programing the ROM has been obtained by *fusing* techniques, wherein an applied electrical pulse permanently and irreversibly destroys selected interconnection lines and leaves those connections intact that provide the desired program. This is a one-shot method and the ROM cannot be reprogramed. In addition the multi-interconnection option consumes chip area, thereby impeding achievement of high circuit density.

Another method for alterably programing a ROM relies on charge storage in a portion of the gate dielectric of the MOSFET. Examples of this approach are the metal-nitride-oxide-silicon (MNOS) memory[4] and the metal-aluminum oxide-silicon (MAS) memory.[5] Memory storage is achieved in both these methods by electron injection and trapping in regions of these sandwiched gate dielectrics by suitably applied gate voltages. The MNOS cell can then be reprogramed nondestructively by proper polarity of gate voltage application to the MNOS structure, or by exposure of the MAS structure to X-radiation. Difficulties to date in controlling the electrical characteristics of the storage dielectrics have thus far limited the application of these two approaches.

As a possible improvement to the above methods, a floating gate avalanche injection MOS (FAMOS) structure having particularly appealing memory storage features has been introduced.[6] The device can be reprogramed nondestructively in a physically well-behaved manner. The FAMOS structure takes the form of a p-channel silicon-gate MOS device as shown in Fig. 2.14. The cell is programed by charge transport to the floating gate in the process of avalanche injection of electrons from source-substrate or drain-substrate junctions. The electrically isolated

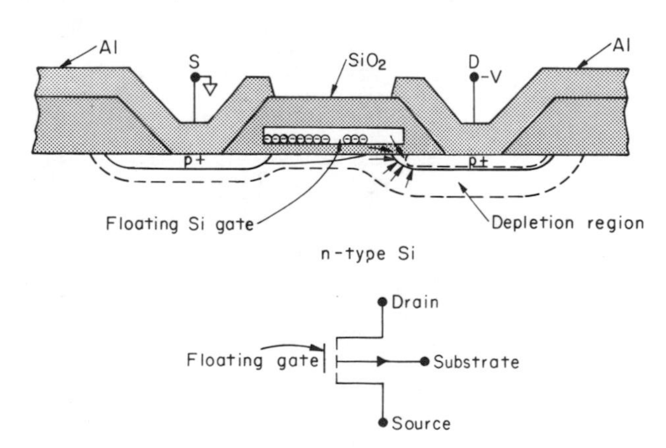

Fig. 2.14. The floating gate FAMOS structure.[6]

gate traps the charge, resulting in a negative gate potential of sufficient magnitude to form a p-type inversion layer between source and drain. The presence or absence of gate charge (logic *0* or *1*) is then sensed by measuring device conductance. The amount of charge transported to the floating gate is a function of the amplitude and duration of avalanche voltage applied to the junction.

To program a FAMOS ROM, x- and y-select lines are simultaneously addressed with a suitable avalanche voltage pulse. It is reported that the *on*-state retention time is greater than 10 years at 125°C.[6] Erasure of the gate charge in preparation for reprograming is accomplished by directing ultraviolet light onto a transparently packaged device. Discharge takes place by photocurrent flow from floating gate back to the silicon substrate. X-radiation can also be used for erasure of the gate charge. The method apparently holds great promise in realizing erasable ROMs that can be reprogramed on command of the system designer.

2.13.2 MOS Approach to Junctionless Devices

It was pointed out at the beginning of this chapter that the motivating factor of circuit economy has stimulated additions to the MOS technology arsenal. This trend is aptly demonstrated in the development of charge-coupled devices (CCDs). These devices require no pre-prepared p-n junctions. Arrays of CCDs consist of a silicon substrate covered by a layer of SiO_2, which in turn is capped with a row of closely spaced (<0.1 mil) aluminum-metal field plates (Fig. 2.15). If all plates are held at a small negative potential and the substrate is n type, holes can be stored in the resulting surface depletion regions (potential wells). If, then, a more negative pulse is applied to a field plate adjacent to one under which charge is stored, the charge will spill over into the deeper potential well (see arrow of Fig. 2.15). In this way, charge can be shifted along the MOS chain. The charge can be read out at the end of the chain by detection of capacitance change in the terminating element. Charge or absence of charge, can be introduced selectively as *1*s or *0*s, respectively, at the input of the chain by radiation-generated electron-hole pairs. Several other methods of charge generation and detection are also possible.

CCDs can provide memory, shift register, delay line, and imaging functions. Circuit density exceeds present forms of MOS/LSI and is greater than 200,000 elements per square inch. If a method of serial input and parallel-mode output

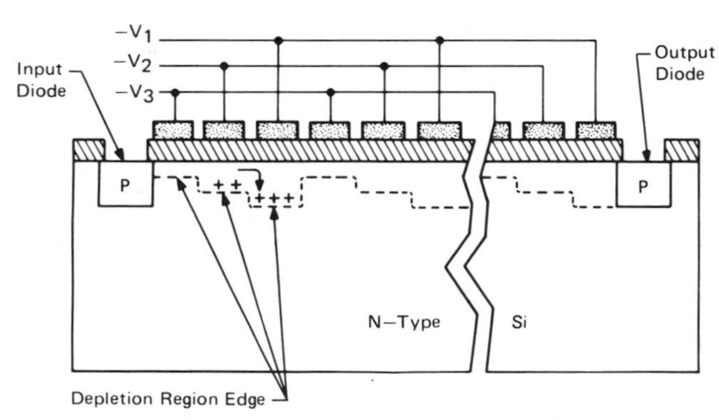

Fig. 2.15. Junctionless MOS charge-coupled-device array.

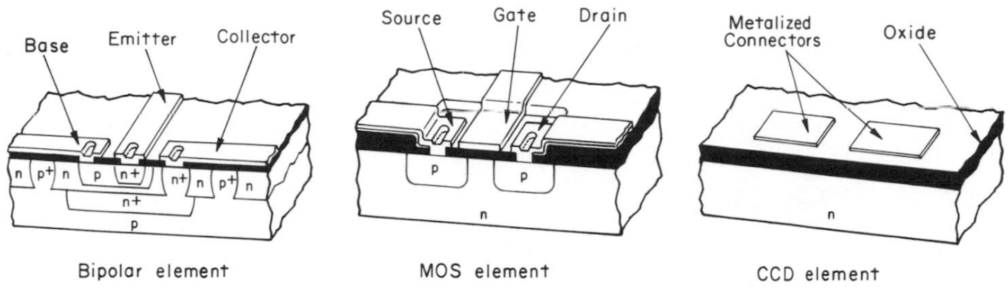

Fig. 2.16. Comparison of bipolar, MOSFET, and charge-coupled devices in cross-sectional view.[7]

signal extraction can be developed, the resulting technique will be of great utility in applications to signal processing such as signal correlation, compression, encoding, and decoding.

The structure of a CCD is compared to MOS and bipolar configurations in Fig. 2.16.[7] The trend is obvious—simplified processing, potentially high yields, and increased circuit density, all resulting in improved economics. These encouraging results further sustain the drive toward new and possibly surprising additions to the MOS technology arsenal.

REFERENCES

1. N. C. Tombs, H. A. R. Wegener, R. Newman, B. T. Kenney, and A. J. Coppola, A New Insulated Gate Silicon Transistor, *Proc. IEEE* (Corresp.), **54:** 87, January 1966.
2. L. L. Vadasz, A. S. Grove, T. A. Rowe, and G. E. Moore, Silicon Gate Technology, *IEEE Spectrum,* **6:** 28 (1969).
3. F. Faggin and T. Klein, A Faster Generation of MOS Devices with Low Thresholds, *Electronics,* **42:** 88, September 29, 1969.
4. D. Froman—Bentchkowsky, An Integrated Metal-Nitride-Oxide-Silicon (MNOS) Memory, *Proc. IEEE* (Letters), **57:** 1, 190, June 1969.
5. S. Nakanuma, T. Tsujide, R. Igarachi, K. Onoda, T. Wada, and M. Makagiri, A Read-only Memory Using MAS Transistors, *ISSCC Digest of Technical Papers,* 68, February 1970.
6. D. Froman—Bentchkowsky, ROM Can Be Electrically Programed and Reprogramed and Reprogramed. . . , *Electronics,* **44:** 91, May 10, 1971.
7. L. Altman, New MOS Technique Points Way to Junctionless Devices, *Electronics,* **43:** 112, May 11, 1970.

3

Reliability Aspects of MOS Integrated Circuits

3.1 INTRODUCTION

The discussion of MOS/LSI reliability is probably best undertaken by first examining the physical characteristics of the devices and then mapping a rational approach to establishing reliability. A physical examination of the product can be made by referring to a photograph of a modern MOS/LSI chip. Fig. 3.1 pictures a 1,024-bit four-phase dynamic shift register which employs $\approx$6,000 MOSFETs on a 150 $\times$ 150 mil chip. As we view this photograph, our question pertaining to reliability takes on a more practical meaning if we ask "after testing and packaging this circuit, will it continue to function satisfactorily, and how long will it remain operable?" It is evident from Fig. 3.1 that, should failure occur, we will not be able to repair the circuit by soldering-in replacement elements or mending failed interconnections, even with the modern techniques of microstitch bonding!

What techniques then are available for establishing reliability assurance? One possibility would be to utilize redundant circuitry on the MOS/LSI chip. For example, information processing by highly parallel redundant methods is effectively employed in the brain, where malfunctioning neurons are bypassed. We are not, however, directing our efforts toward redundancy techniques in integrated circuits at the present time. Circuit redundancy is not included in the design arsenal of MOS/LSI.

As a consequence of the complexity of MOS/LSI, reliability comparisons of one circuit to another cannot be correlated as easily as the corresponding situation was in the discrete transistor era. Furthermore, failure analysis and corrective action in circuitry of the type shown in Fig. 3.1 can be most difficult, in that the specific components or connections that led to the failure can often be almost impossible to identify.

Still another problem in establishing MOS/LSI reliability arises from the multiplicity of MOS technologies and circuit implementations that have evolved during the past few years. It is impractical to establish a numerical value for the reliability expressed in terms of mean time between failures (MTBF)[1] for each of these tech-

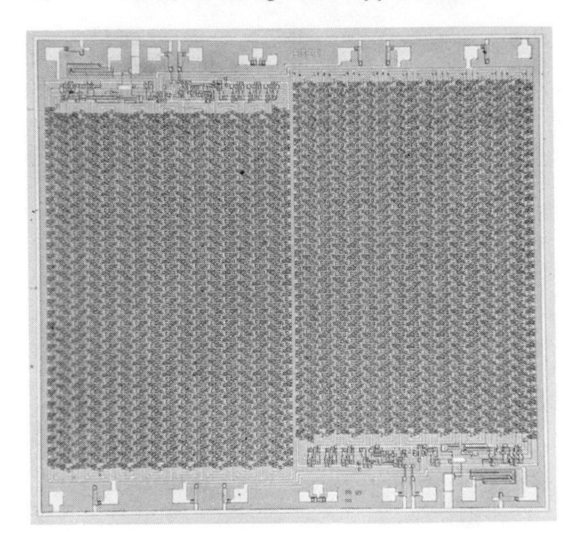

Fig. 3.1. 1,024-bit 4-phase dynamic shift register.

nologies, as they pertain to each of a variety of circuit designs. It is evident that the approach we adopt must be oriented to accommodate a rapidly moving and continually shifting technology.

Economic aspects must also be considered in the establishment of MOS reliability. We should recall that bipolar discrete transistor reliability and integrated-circuit reliability were defined and established through stimulation of the semiconductor industry by the military. Millions of test hours of reliability data for bipolar products were sponsored by the military. In contrast, MOS circuits to date have been used primarily in consumer and industrial systems. Large sums of money have not been made available for generation of MOS life-test data.

The above problems must not, however, deter us from achieving definition of MOS/LSI reliability. This technology which offers low-cost circuitry of high complexity must have assured reliability to give it all a practical meaning. We are thus led to two major approaches for MOS/LSI reliability assurance. The approaches are those of:

1. Building in reliability
2. Testing in reliability

Building in reliability is achieved through circuit design and circuit fabrication with appropriate experimental monitoring. It is inherently an economical approach. Study of its methods will provide aid in advancing our understanding of the reliability aspects of MOS technology and circuits.

Testing in reliability involves the placing of MOS/LSI circuits on operating test. The method is inherently expensive. In the 100 percent burn-in approach all units of a manufactured lot are placed on operating test. Only those units that survive the operating test are shipped to the customer. A second version of the testing in reliability method involves statistical sampling of units from a manufactured lot. The lot sample is placed on operating test, and the mean time between failures of the entire lot is defined by mathematical extrapolation of data obtained from the units sampled.

This chapter is concerned with the details of the two major approaches to achieving MOS/LSI reliability. The building in of reliability has been the primary method of approach to date for MOS/LSI, and its techniques will be considered first.

3.2 THE RELIABILITY OF MOS CIRCUITS OVER OPERATING TEMPERATURE RANGE

3.2.1 Introduction

As a first step in building in reliability for MOS/LSI, let us consider the behavior of the MOSFET as a circuit element over operating temperature range. There are two operating ranges which will be of concern for MOS/LSI circuitry. They are:

Military	$-55°C$ to $+125°C$
Consumer-industrial	$-25°C$ to $+85°C$

To ensure reliability of circuit performance over the above two operating temperature ranges, the circuit must be *designed* to accommodate the basic variations in MOSFET parameters as a function of temperature. Therefore for this case we will *build in* functional reliability through circuit design. To perform the appropriate circuit design (Chap. 4), we must first define the temperature dependence of MOSFET parameters. The parameters of concern are:

1. Threshold voltage V_T
2. Carrier mobility μ

3.2.2 Threshold Voltage as a Function of Temperature

To minimize confusion in sign values, let us specifically consider the variation of threshold voltage with temperature for the p-channel enhancement-mode MOSFET. Recall from Chap. 1 that threshold voltage for the MOSFET is given by

$$V_T = \Phi_{MS} - \frac{Q_{SS}}{C_o} + 2\phi_f - \frac{Q_B}{C_o} \tag{1-79}$$

where

$$\Phi_{MS} = \phi_{mo} - \left(\phi_{so} + \frac{E_g}{2q} + \phi_f\right) \tag{1-75}$$

ϕ_f is temperature dependent through the relation

$$\phi_f = -\frac{KT}{q} \ln \frac{N_D}{n_i} \tag{3-1}$$

which is derivable from Eq. (1-2) and the definition of ϕ_f where

$$n_i = 3.9 \cdot 10^{16} T^{3/2} e^{-E_{GO}/2KT} \tag{1-1}$$

with $E_{GO} = 1.21$ eV

Also

$$Q_B = \sqrt{2\epsilon_{\text{Si}}\epsilon_o q N_D |2\phi_f|} \tag{1-71}$$

and is temperature dependent through ϕ_f. Q_{SS} is assumed to be temperature independent.[2]

Differentiation of Eq. (1-79) with respect to temperature, and employing Eqs. (1-1), (1-71), and (3-1) with the assumption that Φ_{MS} is temperature independent yields

$$\frac{dV_T}{dT} \approx \frac{d\phi_f}{dT}\left[2 + \frac{Q_B}{2C_o|\phi_f|}\right] \tag{3-2}^*$$

where

$$\frac{d\phi_f}{dT} \approx \frac{1}{T}\left[\frac{E_{GO}}{2q} + \phi_f\right] \tag{3-3}^*$$

As an example, consider a p-channel device with:

$$T = 300\,°\text{K} \qquad \phi_f = -0.29\ \text{V}$$
$$N_D = 10^{15}/\text{cm}^3 \qquad E_{GO} = 1.21\ \text{eV}$$
$$t_{\text{ox}} = 1000\ \text{Å} \qquad C_o = 35 \cdot 10^{-9}\ \text{F}$$

Thus from Eq. (3-3)

$$\frac{d\phi_f}{dT} \approx +0.001\ \text{V/°C}$$

and from Eq. (3-2) with $Q_B/2C_o|\phi_f| \approx 0.5$ for this example,

$$\frac{dV_T}{dT} \approx +0.0025\ \text{V/°C} \tag{3-4}$$

The calculated coefficient of threshold voltage variation with temperature is in good agreement with experimentally determined values.[3] Thus for the above example if operating temperature swings through a $\Delta T = +100°\text{C}$, then $\Delta V_T \approx +0.25$ V. If V_T were -3.0 V at 25°C, then at 125°C, V_T will be ≈ -2.7 V. The situation is summarized in Fig. 3.2. We therefore note that an increase in temperature results in a decrease in the absolute value of V_T, and the lowering of threshold is accompanied by an increase in MOSFET conduction current.

*The derivation of Eqs. (3-2) and (3-3) follows closely that given by Vadasz and Grove.[3]

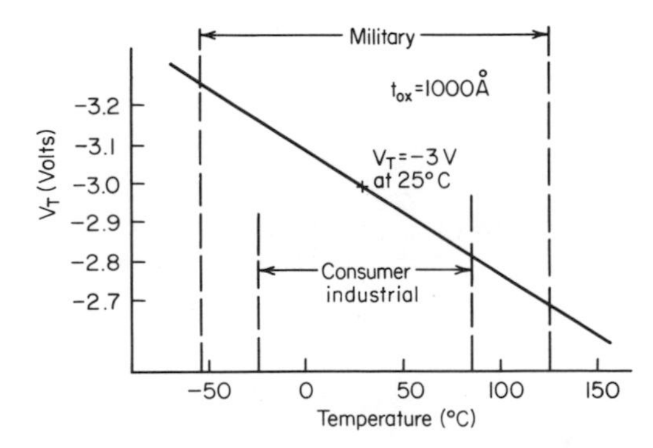

Fig. 3.2. V_T as a function of temperature.

3.2.3 Carrier Mobility as a Function of Temperature

Inversion layer carrier mobility μ is a second factor which plays an important role in MOSFET operation as a function of temperature. In the saturation region of the MOSFET operation:

$$I_D = \frac{\mu\epsilon_{\text{ox}}\epsilon_o W}{t_{\text{ox}} L} \cdot \frac{(V_G - V_T)^2}{2} \tag{1-108}$$

Eq. (1-108) is frequently employed in the form

$$I_D = K' \frac{W}{L}(V_G - V_T)^2 \tag{3-5}$$

where

$$K' = \frac{\mu\epsilon_o\epsilon_{\text{ox}}}{2t_{\text{ox}}} \tag{3-6}$$

K' is a function of measurable factors and is temperature dependent through the mobility term. Since K' is often used by circuit designers in describing MOSFET electrical behavior, the analysis will be related to the temperature dependence of K'. Carrier mobility can be expressed as the inverse first power of the operating temperature,[3] i.e.,

$$\mu \approx \frac{1}{T} \tag{3-7}$$

K' will therefore vary with temperature as shown in Fig. 3.3. An inversion-layer hole-mobility value at 300°K of 170 cm^2 per V-sec typical of (111) silicon orientation,[4] and gate-oxide thickness of 1200 Å, has been utilized in constructing Fig. 3.3. Since surface carrier mobility decreases with increasing temperature, K' decreases with temperature, and this in turn is accompanied by a decrease in MOSFET conduction current.

3.2.4 MOSFET Electrical Characteristics as a Function of Temperature

Carrier mobility and threshold voltage variation with operating temperature affect the conductivity of the MOSFET in opposite senses. We must define which effect dominates the other. To determine the dominant factor, differentiate Eq. (3-5) with

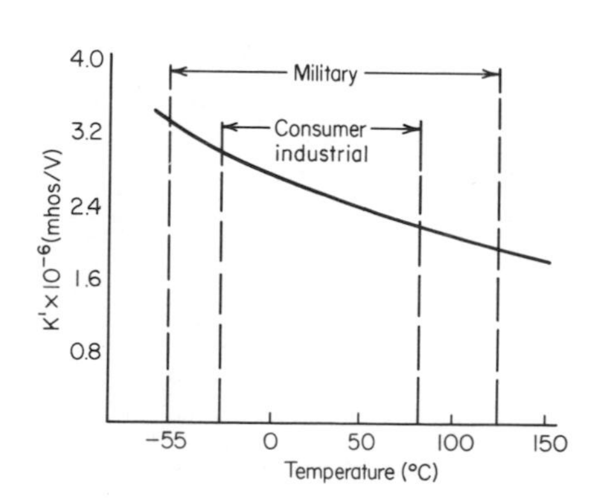

Fig. 3.3. K' as a function of temperature.

respect to temperature, i.e.,

$$\frac{dI_D}{dT} = \frac{dK'}{dT}\left[\frac{W}{L}(V_G - V_T)^2\right] - \left[\frac{K'W}{L}2(V_G - V_T)\right]\frac{dV_T}{dT} \tag{3-8}$$

Combining Eqs. (3-5), (3-6), (3-7), and (3-8) yields

$$\frac{dI_D}{dT} = I_D\left(-\frac{1}{T} - \frac{2}{V_G - V_T}\cdot\frac{dV_T}{dT}\right) \tag{3-9}$$

For the specific example of a p-channel enhancement-mode device with $V_G = -10$ V, $N_D = 10^{15}/\text{cm}^3$, $V_T = -3$ V, $T = 300°\text{K}$, and $t_{\text{ox}} = 1000$ Å, Eq. (3-9) yields with the aid of Eq. (3-4)

$$\frac{dI_D}{dT} = I_D(\underbrace{-0.0033}_{\text{Mobility contribution}} + \underbrace{0.0007}_{\text{Threshold voltage contribution}}) \tag{3-10}$$

The mobility variation with temperature thus plays the dominant role in determining electrical characteristics of the MOSFET as a function of temperature. *The MOSFET therefore becomes less active electrically as the temperature increases.* This is just the opposite of the effect observed for the bipolar transistor, which becomes more active electrically (beta increases) with increasing temperature. The temperature dependence of MOSFET characteristics is far more easily detailed analytically than is that of the bipolar transistor. This is a general trend that prevails in most of the theoretical descriptions of the two devices. The comparatively simple yet precise relationship between the physical and electrical parameters of the MOSFET is one of the advantages to be enjoyed when designing with the device.

It was demonstrated above that the temperature dependence of K' (through mobility) dominates the threshold voltage variation with temperature by a factor of approximately five in the MOSFET electrical characteristics. To first order then, if threshold voltage variation with temperature is neglected, relative drain current can be plotted as a function of temperature as shown in Fig. 3.4. The temperature

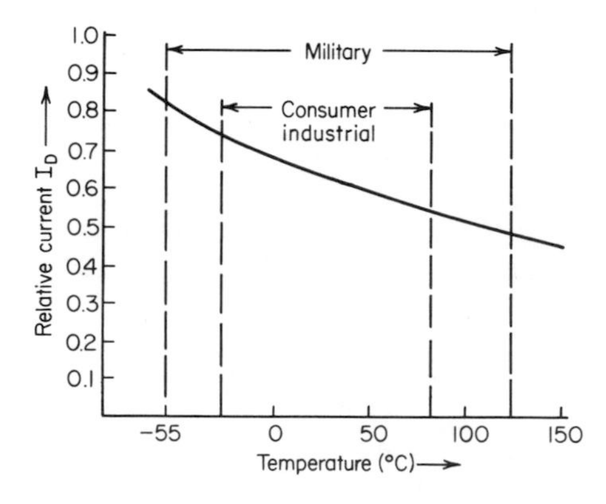

Fig. 3.4. Drain current as a function of temperature.

dependent K' value and the expression

$$I_D = K'(T)\frac{W}{L}(V_G - V_T)^2 \tag{3-11}$$

have been utilized to obtain Fig. 3.4.

3.2.5 MOSFET Switching Speed as a Function of Temperature

Let us assume that a capacitance C charged to voltage V shunts the drain-source terminals of the MOSFET. Then the switching time of the device (the time taken to discharge the capacitance through the MOSFET operating in the saturation region) can be calculated as follows:

$$C \equiv \frac{\Delta Q}{\Delta V} \tag{3-12}$$

$$I_D \equiv \frac{\Delta Q}{\Delta T} \tag{3-13}$$

From Eqs. (3-12) and (3-13)

$$T \approx \frac{CV}{I_D} \tag{3-14}$$

The switching speed S is therefore obtained from Eq. (3-14) as

$$S = \frac{1}{T} \approx \frac{I_D}{CV} \tag{3-15}$$

Since MOSFET conduction current I_D decreases with temperature increase as shown in Fig. 3.4, switching speed will also decrease as temperature increases.

The effects of operating temperature on MOSFET performance considered above are summarized in Table 3.1.

3.2.6 Thermal Resistance Considerations

The foregoing analyses have been derived for arbitrary temperature values. Since the operating MOSFET dissipates power, ensuing heat generation will cause the temperature of the device to rise. If the power dissipated in the device is constant and within the power-handling capabilities of the device, then after a certain time

Table 3.1. Summary of MOSFET Parameter Variations with Temperature

As temperature increases:	
Mobility	decreases
$\lvert V_T \rvert$	decreases
I_D	decreases
Switching speed	decreases
Propagation delay	increases

has elapsed, thermal equilibrium will prevail. The temperature rise will be proportional to the power dissipated in the device. The constant of proportionality is called thermal resistance, θ. For a given power dissipation P, ambient temperature T_a, and thermal resistance between junction and ambient θ_{ja}, the temperature of the junction T_j is

$$T_j = T_a + \theta_{ja}P \tag{3-16}$$

For a typical 40-lead dual in-line package used in MOS/LSI, $\theta_{ja} \approx 50°\text{C}$ per W. If the circuit of interest were, for example, a 200-bit shift register dissipating a characteristic power of 300 mW, and if ambient temperature were 125°C, then the junction temperature would be calculated as

$$T_j = 125 + (0.3)(50) = 140°\text{C}$$

This calculated temperature value can then be employed in the thermal considerations of Secs. 3.2.2 through 3.2.5.

3.2.7 Summary

Comparatively simple but precise relationships exist between physical and electrical parameters of the MOSFET. When these relationships are combined with known temperature dependencies of MOSFET parameters, sufficient safety margins can be designed into the circuit (Chap. 4) to meet electrical specifications over operating temperature range. Reliability is thus being *built into* the MOS/LSI chip by means of circuit design to ensure the required circuit performance over operating temperature range.

3.3 FABRICATION METHODS FOR ENHANCED CIRCUIT RELIABILITY

3.3.1 Introduction

Having defined the temperature variation of MOSFET parameters and having approached the attendant reliability problem through circuit design, let us now examine methods being used to build in reliability through topological layout and fabrication of the MOS/LSI chip.

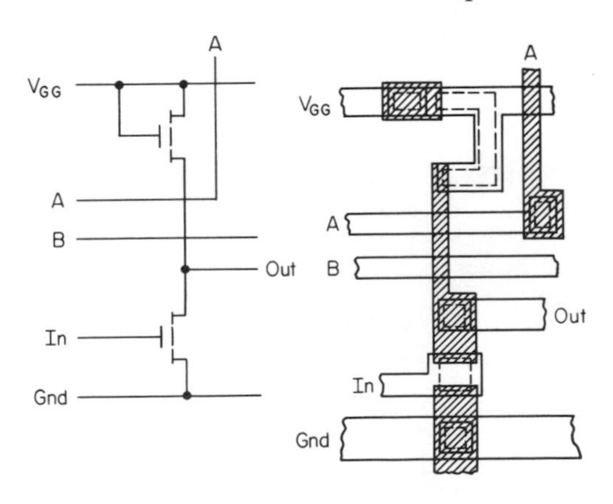

Fig. 3.5. Basic inverter circuit. Cross-hatched regions denote p-diffused areas. Dotted lines denote thin-oxide delineation. Solid lines surrounding dotted lines denote thick-oxide delineation. Clear regions denote aluminum lead patterns.

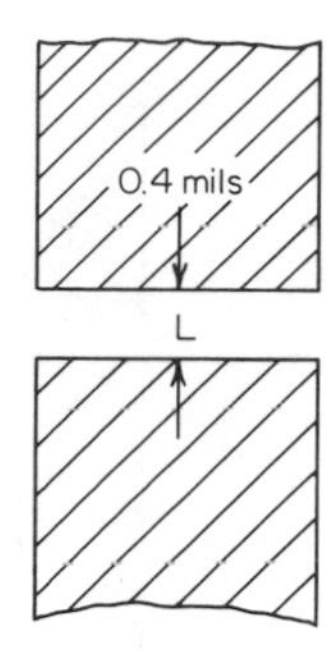

Fig. 3.6. Source-to-drain spacing (channel length L).

3.3.2 The Layout Rules

The layout rules provide a prescription for preparing the photomasks used in the fabrication of MOS integrated circuits. The rules provide an effective communications link between circuit designer and process engineer in the manufacture of MOS/LSI. The goal to be achieved with the layout rules is to obtain the circuit with optimum yield in as small a geometry as possible without compromising reliability of the circuit. The rules are numerous and we will examine only nine of the more important ones. An inverter circuit will be taken as the design vehicle, Fig. 3.5. The layout rules will be given with minimum dimensions in all cases. Included are the constraints which form the basis for rule definition.

1. Source-to-drain spacing (channel length L) defined prior to source-drain diffusion, Fig. 3.6.
   $L = 0.4$ mil
   Constraints:
   a. Depletion-layer "punch through" between source and drain.
   b. Yield.
2. Gate width (channel width W), Fig. 3.7.
   $W = 0.4$ mil when $L = 0.4$ mil
   $W = 0.3$ mil when $L > 0.4$ mil
   Constraints:
   a. Definition of hole etched through 12,000 to 15,000 Å oxide.
   b. Yield.

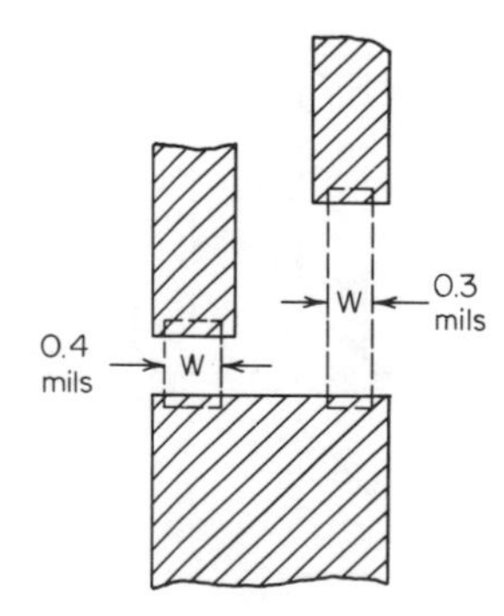

Fig. 3.7. Gate width (channel width W).

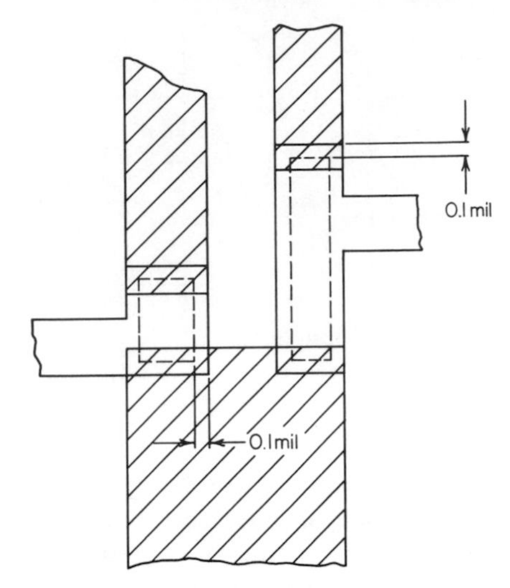

Fig. 3.8. Metal overlap over thin oxide.

3. Metal overlap over thin oxide, Fig. 3.8.
   0.1 mil in W direction.
   0.1 mil in L direction.
   Constraints:
   a. Must cover all thin-oxide areas with metal to eliminate parasitic MOS action caused by possible presence of surface ions.
   b. Yield.
4. p-diffusion width and spacing of p diffusions, Fig. 3.9.
   diffusion width = 0.3 mil.
   diffusion spacing = 0.5 mil.
   Constraints:
   a. Yield in both cases.
   b. Parasitic resistance of "tunnel" interconnections is dependent on diffusion width.

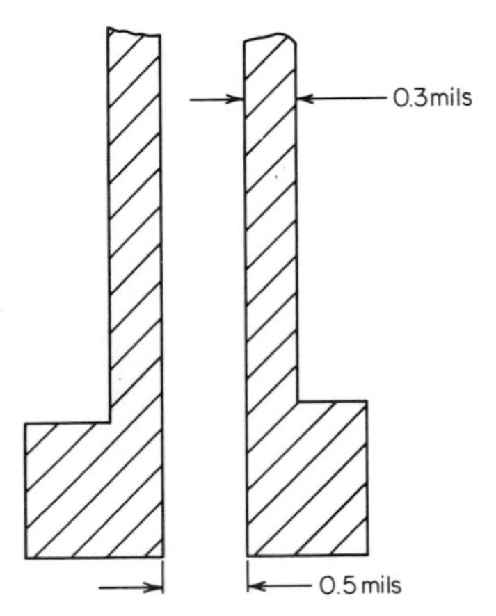

Fig. 3.9. p-diffusion width and spacing of p diffusions.

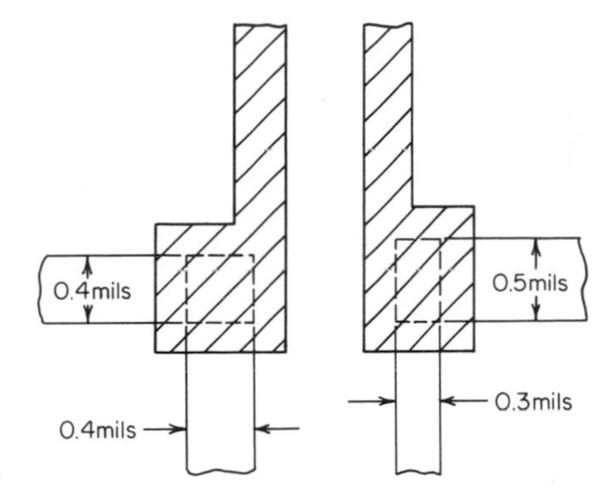

Fig. 3.10. Contact hole length and width.

5. Contact hole length and width, Fig. 3.10.
   0.4 × 0.4 mil (minimum)
   0.3 × 0.5 mil (minimum)
   Constraints:
   a. Definition of hole etched through 12,000 to 15,000 Å oxide.
   b. Yield.
6. p^+ margin around contact, Fig. 3.11.
   0.2 mil total
   Constraints:
   a. Shorting or arcing to ground.
   b. The step over which metal is placed can be the weak link for electrical conduction if metal migration takes place. The metal interconnections must be brought down over the 12,000 to 15,000 Å thick oxide in two steps.

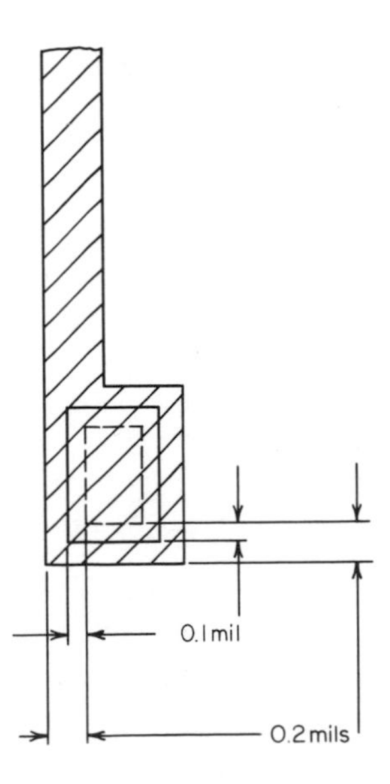

Fig. 3.11. p^+ margin around contact.

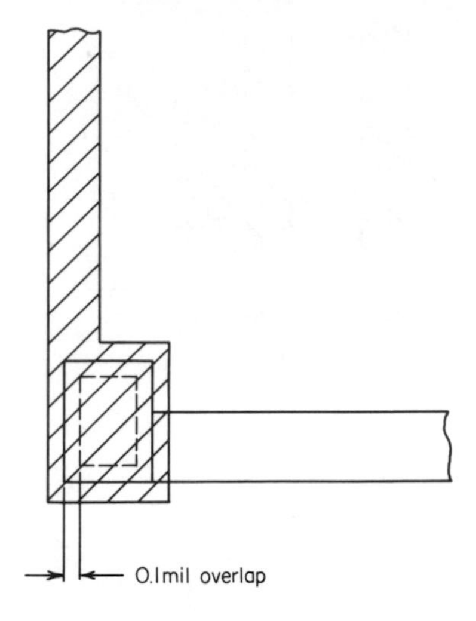

Fig. 3.12. Metal overlap around contact.

7. Metal overlap around silicon contact (at least two adjacent sides), Fig. 3.12.
   0.1 mil total.

   Constraints:
   a. Must ensure good ohmic contact in case of minor misalignment of photomasks during fabrication.
   b. Yield.
8. Metal width and metal spacing, Fig. 3.13.
   width = 0.4 mil
   spacing = 0.4 mil

   Constraints:
   a. Alignment and hence yield in both cases.
   b. Rule: 0.1-mil metal width per 1-mA current. (use >0.4-mil width for power-supply bus lines)
9. Contact-to-gate separation, Fig. 3.14.

   Constraints:
   a. Yield.
   b. Reliability can be marginal. Effect becomes evident after long burn-in.

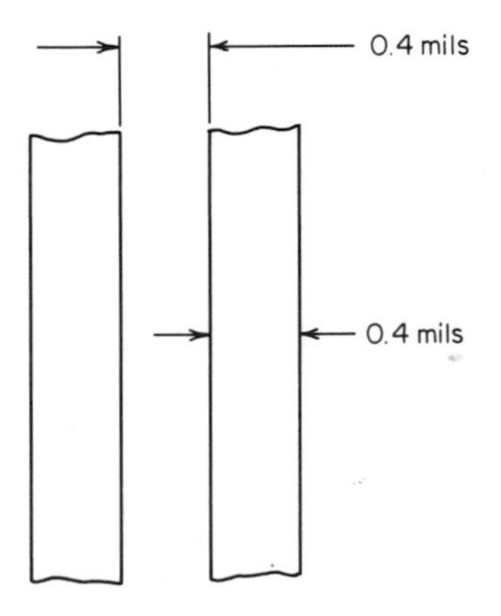

Fig. 3.13. Metal width and metal spacing.

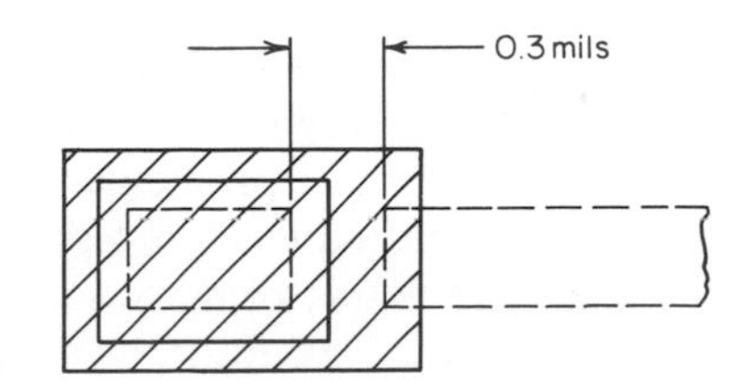

Fig. 3.14. Contact-to-gate separation.

Summary of the layout rules:

1. The layout rules that seriously affect yield will undergo changes to increase circuit density without degrading reliability, as improvements in technology and process equipment are made.
2. Layout rules provide for minimum dimensions. Use larger dimensions when possible.
3. Power dissipation in any case must not exceed 50 mW per mil^2 within the MOS/LSI chip. Thermal resistance and upper operating temperature limit of 200°C establish this rule.

3.3.3 Threshold Stability

The stability of threshold voltage V_T is of major importance to the reliability of MOS integrated circuits. Since

$$I_D = \frac{\beta(V_G - V_T)^2}{2} \tag{1-116}$$

circuit performance is dependent on V_T. It is therefore apparent that drift or instability in V_T during circuit operation can prove disastrous to the reliability of MOS/LSI.

To define basically how V_T instability can arise, note that Eq. (1-79) can be modified by the term Q_{Ox}/C_o to yield

$$V_T = \Phi_{MS} - \frac{Q_{SS}}{C_o} + 2\phi_f - \frac{Q_B}{C_o} - \frac{Q_{Ox}}{C_o} \tag{3-17}$$

The term Q_{Ox}/C_o analytically describes contaminant charge in the gate oxide and hence relates this charge to threshold voltage. The Q_{Ox}/C_o term detailed a very serious instability in V_T which arose in the early development of MOS circuitry (1962 to 1964). The major source of the instability proved to be positive sodium ions present in the gate dielectric, silicon dioxide. Sodium ions are highly mobile in silicon dioxide in the presence of an electric field.[5] Their mobility in silicon dioxide increases as temperature increases. A positive bias on the gate of the MOSFET will drive sodium ions toward the silicon–silicon dioxide interface. As the positive sodium ions approach the silicon surface, negative charges in the bulk silicon will be drawn to the surface, which in turn will make the surface of n-type silicon more strongly n type, or will tend to invert the surface of p-type silicon. This effect will result in increasing the absolute value of the threshold voltage for the p-channel enhancement-mode device. A negative bias applied to the gate under these conditions will produce, with time, just the opposite effect. As a result, threshold voltage is unstable and the situation of course proved intolerable for circuit operation.

Fortunately these problems are now behind the industry, as it was discovered that sodium was introduced onto the gate oxide during the process of aluminum metalization. Once this primary source of sodium contamination was identified, it was soon eliminated by utilizing a high-energy electron beam method for gate metal deposition in an ultrahigh vacuum. MOSFETs now exhibit essentially absolute stability of threshold voltage when subjected to electric fields in the gate oxide of 10^6 V per cm at ambient temperatures of 300°C. These stress conditions far exceed practical values experienced in operation of MOS circuitry. Thus it again becomes evident how *reliability can be built into* MOS/LSI, in this case through processing techniques.

3.3.4 Protective Glass

Reliability can be enhanced by depositing a thin layer ($\approx$5000 Å) of silicon dioxide over the entire surface of the MOS/LSI chip as depicted in Fig. 3.15. This process step:

1. Prevents shorting of circuit elements by encapsulated particulate matter, should it exist after, say, a rugged mechanical stress. Such particulate matter, composed of microsize conducting particles, can bridge unprotected metal lines in the integrated circuit and thereby cause shorting.
2. Provides higher yield since tweezer scratches on oxide or metalization, which can normally occur in wafer processing, are unable to penetrate the protective glass.

In summary: We are again building in reliability on our MOS chip, in this example through processing technology.

3.3.5 Process Monitoring Tools

Reliability of MOS products can be further ensured by process monitoring. The major tools available are:

1. The C-V method, Fig. 3.16*a*, *b*. This technique was discussed in Chap. 1. It is used as a monitor of Q_{SS}, t_{ox}, and substrate doping level.
2. The scanning electron microscope. Photographs obtained with this instru-

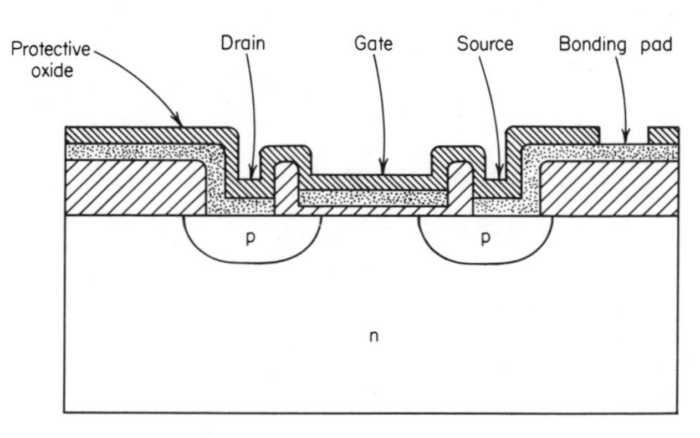

Fig. 3.15. Protective oxide coating.

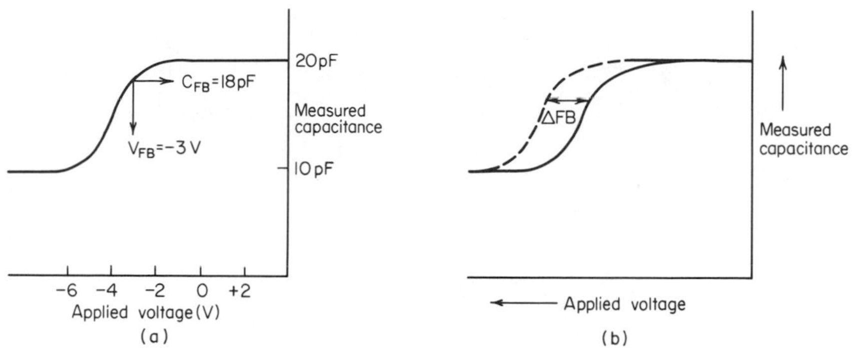

Fig. 3.16. (*a*) MOS capacitance as a function of voltage; (*b*) flat-band voltage shift as measured with MOS capacitor.

ment are shown in Fig. 3.17*a*, *b*, *c*. Magnification is approximately 8,000X. The scanning electron microscope provides photographs of high information content that have a three-dimensional appearance.

3. The optical microscope. A 200X visual inspection with a conventional microscope is made on selected wafers at each process step. An evaluation of alignment, etching, oxide uniformity, metalization quality, etc. can thereby be obtained. Each wafer receives a final visual inspection.

3.4 PROTECTIVE INPUT CIRCUITRY

One of the most publicized reliability problems encountered with MOS/LSI is the susceptibility of these circuits to static charge. The development of static charge in handling (shipping, testing, circuit board assembly, etc.) can rupture the gate oxide and cause catastrophic failure of the device. Design techniques that obviate this possible failure mechanism will be presented in this section.

The problem of gate-oxide failure can be defined by first noting that the dielectric strength of silicon dioxide is approximately $7 \cdot 10^6$ V per cm. Thus, for example, if 85 V is applied across a 1200-Å-thick gate oxide, dielectric breakdown will occur. The breakdown usually results in a permanent electrical short between metal gate and silicon substrate, thereby destroying MOSFET action. The situation is further aggravated by the extremely small value of gate-oxide capacitance C. Since $V = Q/C$ and C is only a fraction of a picofarad, very little static charge is required to produce a potential buildup and resulting electric field across the gate oxide which exceeds the dielectric strength of the oxide. This potential failure mode can be essentially eliminated by incorporating protective input circuitry on the MOS/LSI chip.[6] In this situation, we are building in reliability through circuit design.

Protective input circuitry which provides a voltage-limiting function is placed at each of the signal input pads on the MOS/LSI chip. Several approaches exist for design of these circuits, the simplest being that of a Zener diode. Following are the three methods extensively used in the design and fabrication of protective input circuitry:

(a) Zener-diode chain, Fig. 3.18. The avalanche breakdown voltage of the three Zener diodes arranged in series affords negative 18- to 21-V input protection.

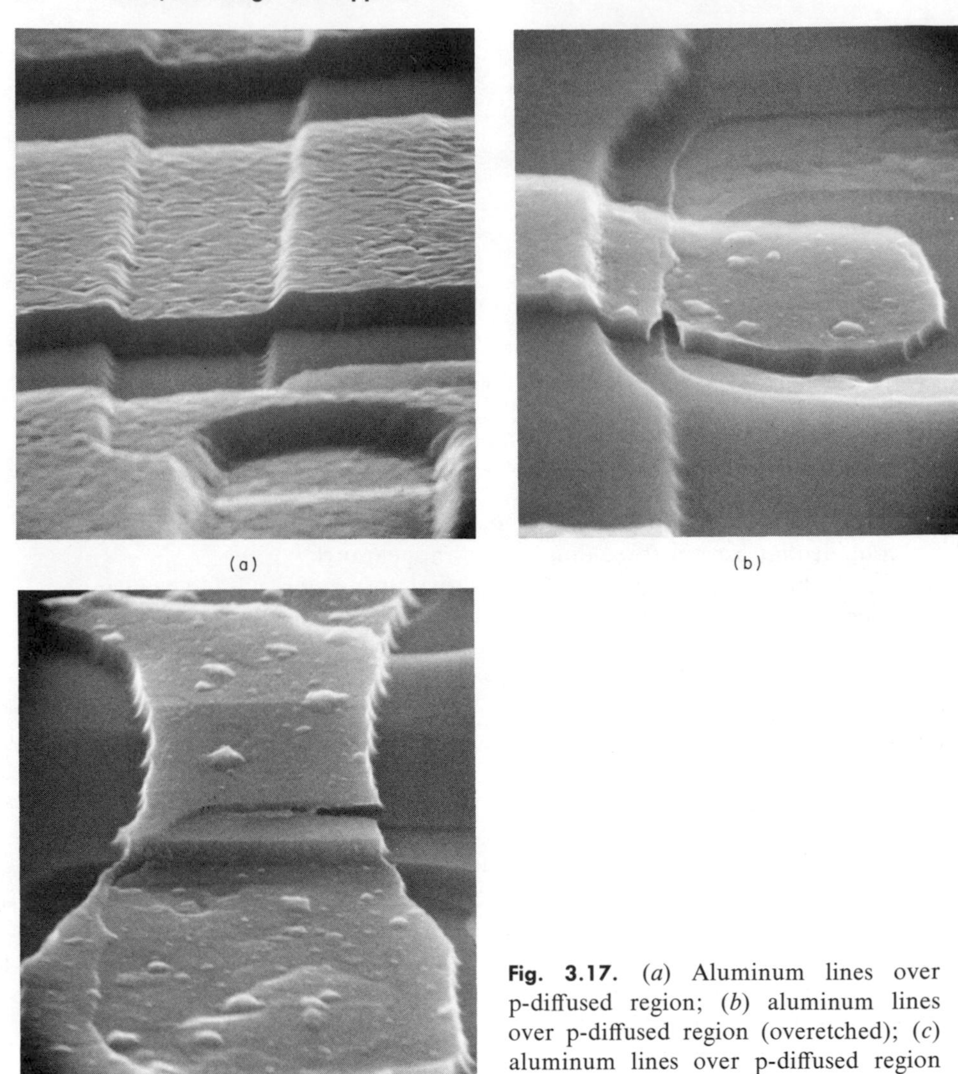

Fig. 3.17. (*a*) Aluminum lines over p-diffused region; (*b*) aluminum lines over p-diffused region (overetched); (*c*) aluminum lines over p-diffused region (oxide step has poor contour).

Positive input voltage protection is realized by the characteristic *on* voltage of the diodes (0.7 V for each diode). Resistor R is utilized to protect the Zener chain from excessive current surges. The resistor also serves to spoil the rise time of the input voltage transient in order that the Zener chain will clip the transient prior to its arrival at the input MOSFET. This method is used when the process includes a bipolar output stage (additional n^+ diffusion).

(b) p^+-n diode and field plate, Fig. 3.19. This circuit provides negative input voltage protection of 40 to 60 V. The crowding of field lines in the vicinity of the field plate initiates the avalanche condition.[7] Positive input voltage protection is provided by the 0.7-V forward diode characteristic. Resistor R protects the input

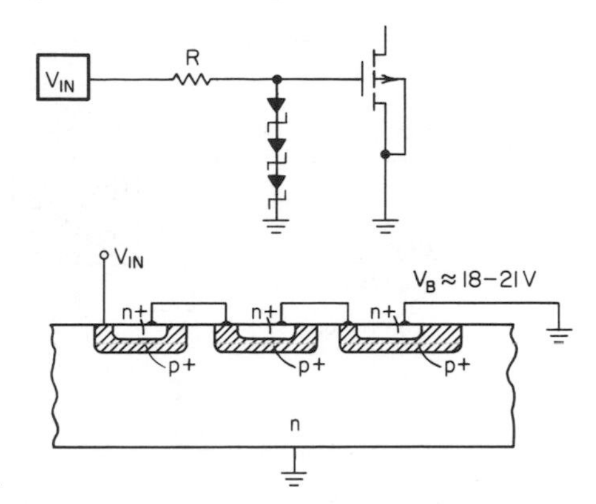

Fig. 3.18. Zener chain input protection.

diode from excessive current surges and also serves to reduce the rise time of the input voltage transient in order that the diode will clip the transient prior to its arrival at the input MOSFET.

(c) Large peripheral p⁺-n diode field plate and thick-oxide MOSFET, Fig. 3.20. This circuit provides negative input voltage protection of 40 to 50 V. It employs the p⁺-n diode and field plate of method b, as well as a thick-oxide MOSFET which has a field threshold of −40 to −50 V. When the thick-oxide MOSFET is turned on by the voltage transient, it will in turn short the transient signal to ground. Resistors R_1 and R_2 function in the same manner as described in methods *a* and *b*, above.

Any of these three protective circuits will provide enhanced reliability of MOS/LSI with respect to excessive input voltage transients. Method *c* is probably more frequently employed than are methods *a* and *b*.

3.5 HANDLING PRECAUTIONS FOR MOS/LSI

As with most electronic products, MOS/LSI requires certain recommended handling procedures. Following are several guidelines which, if observed, will further ensure reliability of MOS products.

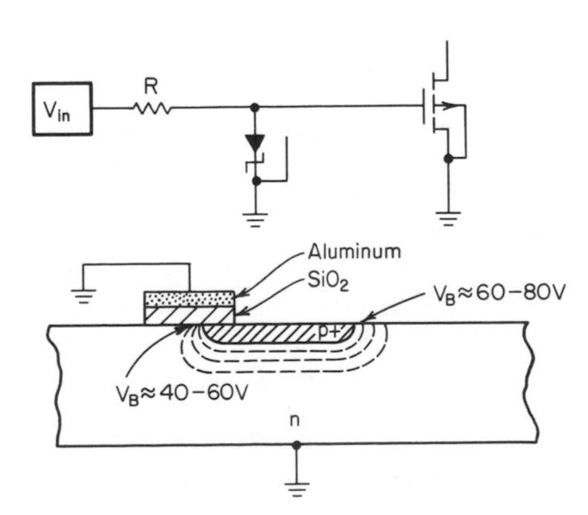

Fig. 3.19. p⁺-n diode and field plate input protection.

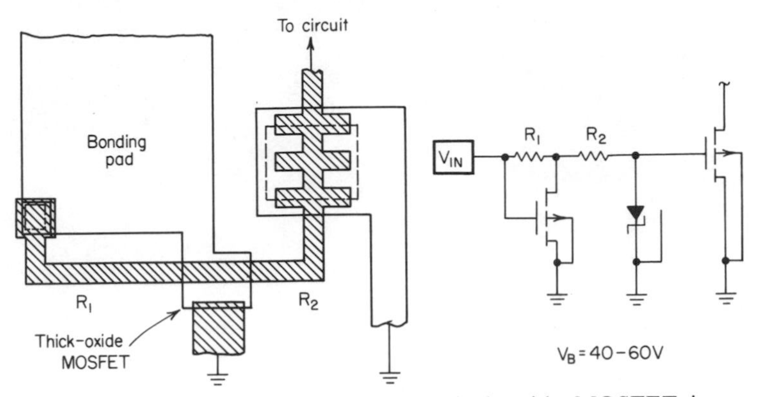

Fig. 3.20. p⁺-n diode, field plate and thick-oxide MOSFET input protection.

1. To minimize static charge at the input terminals, store devices in conductive foam so that all exposed leads are shorted together. To avoid static charge build-up, it is recommended that MOS products not be stored in conventional plastic snow or in plastic trays.
2. Ground the tips of soldering irons.
3. Any unused input lead to the device should be connected to circuit ground.
4. Never forward-bias a lead with respect to case and substrate, since large currents can then flow and damage the device.
5. Use of conductive foam is recommended for bench or desk tops if much handling or breadboarding of MOS circuits is to be performed.
6. Avoid subjecting devices to voltage transients greater than the maximum ratings shown on the data sheet. Do not insert or remove devices from circuit board with power on.
7. In regard to package and lead integrity, avoid excessive heating when soldering ($T < 300°C$). Since the device will be in a multilead package, care should be taken so as not to unnecessarily bend or distort leads. Do not bend leads directly at the lead-to-package-surface interface.

3.6 THE TESTING-IN OF RELIABILITY

3.6.1 The 100 Percent Burn-in Approach

The extreme approach to "testing in" of MOS/LSI reliability is that afforded by the 100 percent burn-in method. As previously described, the method involves placing all units of a customer's order on an operating test. Only those units that survive the tests are delivered to the customer, and thereby substandard components are eliminated. Although this method is frequently employed throughout the semiconductor industry, it does have the following disadvantages:

1. *Costly.* Burn-in test racks and associated electronics must be assembled.
2. *Time consuming.* Testing time can be of long duration unless accelerated

testing methods are used. Accelerated testing is a technique whereby device failure is accelerated by subjecting the devices to stresses in excess of their use or reference condition. The independent variables of accelerated testing are the applied stresses, which may be either mechanical, electrical, environmental, or combinations of these. The choice of an accelerated test program is difficult since it must be stringent enough to detect meaningful changes but not so stringent as to cause degradation of the inherent reliability of the product.

3.6.2 Sample Qualification Tests

The sample qualification tests for establishing MOS/LSI reliability involve statistical sampling of units from a manufactured lot. The lot sample is placed on operating test, under either accelerated or reference operating conditions. The mean time between failures[1] for the given lot is defined by mathematical analysis and extrapolation of the results to the manufactured lot. The disadvantages associated with this method are:

1. *Costly.* Test racks and personnel must be made available for data acquisition. Human error in testing procedures can, in practice, lead to erroneous conclusions.
2. *Time consuming.* Many processes and circuits are candidates for evaluation.

3.6.3 Corrective Action Programs

It is important that reliability test data be directed back to product engineering personnel, to determine the effectiveness of efforts expended in building in reliability. The flow of reliability data to product engineering at Texas Instruments is shown as an example in Fig. 3.21. The following five basic programs appear in Fig. 3.21:

1. In-process V_T stability tests
2. Stability tests
3. Process monitor
4. New process evaluation
5. Quarterly package monitor

MOS circuits are manufactured in the front end or wafer fabrication area. *In-process threshold stability tests* are performed on completed MOS circuits in wafer form. The tests are carried out at elevated temperature with positive and negative voltages applied to the gate. The purpose of these tests is to detect possible ion contamination or other process imperfections which might lead to instability.

After assembly, test transistors are bonded-out and stressed at high temperature and voltage to give additional indications of device stability (threshold voltage and junction leakage current, primarily). These measurements are obtained in the *stability testing* section.

In the *process monitor tests,* representative samples of each process are placed on 1,000-hour high-temperature operating test and 1,000-hour high-temperature life test, to serve as monitor of the manufacturing process. Samples are collected each week. Failure analysis is performed and corrective action is initiated where appropriate.

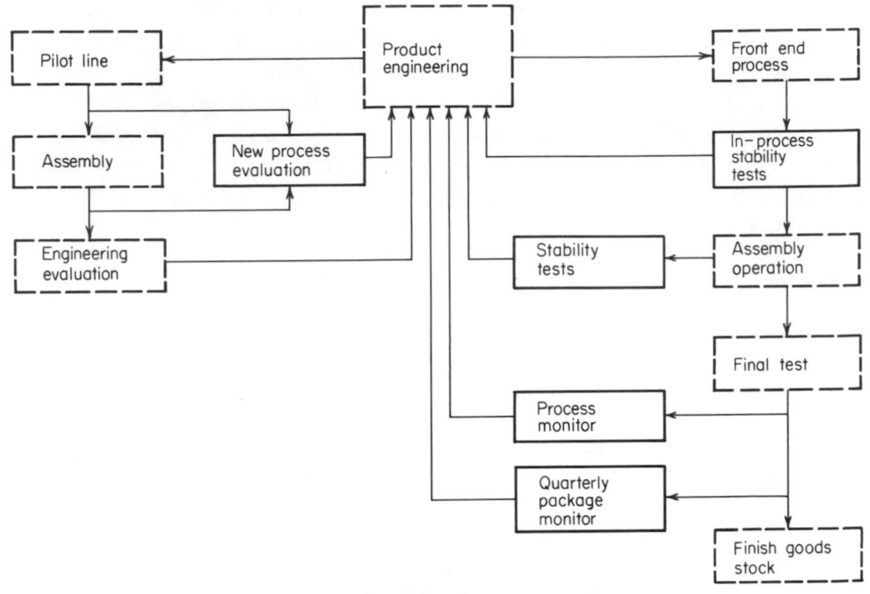

Fig. 3.21. Qualification system chart.

In the pilot line, *new process evaluation* is performed both at the wafer level and assembly level. Failure mechanisms are analysed and corrective action is taken to improve the process prior to release to the manufacturing group.

A *package monitoring program* has been established wherein various package and device types are subjected to electrical and environmental tests four times a year. As an example of how involved programs of this type can become, consider a recently completed reliability study at TI of plastic packages (Fig. 3.22) for MOS/LSI. Over 200 different plastics were evaluated and the choice of one for MOS was governed by the following considerations:

1. Does it contaminate the gate oxide?
2. Does the plastic contain mobile sodium?

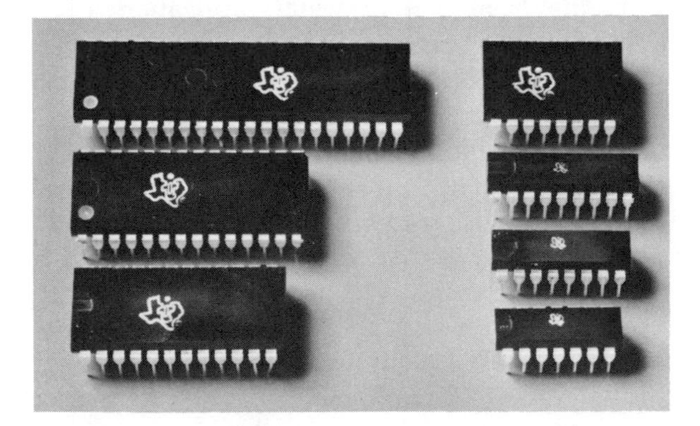

Fig. 3.22. Plastic packages for MOS/LSI.

3. What is the permeability of the plastic to sodium?
4. Does the plastic have good resistance to the ingress of moisture?
5. Is the expansion match of plastic, lead frame, and silicon chip satisfactory?
6. Is the mechanical strength adequate for 40-pin packages?
7. Are the resulting products reliable?

The two approaches adopted to determine whether sodium was mobile in the chosen epoxy plastic were radiotracer and neutron-activation analysis. In the radiotracer studies, sodium chloride labeled with Na^{22} was used, and diffusion at 150°C was performed for various durations. After 600 hours no buildup of sodium was evident in the plastic. Neutron activation supported the above findings. In the investigation silicone plastics were found to be permeable to sodium and hence were discontinued for use with MOS devices. Operating life tests and storage life tests were performed at 125°C and 150°C on MOS/LSI circuitry encapsulated with the chosen epoxy plastic. The results of the tests are shown in Table 3.2.

3.7 CONCLUSIONS

Fundamental conclusions pertaining to MOS/LSI reliability can be summarized as follows:

1. High reliability of MOS/LSI circuitry is basically established by *building in* reliability through fabrication and design. Extensive device testing is used to validate the reliability achieved in the resulting circuits.
2. Precise definition of reliability for MOS/LSI products has been made difficult by the rapidity of developments in fabrication methods and circuit innovations that have characterized this technology. Specific reliability data does exist, however, and Table 3.3, for example, summarizes operating life and storage life tests taken on groups of static MOS shift registers. These results

Table 3.2. Reliability Data for MOS/LSI Plastic Packaging

Operating life at 125°C		Storage life at 150°C	
Device hours at temperature	134,000	Device hours at temperature	358,000
Failures	1	Failures	1
Derated to 55°C		Derated to 55°C*	
Equivalent device hours	910,000	Equivalent device hours	2,800,000
Failure rate (%/1,000 hr)		Failure rate (%/1,000 hr)	
60% confidence level	0.22	60% confidence level	0.073
Derated to 25°C		Derated to 25°C	
Equivalent device hours	2,400,000	Equivalent device hours	8,650,000
Failure rate (%/1,000 hr)		Failure rate (%/1,000 hr)	
60% confidence level	0.083	60% confidence level	0.023
MTTF for 25°C operation	1,200,000	MTTF for 25°C operation	4,300,000

*Acceleration factors obtained from NASA Report No. CR 1349, vol. 4, May 1969.

Table 3.3. Summary of Life Test Data for MOS Static Shift Registers Series 3000, TO-100 Package, (111) Orientation

Test description	Number of failures		Length of longest test hours	Actual device hours at temp extremes	Equivalent hours at		60% upper confidence level equivalent % failure rate /1,000 hr	
	Deg.	Cat.			25°C	55°C	25°C	55°C
+125°C Operating life	1	0	6,000	346,000	6,230,000	2,350,000	0.034	0.084
+150°C Storage life	0	1*	6,000	387,000	9,560,000	3,100,000	0.021	0.065
Totals	1	1	6,000	733,000	15,790,000	5,450,000	0.020	.057

*Catastrophic failure due to conducting particle on chip.
Note: Acceleration factors obtained from NASA Report CR1349, vol. 4, May 1969.

coupled with previously obtained data indicate that the reliability of MOS products is improving with time. It can be said that we are moving down the slope of the reliability learning curve and have, in general, reached a position in 1972 where bipolar integrated circuits were in 1967, cf. Fig. 3.23. Although no impediment to further MOS/LSI reliability improvement is anticipated, certainly a great deal more reliability data must be assembled in order to mathematically establish indicated improvement.

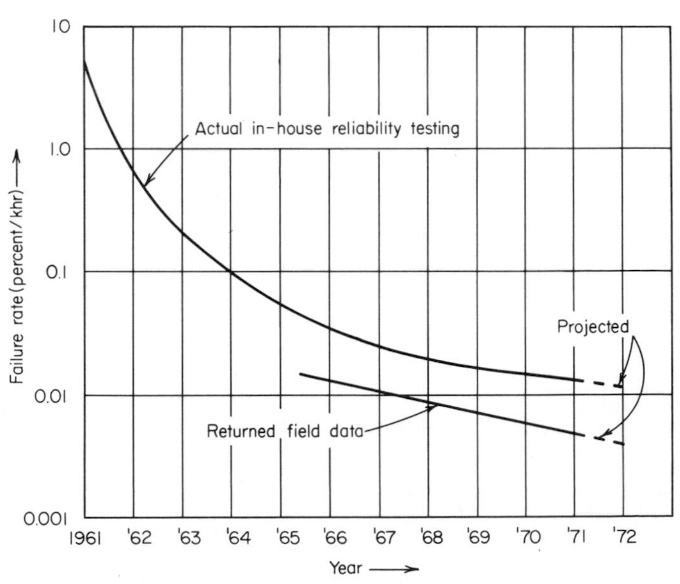

Fig. 3.23. Texas Instruments' bipolar integrated circuit failure-rate history.

3. With the existing complexity advantage of MOS over bipolar (approximately a factor of five) and the predicted future increase of MOS complexity, improved *system* reliability should be enjoyed when using MOS/LSI. (Complexity is defined here as the number of effective or equivalent gates per package pin.)

REFERENCES

1. G. W. A. Dummer and N. B. Griffin, "Elcctronics Reliability-calculation and Design," p. 40, Pergamon Press, London, 1966.
2. A. S. Grove, B. E. Deal, E. H. Snow, and C. T. Sah, Investigation of Thermally Oxidized Silicon Surfaces Using Metal-Oxide Semiconductor Structures, *Solid State Electronics,* **8:** 145, February 1965.
3. L. Vadasz and A. S. Grove, Temperature Dependence of MOS Transistor Characteristics below Saturation, *IEEE Trans. on Electron Devices,* **ED-13:** 863(December 1966).
4. D. Colman, R. T. Bate, and J. P. Mize, Mobility Anisotropy and Piezoresistance in Silicon P-type Inversion Layers, *J. Appl. Phys.,* **39:** 4-1923(1968).
5. E. H. Snow, A. S. Grove, B. E. Deal, and C. T. Sah, Ion Transport Phenomena in Insulating Films, *J. Appl. Phys.,* **36:** 1664, May 1965.
6. R. S. C. Cobbold, "Theory and Applications of Field-effect Transistors," p. 267, Wiley Interscience, New York, 1970.
7. A. S. Grove, O. Leistiko, and W. W. Hooper, Effect of Surface Fields on the Breakdown Voltage of Planar Silicon p-n Junctions, *IEEE Trans. Electron Devices,* **ED-14:** 157, March 1967.

4

Inverters, Static Logic, and Flip-flops

4.1 INTRODUCTION

The most basic MOS logic circuit is the gate that performs the logical NOT function. We will refer to this circuit, in which the differential signal polarities reverse between input and output, as an *inverter*. Each inverter contains a load device and a driver or active device, as shown in Fig. 4.1. Here the EIA graphic symbols denote that both the load (upper device) and the driver (lower device) are p-channel, enhancement-type transistors.

Of course, there are many other technologies available for the MOS inverter circuits. One can design inverters using n-channel or p-channel, each with depletion-mode or enhancement-mode options. The practical circuit designs are presently limited to enhancement-type driver devices because the alternate selection of a depletion driver will not permit voltage polarity matching between direct-coupled stages. Since the enhancement-type driver device permits circuit operation using gate and drain bias voltages of the same polarity, it offers circuit compatibility for direct-coupled stages. We will confine our discussion in this chapter to inverters with enhancement-mode drivers.

The MOS inverter invariably uses an MOS device functioning as a load resistance. The small size of the MOSFET is the basic motivation for its use as a load equivalent resistor. An MOS load device of 100K Ohms occupies approximately 1 mil^2. The same resistance value obtained using standard diffusion methods and sheet resistances in the range of 20 to 200 Ohms per square (as are typically used in processing) would require a much larger area for the load device. Other important advantages accruing from the use of the MOSFET as the load device result from the gate control which permits the load device to be turned off in cases where power dissipation at reduced levels is desired. When the load device is clocked, the power dissipation becomes a function of the duty cycle of the clock.

A partial list of inverter combinations is given in Table 4.1. Here p- and n-channel driver devices, all operating in the triode region when they are turned on, are coupled within the inverter to load devices of several combinations. A designation system

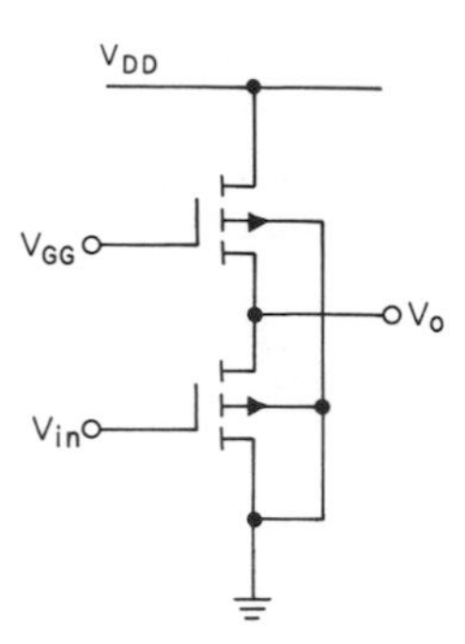

Fig. 4.1. Basic inverter circuit with common substrate.

is used in Table 4.1 to denote various inverter combinations: for instance, PELT, for p-channel enhancement-mode load, triode region of operation. Another common inverter combination using p-channel devices is the PELS configuration. Here the driver device is again a p-channel enhancement device and PELS uniquely specifies the load. PELS in this case denotes p-channel enhancement-type load operating in a saturation region.

If a depletion load is used with p-channel devices, we have an inverter combination which we will designate PDLS or PDLT, with the load biased to operate in a saturation and triode region, respectively. For n-channel devices the designator is similar. For instance, an inverter with n-channel driver and load matching that of the PELT case would be the NELT, as shown in Table 4.1. A simple designator for the inverter combination in which the load and the driver device are of differing channel types is simply designated as CMOS—C for complementary. The CMOS has become standardized using an n-type silicon substrate with the n- and p-channel devices fabricated within. Other CMOS circuitry uses insulating substrates.

The various MOS inverters can be processed with silicon substrates or with insulating substrates such as sapphire. The distributed capacitances and, therefore, the speed power parameters describing a particular inverter depend critically upon the device type, device mode of operation, substrate material, and layout topology detail.

Let us analyze several of the common inverter designs separately for dc static transfer characteristics, layout geometry, and operating speed. We will be able to make rough comparisons for design decisions relating to the various inverters. One

Table 4.1. Partial List of Inverter Combinations

Driver	Load	Designation
p-channel, enhancement (triode)	p-channel, enhancement (triode)	PELT
p-channel, enhancement (triode)	p-channel, enhancement (saturation)	PELS
p-channel, enhancement (triode)	p-channel, depletion (saturation)	PDLS
n-channel, enhancement (triode)	n-channel, enhancement (triode)	NELT
n-channel, enhancement (triode)	n-channel, enhancement (saturation)	NELS
n-channel, enhancement (triode)	n-channel, depletion (saturation)	NDLS
n-channel, enhancement (triode)	p-channel, enhancement (triode)	CMOS

must recall at all times that as technologies change and layout innovations occur, they may change the processing-technology/circuit-layout combination that provides the most effective solution to a given design problem. There is a great flux in both design innovation and process technology at this writing, and one must be very careful to avoid broad comparisons relating to relative value, in which the data used are not of a fundamental nature.

4.2 INVERTER OF P-CHANNEL ENHANCEMENT-TYPE LOAD DEVICE IN TRIODE REGION (PELT)

A simple inverter circuit that can be integrated onto a monolithic substrate is the PELT, which uses all p-channel devices of the enhancement type. A second power supply, V_{GG}, is required to maintain the load device in the specified triode region. Therefore, the magnitude of the power-supply voltage V_{GG} is larger than the supply V_{DD} which supplies static power to the inverter. The power requirements from the V_{GG} supply, although negligible for the dc case, do involve capacitive power-consuming transients for dynamic switching operation. The PELT inverter circuit shown in Fig. 4.2*a* requires an n-type common substrate connected to the most positive voltage level. Representative threshold voltages for both the load and driver device are selected as -2.3 V together with V_{DD} and V_{GG} values of -5 and -17 V. The transfer characteristics for this choice of threshold and voltage supply parameters are shown in Fig. 4.2*b*. These curves are obtained from the basic MOS equations, and the effect of substrate bias on the threshold voltage is considered. For these computer calculations, an infinitely large saturation resistance is used for the devices.

The transfer characteristics for the inverters are critically dependent upon a parameter called the *beta ratio,* which we designate β_R. The beta ratio is an electrical parameter and is the ratio of transconductances for the driver to that of the load device in the inverter. One can specify separately a geometrical beta ratio, which refers to the apparent gain ratio that one observes in calculating only from the geometrical photomask layout. The more important beta ratio is the electrical beta ratio β_R; this is the one normally referred to in our discussion. It is the actual ratio of transconductances between the driver and the load devices. This electrical β_R is used to determine the transfer characteristics of Fig. 4.2*b* where substrate-biasing effects on the load device threshold are included. The mask geometry for the PELT inverter is given in Fig. 4.2*c* with width-to-length (W/L) ratio of 5 for the driver and 0.25 for the load device. As the beta ratio increases, the difference in transconductance values between the load and driver devices increases, and the incremental dc gain as shown in Fig. 4.2*b* is seen to increase in the transition region. Conversely, you will note that as β_R decreases below a critical value of approximately 10, the slope of the transfer characteristic in the transition region is reduced below unity and the logical gain of the device is generally not useful for digital applications.

The layout selected for the PELT inverter of Fig. 4.2*c* requires approximately 6.5×3.2 mils of area on the chip. Standard design rules are applied here, including 0.1-mil gate overlap onto the source-drain diffusion mask regions. The metalization stripes have a minimum of 0.3-mil width. The connection to the substrate is

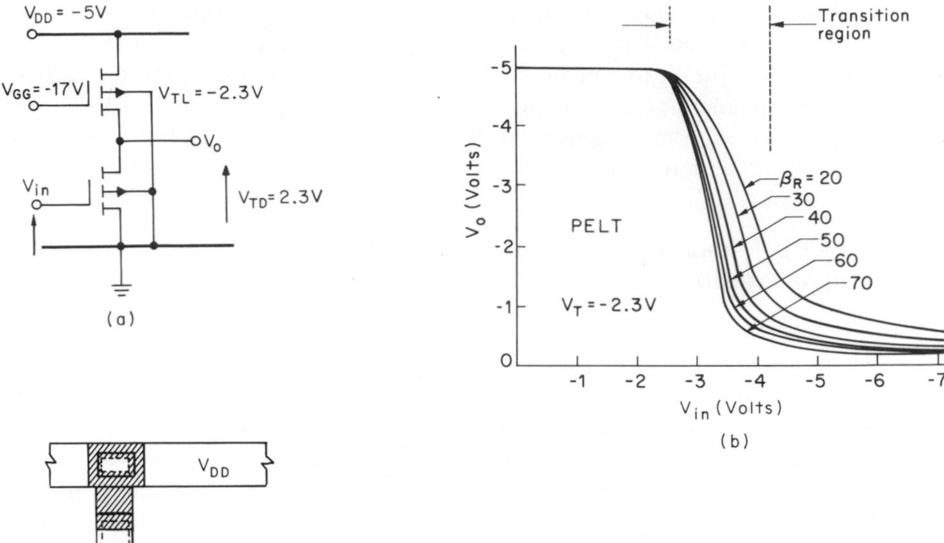

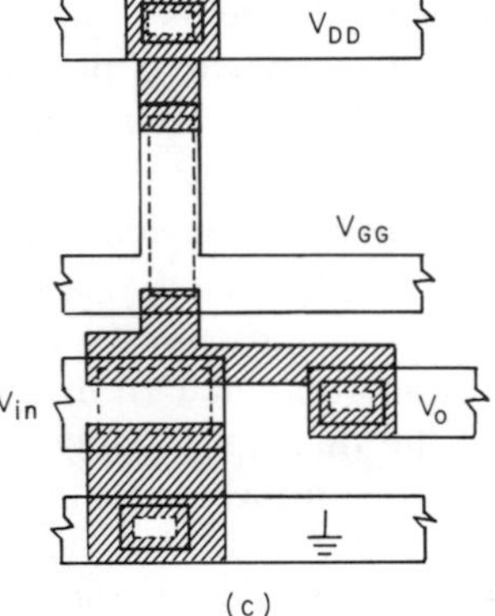

Fig. 4.2. Inverter, p-channel, enhancement-type load in triode region (PELT): (*a*) circuit schematic; (*b*) transfer characteristic curves; (*c*) mask geometry for a masking $\beta_R = 20$.

normally made to the back of the monolithic chip. During processing, a geometrical beta ratio of 20 in a design layout here will increase to approximately 35 and higher, because of the side-diffusion of impurities that occurs in diffusion.

4.3 PELS INVERTER

Here the load device is p-channel enhancement-type operated in a saturation, and is denoted as PELS. This configuration permits the use of a single power supply. Figure 4.3*a* shows a schematic circuit diagram for the PELS inverter where the threshold voltage is -2.3 V and the power supply voltage is -17 V. The transfer characteristics for this circuit, including the effect of back-gate bias and with the power supply voltages shown, have been calculated by computer and are plotted in Fig. 4.3*b*. Note that the portion of the transfer characteristic within the transition region is more linear than in the PELT case. This improved linearity results from the fact that both the driver and load device are in saturation through the voltage range specified by the transition region. Since the nonlinearities are similar for both the load and driver device in the transition region, the net effect tends to be a linearization of the transfer curve. The PELS configuration, in circuitry where the quiescent operating point can be maintained within the transition region, provides a small-signal gain of approximately $\sqrt{\beta_R}$.

The PELS inverter for digital applications contains an effective load resistance value which is larger than that for a comparable PELT inverter. The resulting effect is that the PELS inverter has less capability for charging load capacitors through the load device than does the PELT inverter. Therefore, the PELS inverter tends to switch slower than the PELT circuit during the portion of the transient in which the output voltage magnitude is increasing. The discharge portion of the switching transient is similar for both circuits.

4.4 PDLT INVERTER

When a depletion-type device operating in the triode region is substituted into the load position in a p-channel inverter, one obtains the PDLT configuration. A configuration using p-channel devices in PDLT operation is shown in the circuit schematic of Fig. 4.4*a*. Here only a single power supply is required. The gate of

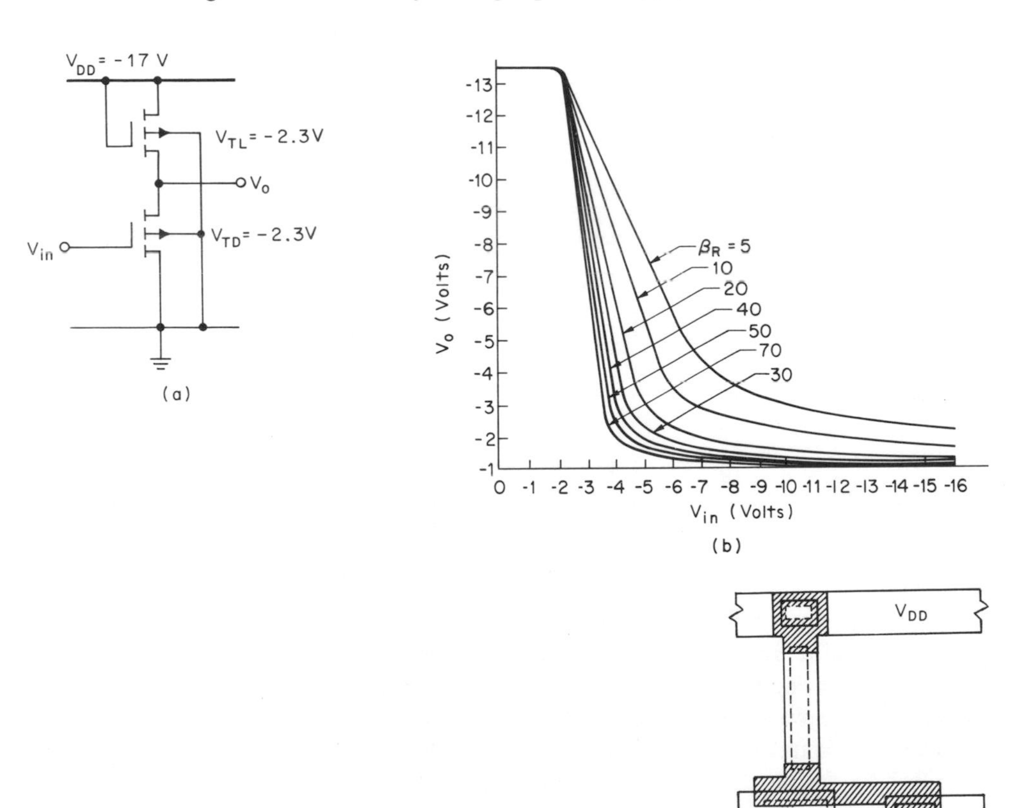

Fig. 4.3. Inverter, p-channel, enhancement-type load in saturation region (PELS): (*a*) circuit schematic; (*b*) transfer characteristic curves; (*c*) mask geometry for a masking $\beta_R = 20$.

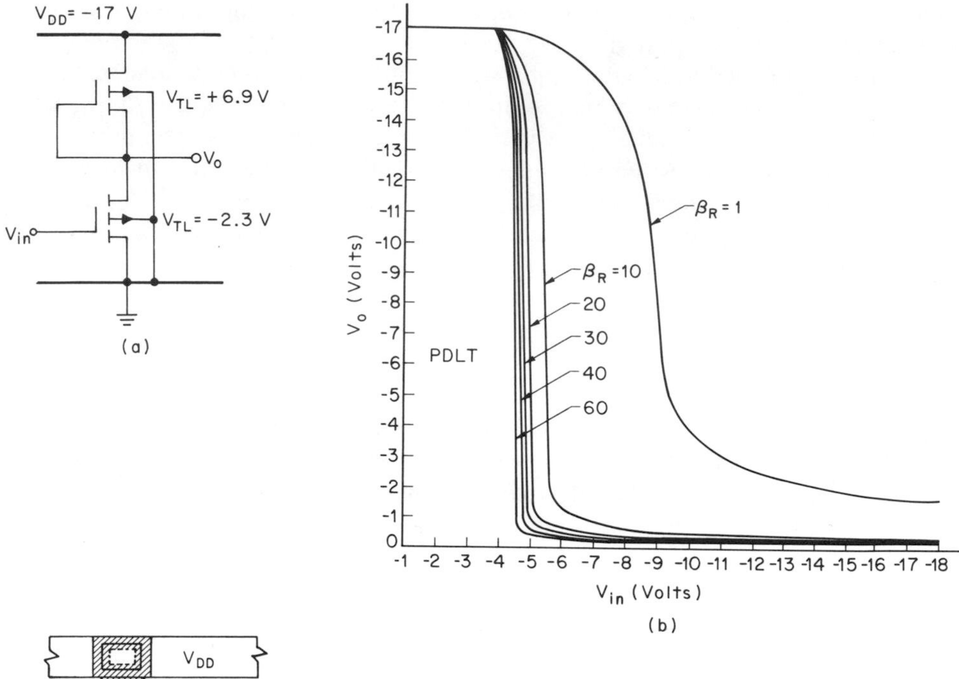

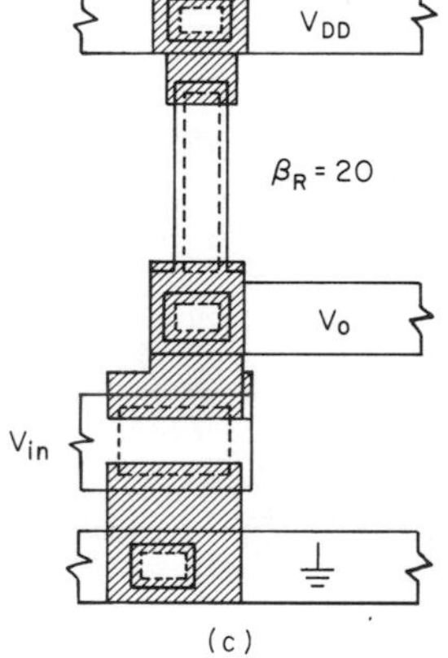

Fig. 4.4. Inverter, p-channel, depletion-type load in triode region (PDLT): (*a*) circuit schematic; (*b*) transfer characteristic curves; (*c*) mask geometry for a masking $\beta_R = 20$.

the load device is connected internally into the inverter signal output node. The minimum value for effective load resistance is generally desirable, and this is obtained by shorting the gate of the load device to the output signal node within the inverter. A power-supply voltage of -17 V has been selected for the calculation of transfer characteristics. The threshold voltage value is 6.9 V for the load, and -2.3 V for the active device. The transfer characteristics have been calculated and plotted with the electrical beta ratio shown as a parameter in Fig. 4.4*b*. These calculations include the effect of back-gate bias on the load device. The transfer characteristics within the transition region provide greater than unity gain over the entire range of β_R shown (values as low as one). The small-signal gain within the transition region is higher for this inverter compared with the PELT and PELS configurations of the same beta ratio. The small-signal gain for β_R ratios in excess of 30 will definitely be limited by the saturation resistance of the MOS transistor. Since this saturation resistance has been assumed for calculation purposes to be infinity, the transfer characteristic of Fig. 4.4*b* shows a gain at high beta ratios to be somewhat higher than is actually observed in practice.

The typical mask geometrical layout shown in Fig. 4.4*c* for the PDLT inverter is smaller than that for the previous inverters of Figs. 4.2 and 4.3, and requires approximately 5.5 × 2.5 mils of chip area. The advantages of the reduced area requirement for the PDLT inverter are offset by the increased difficulty in processing for the fabrication of the depletion-type load device.

Each of the three inverter circuits considered, PELT, PELS, and PDLT, has an output voltage level which depends upon β_R. These inverters are called *ratio-type* inverters. The higher β_R values result in increased logic swing when used with a given power supply. The PELT and PDLT inverters provide a maximum output voltage magnitude under no-load conditions corresponding to that of the V_{DD} supply value. The PELS inverter differs in that the maximum output voltage level that the inverter will provide to an external gate is limited to a magnitude equal to the supply V_{DD} value reduced by the threshold voltage of the load device. This minimum voltage drop across the load device equal to the threshold value is basic to the operation of the PELS inverter and results from the fact that we have wired the load for a saturation bias condition.

4.5 CMOS INVERTER

An inverter circuit in which the output voltages do not depend upon β_R can be designed using complementary MOS active devices for both the pullup and pulldown devices. Circuits in which the logic swing is independent of β_R are termed *ratioless*. The circuit we call *complementary MOS* is illustrated in Fig. 4.5*a*. The circuit is symmetrical because the load and active device may be interchanged in the circuit schematic, with the power supply polarity reversed, resulting in no particular difference in operation of the circuit. Each device has its respective substrate connected to the appropriate power-supply level.

When the driver and load device are complementary, as in the CMOS inverter, the power dissipation at quiescent operating points is determined by the leakage of the p-n junctions, and therefore provides micropower circuit operation in the nanowatt range for dc static conditions. For CMOS inverter operation the gate of both devices is connected to the data source node. Therefore, each device operates in a triode or cutoff region under normal static conditions. The power dissipation during switching transients includes the important dissipative component of charging and discharging the capacitors. The maximum switching speed for each of the various designs is determined by the gate-to-source, gate-to-drain, diffusion-to-substrate, and other capacitance values. The inverter does maintain its output signal node at static voltage levels, the same as the previous three inverters, and operates at frequencies down to dc.

The power dissipation of the CMOS circuit is approximately proportional to frequency, up to the maximum operating frequency. Whenever one device is

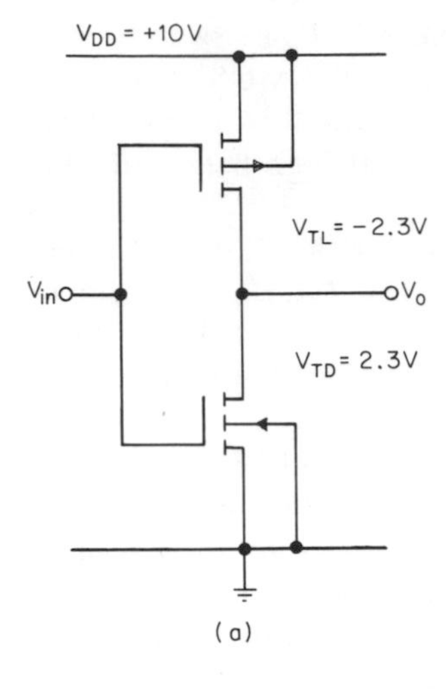

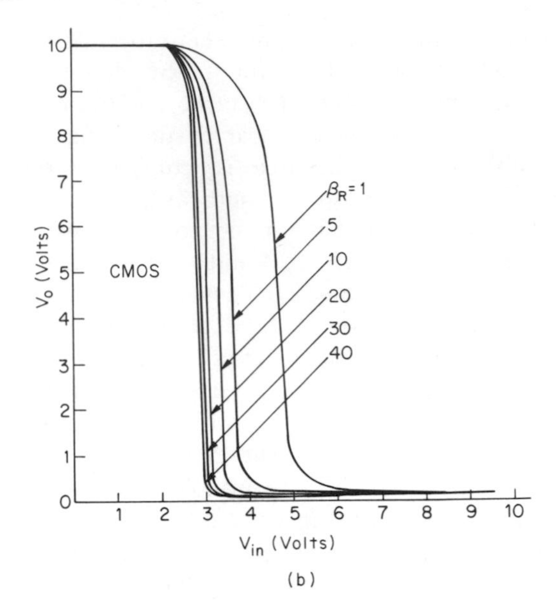

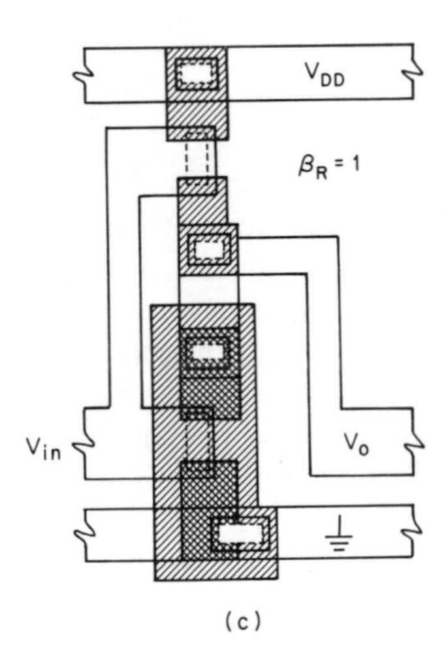

Fig. 4.5. Complementary MOS inverter (CMOS): (*a*) circuit schematic; (*b*) transfer characteristic curves; (*c*) geometry for a masking $\beta_R = 1$.

conducting *on,* at least one complementary device is held *off,* and the net circuit quiescent current is determined only by the leakage current of the *off* device.

If devices of 2.3-V threshold magnitude are used with a 10-V power supply, the transfer curves of Fig. 4.5*b* result. The large small-signal gain within the transition region, similar to the PDLT inverter, is characteristic of CMOS. Approximate equality between transconductance values for the two devices within the inverter is desirable. Since the carrier mobility is larger for the n-channel device, the β_R for optimum operation is greater than unity.

The complementary circuitry does require additional fabrication steps, as required by the illustrative layout of Fig. 4.5*c*. The example uses two diffusions into the p-type substrate. The n-type diffusion has been used to surround the active region of the driver transistor permitting an isolation of that device from the p-type substrate. Notice that the p-type monolithic substrate is connected to the most positive supply voltage point, and the p-n junction existing between the n-type diffused region and that substrate is reverse biased. It is the leakage current existing between the substrate and the n-type isolation diffusion which accounts for most of the dc power dissipation in this particular configuration. An alternate fabrication process for the CMOS circuitry involves epitaxial deposition techniques to define the two types of MOS transistors. High-performance CMOS devices can be made by the silicon-gate process using an n-type epitaxial deposition. This technique would of course require a different layout than that shown in Fig. 4.5*c*.

With comparable design tolerances, the CMOS inverter will require additional area compared to the other inverters. The CMOS technology generally permits operation to higher frequencies than we observe for the ratio-type inverters. The CMOS voltage supply levels can be quite low compared with other gate configurations due to the symmetry in the transfer characteristics. The representative supply and threshold voltage levels selected for these examples are summarized in Table 4.2.

4.6 STATIC NOISE MARGINS

The input and output characteristics of an MOS inverter can be used to determine the operating points for that gate when it is driven by similar gates. These operating points determine the logical 1 and logical 0 levels for implementing digital logic. One can define convenient parameters on the transfer curve for purposes of discussing such things as the ability of a circuit to switch accurately in spite of a high level of noise superimposed on logic signals. In Fig. 4.6 a representative transfer curve is shown in which the total logic swing is defined as being the voltage difference between the two quiescent static voltage levels at A and B, as shown. Generally an increased logic swing will result in increased noise immunity with a particular power supply. There are two points on the transfer characteristic at which the magnitude of the differential gain is unity. These are termed *unity gain points* and are labeled V_{UGA} and V_{UGB}. The transition region lies between these two points. We can define a transition width voltage as the change in input voltage necessary to shift between operating points corresponding to the first and the second unity

Table 4.2. Supply and Threshold Voltages Selected for Inverter Samples

Configuration	V_{DD}, Volts	V_{GG}, Volts	V_{TL}, Volts	V_{TD}, Volts
PELT	−5	−17	−2.3	−2.3
PELS	−17	n.u.*	−2.3	−2.3
PDLT	−17	n.u.	6.9	−2.3
CMOS	10	n.u.	−2.3	2.3

*n.u. = V_{GG} not used.

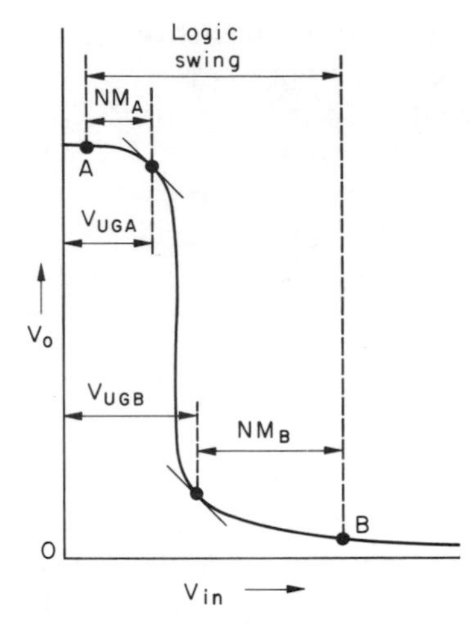

Fig. 4.6. Transfer curve with noise margins, logic swing, and unity gain points defined.

gain points (V_{UGA}—V_{UGB}). Static operating points *A* and *B* may correspond to a logical 0 or a logical 1, depending upon whether one is considering positive or negative logic, and upon the polarity of the power supply voltages. The transfer characteristics for which representative examples have been given previously in this chapter show the magnitude of the input voltage V_{in} increasing toward the right irrespective of its polarity.

The transfer characteristic furnishes us with information which relates directly to the magnitude of the noise voltage, superimposed on the static logic levels, that the gate can tolerate before changing states. To be specific, we define the noise margin *NM* of the inverter to be the voltage difference, measured on the input voltage axis of the transfer characteristic, between the operating point and the nearest unity gain point.

$$NM_A = V_{UGA} - V_A \tag{4-1}$$

$$NM_B = V_B - V_{UGB} \tag{4-2}$$

where the subscripts *A* and *B* denote the two separate noise margins. The unity gain voltages generally represent the threshold for voltage switching between states. Each inverter is characterized by two separate noise margins. Either noise margin can be the limiting factor in determining noise immunity for a particular circuit. It should be pointed out that the definition of noise margin used is a conservative estimate in practical circuits. In many cases noise signals can move the operating point somewhat into the transition region without causing a false output, thereby providing an effectively higher noise margin.

Sources of noise that may cause an inverter to indicate a false logical output pose a very important concern in design. The sources of noise that result in false signals may be externally or internally derived. The external sources of noise, generally crosstalk from external interconnects, dominate in certain environments. For example, rf interference falls into this category. A second type of noise is internally generated by capacitive or inductive coupling between signal lines on the chip. If a gate is controlling a large logic voltage swing in one signal line and there exists

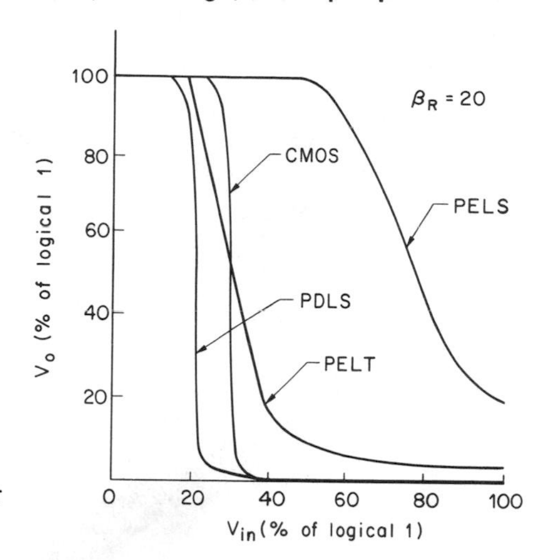

Fig. 4.7. Comparison of static transfer curves.

appreciable capacitive coupling to an adjacent line, then crosstalk, or cross coupling of signal levels, results. If the undesired voltage is transferred to the input of a logic circuit, that circuit may temporarily change states, resulting in an erroneous output. For example, the capacitance between metalization and diffused regions might introduce an error signal during a transient switching operation. Since internally generated noise is proportional to the logic swing, it is often convenient to consider the ratio of noise margin to logic swing as a figure of merit. This dimensionless quantity we will define as noise immunity (NI).

A rough comparison of noise immunity for those examples of inverter circuits we have been considering is shown in Fig. 4.7. A beta ratio of 20 has been selected for purposes of comparison in each case with the exception of the CMOS, where the beta ratio is unity. Notice that the noise immunity for the PELT inverter is the lowest, because the power-supply voltage is only 5 V compared to larger supply voltages used for the other inverters. In Fig. 4.7 a transition region centered at the 50 percent level for the input voltage V_{in} is close to optimum, representing a matching for the noise margins at the high and low ends of the transition.

Noise margins are compared in greater detail in Table 4.3. The noise margins are expressed (1) as a percentage of a power-supply voltage reduced by threshold

Table 4.3. Noise Margin Comparison

Configuration	β_R	V_{DD}, Volts	V_{TD}, Volts	V_{TL}, Volts	NM_B % of $(V_{DD} - V_{TD})$	NM_B Volts	NM_A % of V_{TD}	NM_A Volts
PELT*	20	−5	−2.3	−2.3	19	0.5	74	1.7
PELS	20	−17	−2.3	−2.3	51	7.5	106	2.5
PDLT	20	−17	−2.3	6.9	90	13.3	110	2.6
CMOS	20	10	2.3	−2.3	88	6.8	104	2.4

* $V_{GG} = -17$ V.

value, and (2) by their actual magnitudes in volts. The PELT inverter has its smallest noise margin at operating point *B* of 0.5 V. Fortunately low noise margin for the PELT inverter is offset by the rather low impedance level at operating point *B*, and therefore the inverter provides acceptable immunity to noise. The noise margin and the input characteristic for the gate, as well as the circuit impedance levels, should be considered in predicting the overall susceptibility of the circuit to noise voltages.

For practical designs one must also consider the variation in transfer characteristics from circuit to circuit. In considering the static characteristics of the inverter, we have, up to this point, limited our discussion to nominal characteristics. In actuality these characteristics depart from their normal values when variables such as production tolerances, temperature, specific circuit source and loading conditions, power-supply drift, and aging are taken into account. If the upper and lower limits of the terminal characteristics are obtained by allowing for these variations, the worst operating case among various operating conditions can be determined easily. In other words, instead of being single line relationships, the terminal characteristics now become a spectrum of curves relating to the range of operating parameters. One can do worst-case analyses using limit calculations or Monte Carlo techniques. Most useful calculations involving MOS integrated-circuit design and analysis require a computer. Therefore, we will not concentrate on developing analytical shortcuts in this text for purposes of making these calculations. Instead, we will present a brief summary of computer-aided design techniques.

4.7 COMPUTER MODELING OF TRANSIENT CHARACTERISTICS

The transient performance of an inverter within a logic gate depends not only upon the particular geometry and process technology, but also upon the power-supply voltages and loading conditions. For comparison purposes, the four inverter configurations which we have been considering are modeled, and computer calculations have been obtained for the resulting transient response.

The models used for these inverters are shown in Fig. 4.8*a* for the PELT, PELS, and PDLT inverters and in Fig. 4.8*b* for the CMOS configuration. The standard equations for triode and saturation operation describe the current sources shown as J_1 and J_2 in the equivalent circuits. For illustrative purpose, the input voltage V_{in} is assumed to be much faster than the switching transient within the inverter. The inverter is assumed to drive an identical loading stage (fan-out) for which the loading gate-to-source capacitance is included in the model. The diffused junctions are considered to be voltage-dependent capacitors corresponding to the source and drain diffusions connected into the output node. Capacitance C_6 and capacitance C_{10} refer to the ratio and ratioless inverters with corresponding diffused capacitance values. Capacitance C_9 shunting the gate of the load device to ground represents the metalization-to-substrate value which is in parallel with the power-supply V_{GG} line. The additional voltage-dependent capacitor C_{10} is associated with the diffused junction between the output voltage node and the substrate for the CMOS. The resistors R_1 and R_2 are arbitrarily large values required by the computer-aided-analysis program (SCEPTRE) and do not affect the transient response. Other parameters used for the computer analysis include capacitance values (pF per mil^2):

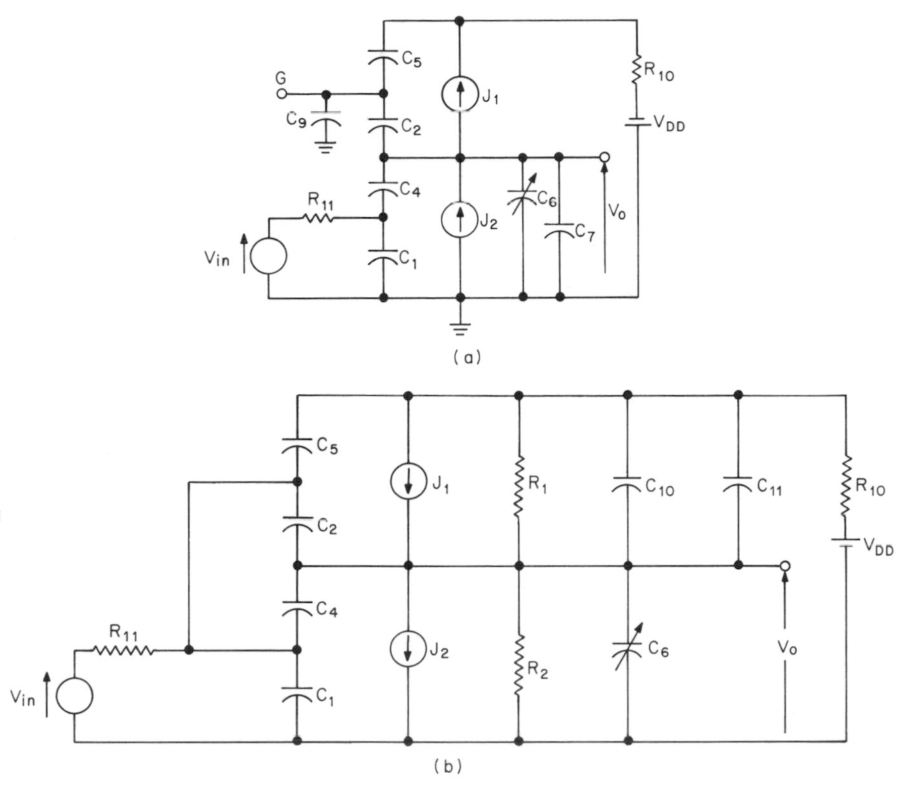

Fig. 4.8. Equivalent circuit for inverters: (*a*) PELT, PELS, and PDLT configurations; (*b*) CMOS configurations.

for the thin oxide, 0.18; thick oxide, 0.018; p-n junction at zero bias, 0.11; and gate-to-substrate oxide, 0.12. The specific capacitance values taken from the mask geometry previously shown for the various inverters have been summarized in Table 4.4. The beta ratio β_R value is selected at 20 for each inverter, with the exception of the CMOS, for which a value of unity is used.

Table 4.4. Capacitance Values Used for Lumped Equivalent Circuits

Configuration	C_1	C_2	C_4	C_5	C_6^*	C_7	C_9	C_{10}^*	C_{11}	R_1	R_2	R_{11}	R_{10}
PELT	0.0659	0.0135	0.0394	0.0124	0.232	0.00144	0.0653	n.a.	n.a.	∞	0	∞	0
PELS	0.0659	0.0125	0.0394		0.232	0.00144	0.0475	n.a.	n.a.	∞	0	∞	0
PDLT	0.0659		0.0394	0.0124	0.308	0.0489	included in C_7	n.a.	n.a.	∞	0	∞	0
CMOS	0.0302	0.0129	0.0129	0.0657	0.489		included in C_1	0.285	0.0245	10 MΩ	100 Ω	100 Ω	100 Ω

All capacitances are in pF.

*Indicates voltage-dependent capacitance $= C_6/(1 + V)^{0.45}$ for PELS, PELT, and PDLS

$$= C_{10}\sqrt{\frac{0.6}{0.6 + V}}$$ for CMOS

where V is the voltage across the capacitor.

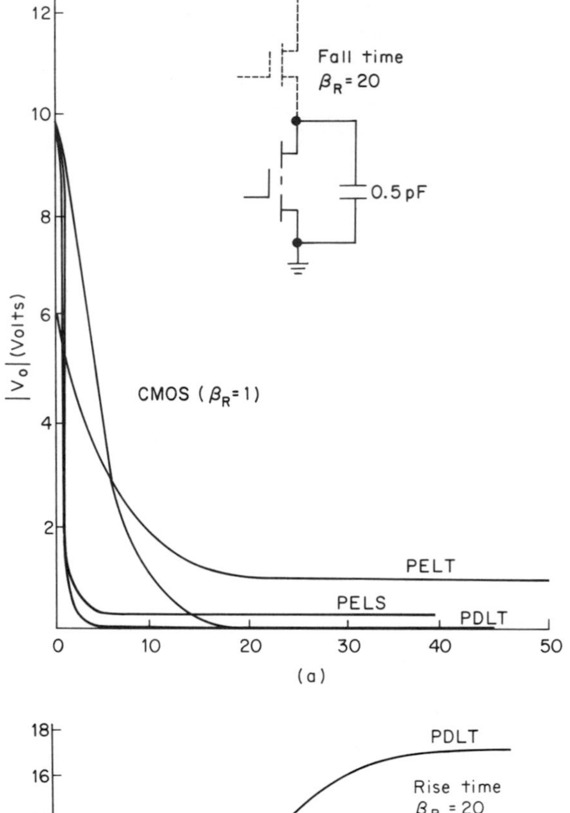

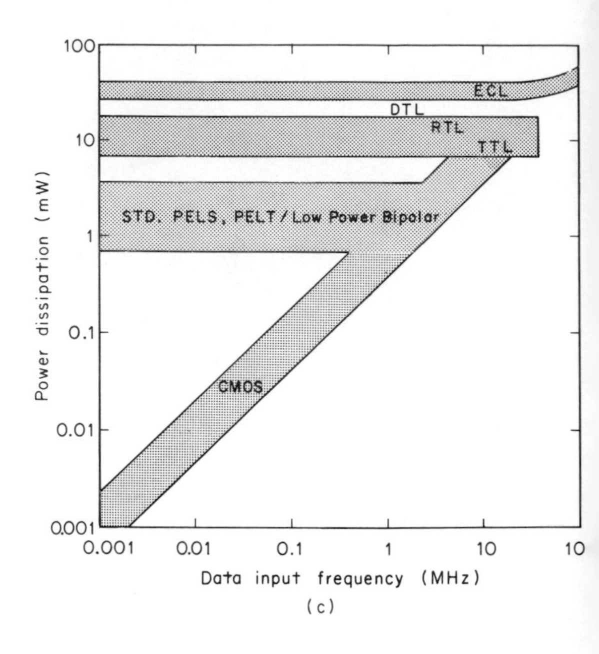

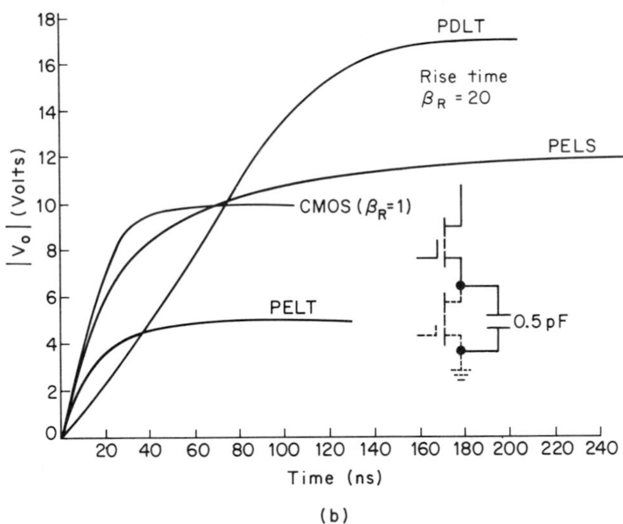

Fig. 4.9. Transient response for inverters: (*a*) lower (driver) device conducting; (*b*) upper (load) device conducting. The electrical beta ratio $\beta_R = 20$, and parameters are given in Tables 4.3 and 4.4; (*c*) representative power dissipation values for static gates as a function of data input frequency.

The most common method of specifying transient performance of a logic gate is to determine the propagation delay and rise and fall times for the voltage waveform from that gate. Two separate output voltage transient waveforms are shown in Fig. 4.9*a* and *b* as determined from the equivalent circuits and parameters above. The response of Fig. 4.9*a* refers to that transient in which the output capacitance is discharged through the lower device within the gate. This discharge path from quiescent operating point *B* to point *A*, referred to in Fig. 4.6, is noted to be faster compared to the other transient in each case except CMOS. These values are, of course, representative only for the very specific examples selected here but do indicate that the discharge transient in Fig. 4.9*a* generally is faster than the charging transient illustrated in Fig. 4.9*b*. In the ratio-type inverter the driver device operated in the triode region permits a fast discharge of the load capacitance. The load device within the ratio-type inverter, having a reduced transconductance, does not permit a match in speed for charging the output capacitance during the transient. In the rise-time transient of the ratio-type inverters in Fig. 4.9*b*, one notes that the PELT, with its load device operated in the triode region, offers the fastest turn-on transient. The PELS inverter shown here provides a much slower rise-time transient because the load device operates in saturation, and not in the low resistance triode region. The PELS inverter has always been slower in circuit applications as compared to the other inverters.

The representative example of the PDLT inverter could have been selected using a load device of increased transconductance. With careful selection of the threshold voltage and the geometry in the PDLT inverter case, one can obtain transfer characteristics that are quite symmetrical and very fast. The PDLT inverter can approach, if not match, the CMOS design for rise- and fall-time speeds, given proper and careful circuit design. The switching time required to move between a quiescent point and a unity-gain point provides a fairly accurate measure of the time required to switch an inverter. These times are summarized in Table 4.5 for the four cases considered. Here the slower rise time as compared with the turn-off time is clearly shown for each circuit. The rise time can be made approximately equal to the fall time with careful circuit design for the PDLT and CMOS cases, but will always be much slower for the remaining ratio-type inverters. It should be pointed out that the transient characteristics for which we have made these illustrative calculations are standard-gate layouts and do not possess the reduced capacitance one can obtain using a silicon-gate, self-aligning process. The self-aligned gate will provide an increase in speed of at least a factor of 2 in most practical circuit designs.

The switching transients of Fig. 4.9*a* refer to p-channel devices in every case except CMOS. It should be noted that the considerable increase of mobility that can be obtained through design with n-channel MOSFET inverters is very desirable. The switching speed, roughly proportional to channel mobility, points toward an obvious

Table 4.5. Transient Response Times for the Four Inverters Selected

Configuration	Time to reach V_{UGB}, ns	Time to reach V_{UGA}, ns
PELT	3.2	38.2
PELS	5.0	250.0
PDLT	8.8	157.5
CMOS	5.5	30.5

design preference for n-channel enhancement-type devices for inverters as production technology permits. In the past, production yields have not provided cost-effective n-channel MOSFETs for LSI, although promising work in this direction is certainly under way.

In addition to the switching transient speed, we are concerned with power dissipation and area required for circuit implementation. A minimum power dissipation is especially important in large-scale integration, where the total circuit power approaches that of the header or package dissipative limit. Also, we prefer a reduced power dissipation to limit the requirements for the power supply, which often involves high-speed clocking drivers. Therefore, another important figure of merit is the propagation-delay power product (generally referred to as the speed power product). A minimum value for this product is, of course, desired.

The propagation-delay time represents a slightly more complex function than the rise and fall times since the source impedance of the driver device must be considered. The propagation delay is normally defined as the time delay between the 50 percent points in the input and output waveforms for a particular gate in serial cascade with similar gates. The delay time depends not only upon the internal response of the gate, but also upon the capacitance at the input node and the load capacitance at the output terminals. Consequently, fan-in and fan-out conditions, as well as the amount of stray capacitance at the input and output nodes, should be specified when quoting delay times. The power dissipation for the ratio-type inverters is approximately independent of data input frequency up to relatively high clocking rates and is proportional to the duty cycle for the power supply V_{GG}. At higher frequencies the power dissipation for the ratio-type inverter will be proportional to clocking rates because of dissipation associated with capacitive loading.

Typical power dissipation levels for the various circuits are shown in Fig. 4.9*c*. The standard PELS inverters will dissipate one to a few milliwatts of power per gate. The PELT inverter has slightly reduced power dissipation compared with the PELS circuit. The ratioless inverters, which include the CMOS inverter, have a power dissipation proportional to the data input frequency. The CMOS power dissipation is comparable to that of the ratio-type inverters at high frequencies but is reduced to nanowatt levels for static holding. It should be pointed out that the p-channel enhancement-type devices can be used in a ratioless type inverter circuit in which multiple-phase clocking is used. These ratioless inverters are introduced in Chap. 5.

Another very important figure of merit is the propagation-delay power area product commonly referred to as the speed power area product. Since the area required for a given circuit function is an important factor in determining the cost of that part, one cannot omit the consideration of *area* from an overall figure of merit that includes economic considerations. We have made computer calculations for this figure of merit using various inverters. Minimum values for speed power area products have occurred for p- and n-channel depletion-load triode inverters (PDLT and NDLT). The speed power area cost product minimizes for that economical inverter for which we are able to use a single power supply, the PELS. It should be emphasized that these general comparisons refer to specific technology at the present time and are subject to revision by both design and technology innovation at any time.

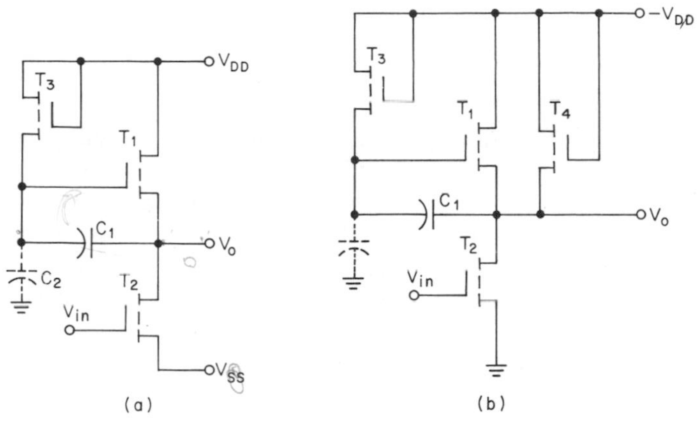

Fig. 4.10. Bootstrap-pullup inverter for increased drive capability: (*a*) dynamic only; (*b*) with reduced drive for dc holding.

4.8 BOOTSTRAP-PULLUP INVERTER

There are novel circuits, useful for special applications, that provide excellent speed power area product values. This figure of merit has a particularly low value for the bootstrap-pullup inverter circuit of Fig. 4.10*a*, useful for fast switching operations in which one desires the ability to drive larger capacitive loads. You will recognize this circuit as being similar to the PELS inverter because a single power supply is used. However, a difference results because an additional device, T_3, is used between the most negative circuit point and the gate of the load device. The additional transistor T_3 coupled to the capacitance C_1 offers very important advantages for fast switching transients. The switching transient can be followed by observing that initially, transistor T_2, operating in a triode region, permits a voltage of $V_{DD} - V_T$ to appear across capacitance C_1. Next, as the input voltage V_{in} switches to a zero voltage level, the output voltage V_o tends to increase as device T_2 ceases to conduct. The large voltage appearing across capacitance C_1 represents the gate bias for the load device T_1. During the initial period, the load device was operating in a saturation region, the driver device was in a triode region, and the output node V_o level is at a low voltage value. As the driver device T_2 ceases to conduct, the larger voltage magnitude across capacitor C_1 remains during the switching transient, provided that the capacitance C_2 in shunt is of small enough magnitude. As the voltage at the output node increases in magnitude, the load device is forced into a strong triode region, resulting in a fast turn-on for V_o. The gate of the load device T_1, in that case where the shunt capacitance C_2 is completely neglected, reaches a maximum magnitude of $2\ V_{DD} - V_T - V_{SS}$, where V_{SS} is the substrate bias value. Therefore, the additional device T_3 acts as a pullup transistor and enables the single power-supply stage to approach the PELT inverter in performance.

Notice that the inverter of Fig. 4.10*a* will not operate under static dc conditions because the charge will drain from capacitance C_1 through the reverse-biased p-n junctions. Specifically, one obtains a minimum holding time, in the range of a few milliseconds for the circuit of Fig. 4.10*a*. The operation of the circuit can be extended, however, to a limited holding condition at dc by the addition of a sustaining transistor T_4 which is placed in parallel with the load device T_1. The inverter which includes the static holding device T_4 is shown in Fig. 4.10*b*. This holding device has a very low transconductance and does not substantially reduce the switching speed of the inverter, but does extend operation of the inverter to lower frequency. The bootstrap-pullup inverter is useful only for driving highly capacitive loads; it will not permit faster operation in a standard cascade circuit situation. The bootstrap-pullup inverter does include additional capacitance in the form of C_1 and C_2. For specific cases when the load capacitance is increased in excess of 1 pF, one finds important advantages in switching speed resulting directly from the use of the bootstrap-pullup. Some specific circuits, with device geometry similar to Fig. 4.4*c*, have been characterized using the pullup device under capacitive load situations. The results summarized in Table 4.6 indicate that a 0.5 pF load permits a rise time of 18 ns when the load device T_1 has a W/L ratio of $1(\beta_R = 20)$. This rise time increases to 78 ns when the W/L ratio is reduced to $\frac{1}{4}$. The output rise time of Table 4.6 is defined as the time required to change V_o following an input at $t = 0$ to 90 percent of the full output voltage swing. These figures refer to the use of a pullup device with a W/L ratio of unity. The inverter circuit with the additional devices discussed is especially important in interfacing from an MOS circuit to external loading devices located off the chip.

4.9 NOR/NAND LOGIC GATES

The inverter circuit performs the logical NOT function in which the input data is complemented by the inverter. The most positive output voltage from the inverter is considered to be a logical 1 by the convention of positive logic. Similarly, the most negative output voltage from the inverter is considered to be a logical 1 according to the negative-logic convention. This definition of positive and negative

Table 4.6. Switching Parameters for a Bootstrap-pullup Inverter Driving a 0.5 pF Capacitive Load

Driver W/L ratio	β_R	Depletion load pinch-off voltage V_T, Volts	Output rise-time, ns (see comment)
4	20	9	395
		12	228
		18	117
		24	78
20	20	9	100
		12	57
		18	28
		24	18
40	20	9	69
		18	19
		24	12
80	20	9	53
		18	15
		24	10

Table 4.7. Positive and Negative Logic Defined

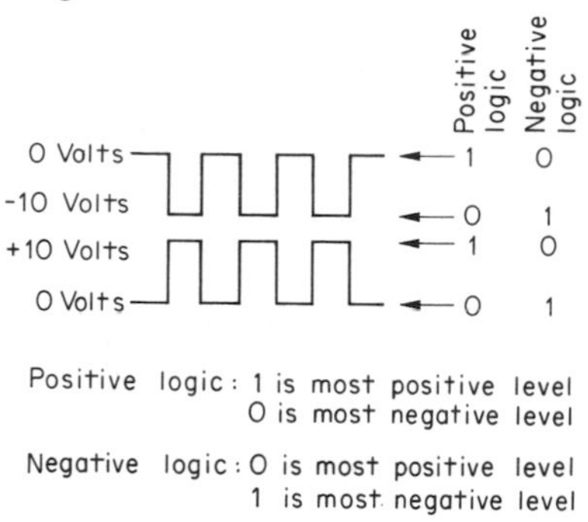

logic is general and may refer to logic derived from any circuit element. The concept is summarized in Table 4.7. For designs with conventional bipolar integrated circuits, one normally uses positive logic to describe the circuit action because the power-supply voltages swing between a zero and a positive value.

With MOS integrated circuits using the p-channel enhancement-type devices, one works with drain voltages which are negative with respect to the source, and a negative-logic convention is sometimes helpful for analyzing circuit operation. The output of a PELS inverter, according to a negative-logic convention, for instance, is a logic high or 1 when the output voltage is at a maximum magnitude with respect to the source. Throughout our discussion in this chapter we observe the negative-logic convention, although some comparison between the two types of logic convention is made. The logic symbols used are taken from MIL-STD-806B.

NOR/NAND logic can be implemented by simple MOS circuits as shown in Fig. 4.11*a*. In Fig. 4.11*b* the PELT inverter containing two parallel devices in the driver position performs the NOR logic function, according to a negative-logic convention,

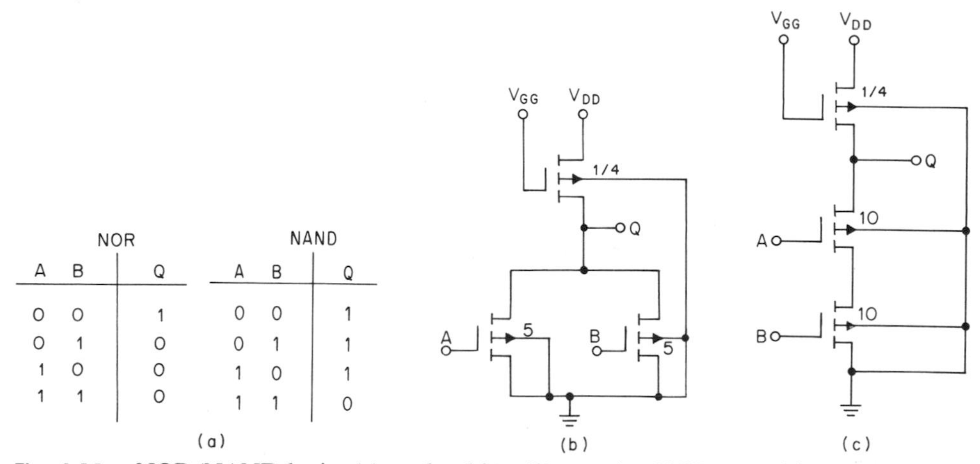

NOR

A	B	Q
0	0	1
0	1	0
1	0	0
1	1	0

NAND

A	B	Q
0	0	1
0	1	1
1	0	1
1	1	0

Fig. 4.11. NOR/NAND logic: (*a*) truth tables; (*b*) negative-NOR or positive-NAND gate; (*c*) negative-NAND or positive-NOR gate. PELT implementation.

with the negative-going power supply. This same PELT gate also performs the NAND function, according to a positive-logic convention. Similarly, in Fig. 4.11*c* the gate provides a positive-NOR or a negative-NAND function. In the gate of Fig. 4.11*b* the worst-case (maximum V_o) effective β_R occurs for the case with a single driver device conducting. This worst-case β_R value is 20 and results from W/L ratios in the load and driver of $\frac{1}{4}$ and 5, respectively. The effective beta ratio increased to 40 when a second driver device conducts. We refer to the gate of Fig. 4.11*b* as having a worst-case beta ratio of 20. The gate of Fig. 4.11*c* also has a beta ratio of 20 since the transconductance of the two series devices is specified as double ($W/L = 10$) that of a single driver device in Fig. 4.11*b*. The transconductance consideration is very important in the design of complex arrays since β_R directly affects the noise margin and noise immunity of the circuit. The static transfer characteristics of the circuits in Fig. 4.11*b* and *c* will be identical for the same power supply and input voltage value, except for a small difference in the back-gate bias effect on the load devices.

The complementary CMOS gates are shown in Fig. 4.12. A positive-NAND function is obtained in Fig. 4.12*a*, and a positive-NOR function is obtained in Fig. 4.12*b*. Here one requires two devices in the gate for each input data line. Each input is connected directly to the gate of a driver and a corresponding load device. This double gate connection increases the device count for CMOS inverters above that for a comparable circuit function using the PELS, PELT, PDLT, and other ratio-type circuits. The circuit action for the gates in Fig. 4.12 is essentially the same as that for the CMOS basic inverter, with the load and driver devices all operating in a triode conducting region and providing the symmetrical circuit switching action. The input data line enables its corresponding driver or load—but not both—for either input data position.

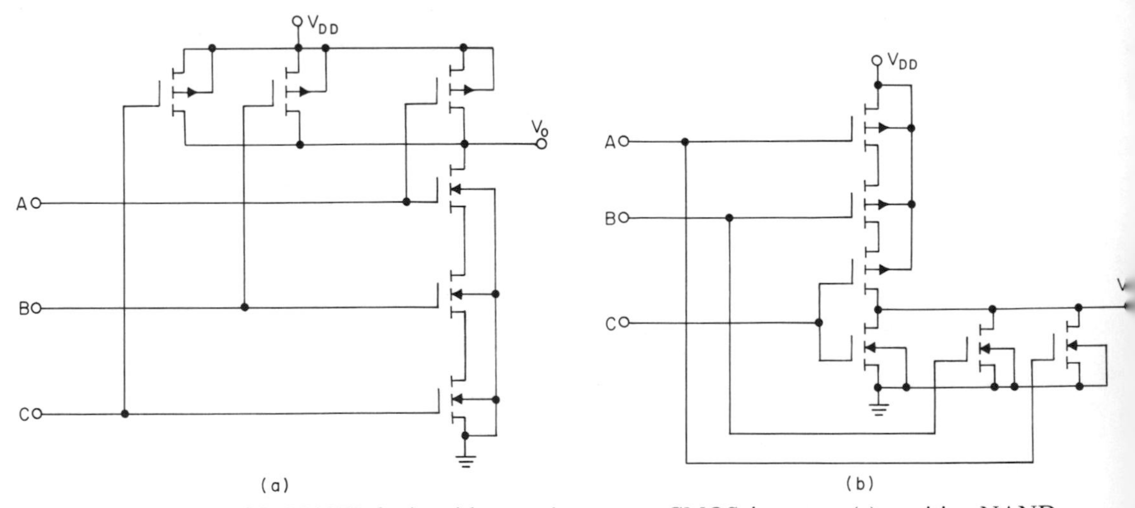

Fig. 4.12. NOR/NAND logic with complementary CMOS inverter: (*a*) positive-NAND gate; (*b*) positive-NOR gate.

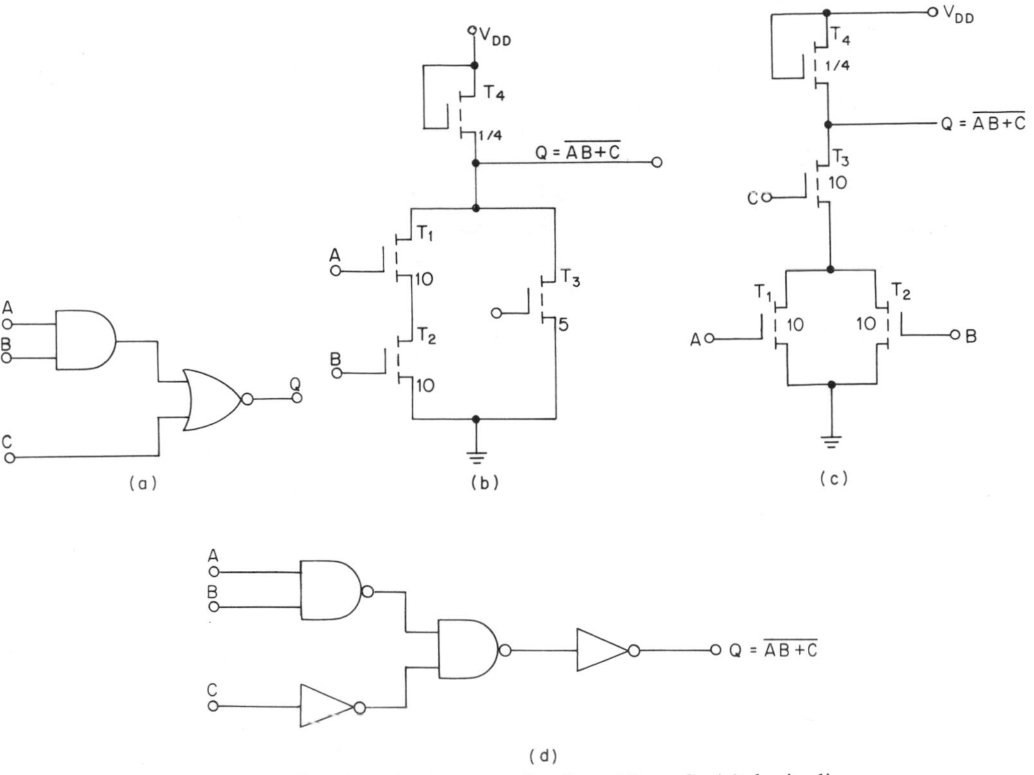

Fig. 4.13. Random logic example $Q = AB + C$: (*a*) logic diagram; (*b*) negative-logic, PELS circuit; (*c*) positive-logic, PELS circuit; (*d*) alternate logic diagram using all NAND and inverter gates.

A random logic example is shown in Fig. 4.13 for implementing the function $Q = \overline{AB + C}$. In Fig. 4.13*a* the logic diagram is given for this Boolean expression, with a PELT implementation shown in Fig. 4.13*b* using negative logic. Note the efficiency with which the MOS circuitry permits implementation of a two-level logic function using only four devices in a circuit which includes a single load transistor. The propagation delay for this two-level logic gate is basically that of a single inverter, with the additional capacitance associated with multiple devices, but not multiple stages. The same device count is required to implement this function using positive logic in PELS stage as shown in Fig. 4.13*c*. The design of Fig. 4.13*b* is more desirable than that of Fig. 4.13*c*, because the transconductance of the individual devices is reduced, providing an improved noise margin and reducing the total area required for the gate. When one is required to use NAND gates throughout, as is often required in bipolar implementation, the logic requires three levels, as shown in Fig. 4.13*d*. Note that the MOS implementation requires the use of a single load device, providing an equivalent triple-level NAND logic. Among the unique abilities of MOS, and generally not available using bipolar circuitry, are the multiple logic levels obtained using a single load device.

4.10 HALF-ADDER CIRCUIT IMPLEMENTATIONS

The flexibility available in MOS circuitry permits the implementation of random logic using a variety of circuit configurations. A half-adder function is implemented in Fig. 4.14 as an example showing five different circuits all based on the PELS basic inverter scheme. In Fig. 4.14*a* a device count of nine provides the sum Q_S and Q_C true values derived from a double-rail input. Two delay levels are present in this circuit and minimum transconductance values are possible, permitting the maintenance of a beta ratio of 20 within each logic level.

In Fig. 4.14*b* an increase in the number of devices permits one to obtain the half-function using complemented inputs only, with minimum-geometry driver devices. In Fig. 4.14*c* all true levels are available, directly permitting matching of logic levels for coupled circuits. The circuit of Fig. 4.14*c* requires implementation of the negative-NAND function and therefore requires driver devices that are not minimum transconductance. The device count in the circuit in Fig. 4.14*c* increased to 16, and therefore the circuit requires an increased area for implementation on the chip. A more efficient circuit is shown in Fig. 4.14*d*. This circuit requires 11 devices and further maintains the true logic levels at both input and output. The

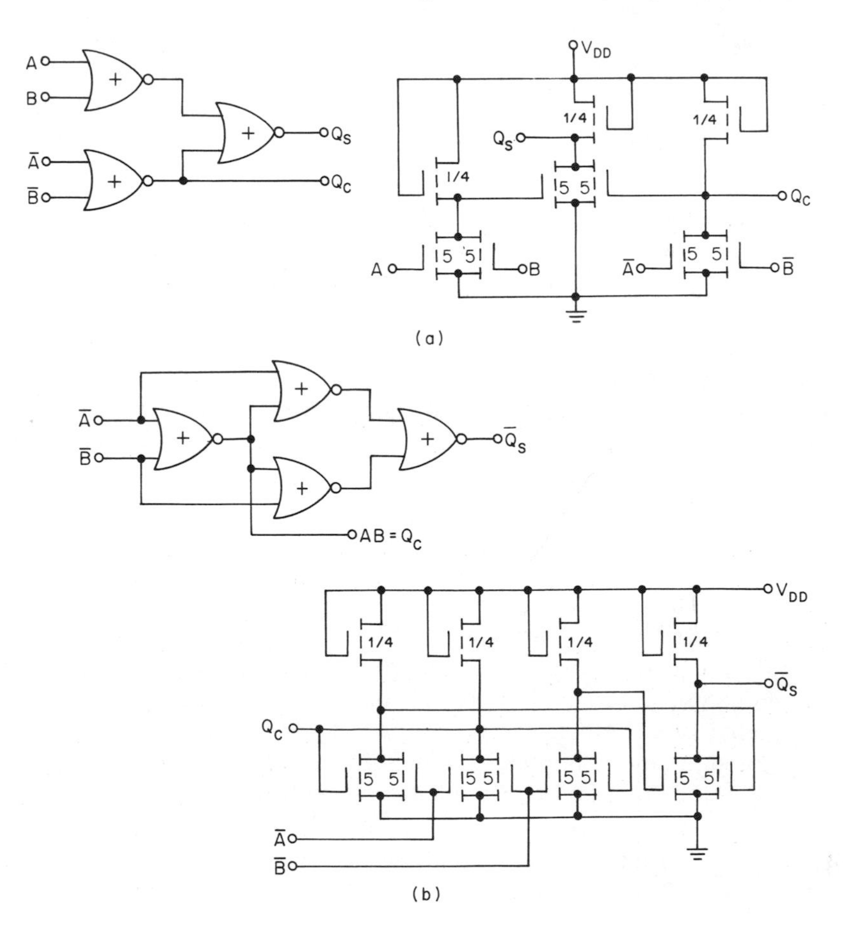

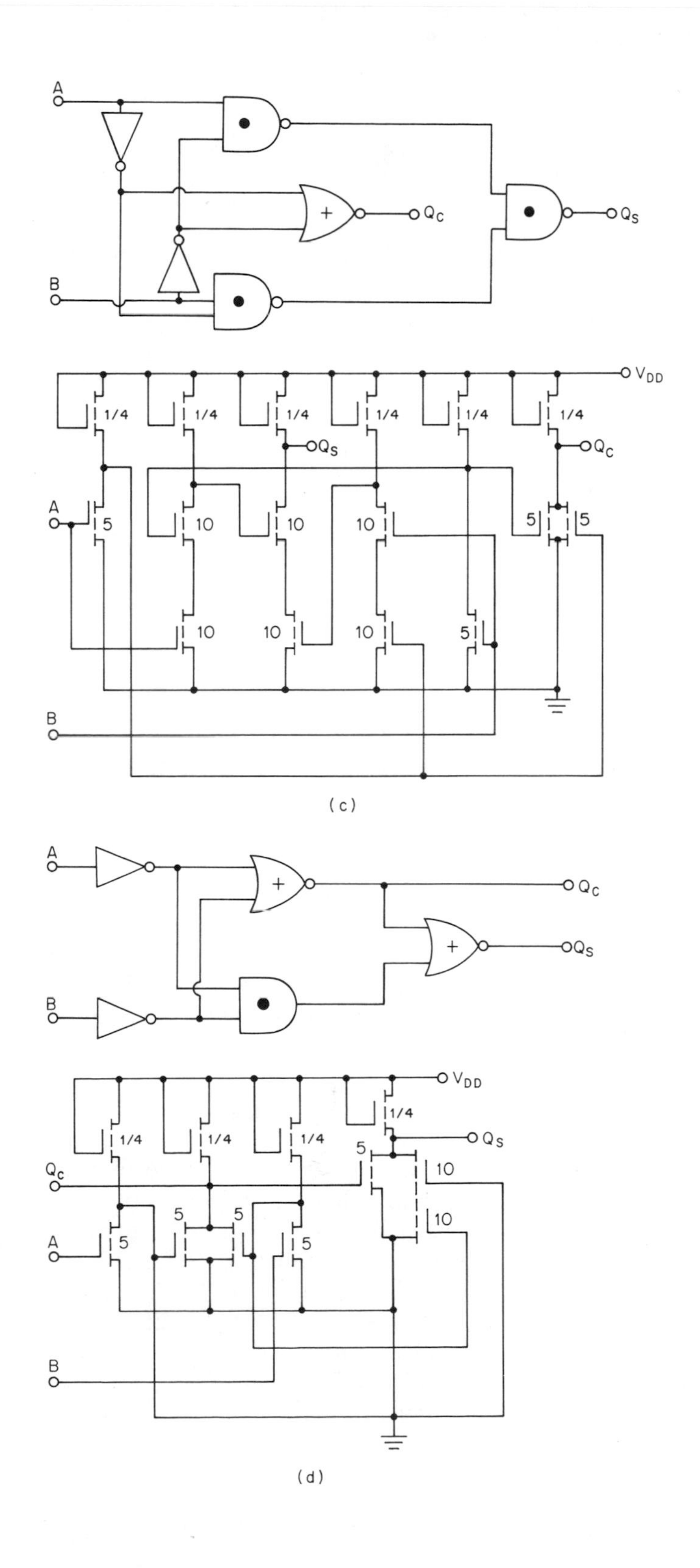
A
B
Q_C
Q_S
V_{DD}
1/4
5
10
(c)
A
B
Q_C
Q_S
V_{DD}
1/4
5
10
(d)

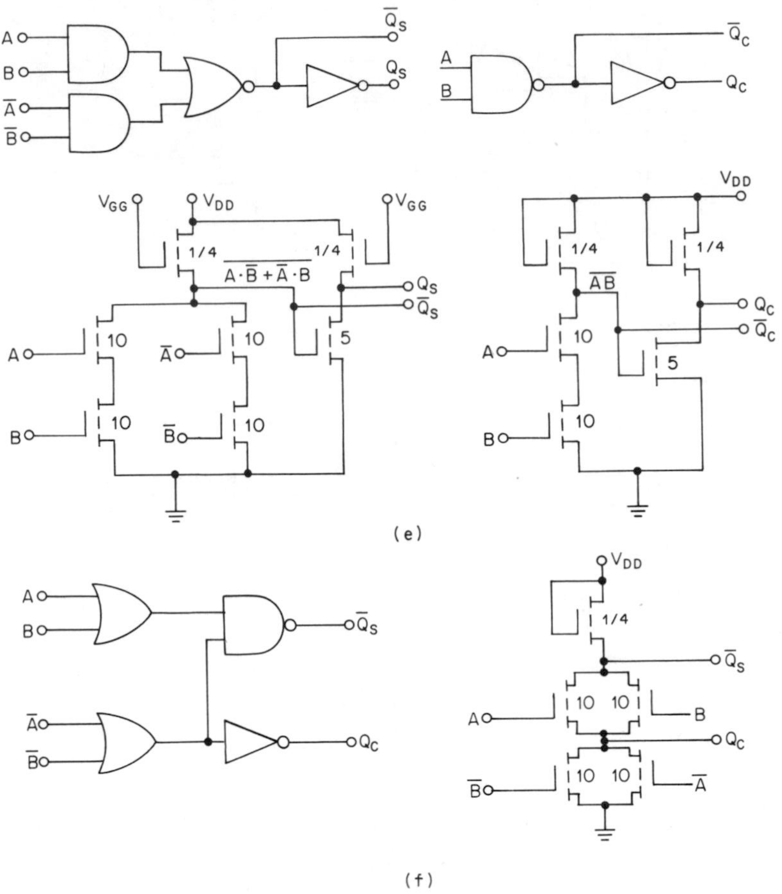

Fig. 4.14. Half-adder circuitry showing negative-logic equivalents; (*a*) all NOR gates; (*b*) all NOR gates with complemented logic inputs only; (*c*) all true levels with 16 devices; (*d*) all true levels with 11 devices; (*e*) both complemented and true logic levels available at the output; (*f*) low-device-count design.

number of logic levels required here is two, with a minimum-transconductance geometry maintained throughout, except for two devices that require increased gain.

One can roughly estimate the area requirement for one of these circuits by summing the transconductances for the component devices within the circuit. For instance, in Fig. 4.14*d* the transconductance, or g_m, count sums to 61. This g_m count may be compared with 104 for the case of circuit Fig. 4.14*c*.

Still another circuit configuration is shown in Fig. 4.14*e* which requires a double-rail input function and uses negative-AND gates in this PELS circuit. The carry signal Q_C is not available from the sum-circuit devices, and therefore it must be implemented with additional devices. A novel circuit configuration is shown in Fig. 4.14*f*. In this circuit the carry Q_C signal is derived from within a summing series-

Table 4.8. Comparison of PELS Half-adder Circuits

Circuit	Device count	g_m count	Compatible levels for fan-out?	Noise margins adequate?	Number delay levels
(a)	9	42	No	Yes	2
(b)	12	52	No	Yes	3
(c)	16	104	Yes	Yes	3
(d)	11	61	Yes	Yes	3
(e)	12	86	Yes	Yes	2
(f)	5	44	No	No	1

combination. A device count of five is shown and a single delay is present, although the logic levels available at the output are not of the correct logic match for driving a similar stage in fan-out. The design of Fig. 4.14*f* is undesirable, however, because the noise margin is not acceptable for the carry signal, and capacitive spiking raises the possibility of spurious output transient levels.

A comparison of these half-adder circuits is made in Table 4.8 using various figures of merit. The g_m count refers to the sum of W/L ratios for all devices in the various circuits. Three of the five circuits shown in Fig. 4.14 provide output levels of the correct logic polarity for driving a similar stage. The other two circuits may be made compatible with similar circuits by the addition of either one or two inverter circuits, with an increase in device count of two and four, respectively. The circuit shown in Fig. 4.14*d* offers a load device count of one and a g_m count of 61 and is optimum compared to the other circuits that provide compatible levels for driving similar circuits in fan-out. There are many MOS circuit configurations that will implement the same logical function. One must consider a range of circuit possibilities to select an optimum implementation.

The sum signal Q_C in the half-adder is identical to the exclusive-OR function and has been implemented with CMOS gates in Fig. 4.15. Here two logic levels are used to obtain the desired function. With CMOS one can also implement a dual-level function using a common set of load devices as is done in the AND-NOR section of this circuit.

4.11 FULL-ADDER CIRCUITRY

An extension in design from the half-adder provides the full-adder function. The full-adder has three inputs: the addend A, the augend B, and the previous stage carry C. It has two outputs: the sum Q_S and the carry Q_C. From the truth table given Fig. 4.16*a* we derive the Boolean equation for the full adder.

$$Q_S = ABC + A\bar{B}\bar{C} + \bar{A}\bar{B}C + \bar{A}B\bar{C} \qquad \text{(4-3)}$$

$$Q_C = AB + AB + AC \qquad \text{(4-4)}$$

The sum Q_S may also be expressed as follows (using exclusive-OR functions):

$$Q_S = (A \oplus B) \oplus \mathrm{C} \qquad \text{(4-5)}$$

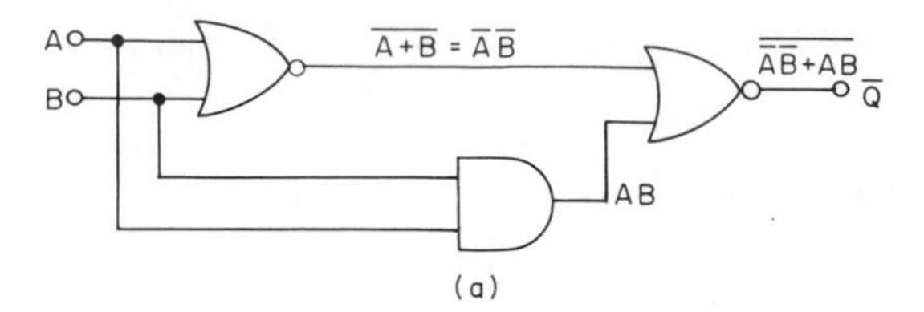

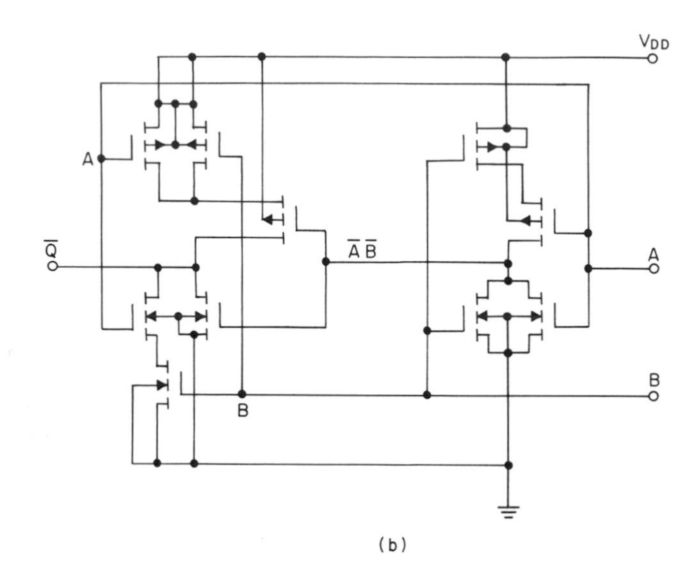

Fig. 4.15. Exclusive-OR (XOR) function implemented with CMOS (positive-logic): (*a*) logic diagram; (*b*) circuit schematic.

and the logic diagram is shown for this function in Fig. 4.16*b*. One may implement the full-adder using the half-adder and an exclusive-OR component with the logic diagram shown in Fig. 4.16*c*. A straightforward approach would be to implement the various full-adders using the half-adder components previously discussed. This, it turns out, results in a large device count and is, therefore, not optimum. The full-adder logic diagram can be rearranged in various ways, one example of which is shown in Fig. 4.17*a*. This diagram suggests a direct implementation of the full-adder function using the Boolean expressions

$$Q_S = Q_C(A + B + C) + ABC \tag{4-6}$$
$$Q_C = AC + BC + AB \tag{4-7}$$

The direct implementation following from these Boolean expressions is given in Fig. 4.17*b* and requires a total of 21 devices. This circuit provides both the true and complemented output sum and carry signals, with only a true input signal level required.

An optimum full-adder circuit implementation is determined after considering the range of circuit possibilities. One can obtain the full-adder function directly from the logic diagram of Fig. 4.18*a* or the Boolean expressions

$$Q_S = \bar{Q}_C(A + B + C) + ABC \tag{4-8}$$
$$Q_C = AB + (A + B)C \tag{4-9}$$

The resulting PELS circuit schematic in Fig. 4.18*b* requires only the true signal levels and a total of 18 devices. One disadvantage of this circuit results from the fact that the series string of three separate devices may degrade the noise immunity of the circuit; but the total area required for this circuit would be less than that for other implementations using ratio-type circuitry.

4.12 THE BISTABLE ELEMENT

One of the static storage elements in MOS circuitry is equivalent to the Eccles-Jordan bistable element in vacuum-tube amplifiers. The monolithic integrated-circuit bistable element functions basically like the vacuum-tube circuit. Figure 4.19 shows the logic diagram, PELS circuit equivalent, and CMOS circuit equivalent for bistable elements in Fig. 4.19*a*, *b*, and *c*, respectively. The bistable element contains inputs permitting one to set or reset the double-rail output nodes from the circuit as desired.

Another bistable element is the latch, a basic element, as described above, in which data can be clocked in through an internal AND gate. The truth table for the latch is shown in Fig. 4.20*a*, together with a logic diagram shown in Fig. 4.20*b*. The PELT implementation for the latch is given in Fig. 4.20*c* in negative-logic convention. Information present at a data input is transferred to the Q output when the clock C is high, and the Q output will follow the data input as long as the clock remains

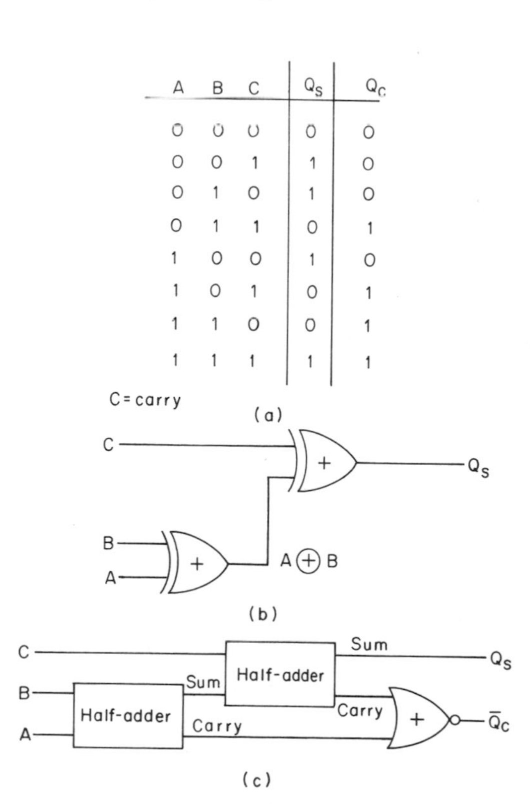

Fig. 4.16. Full-adder design: (*a*) truth table; (*b*) sum derived from dual XOR gates; (*c*) dual half-adders provide complete full-adder function.

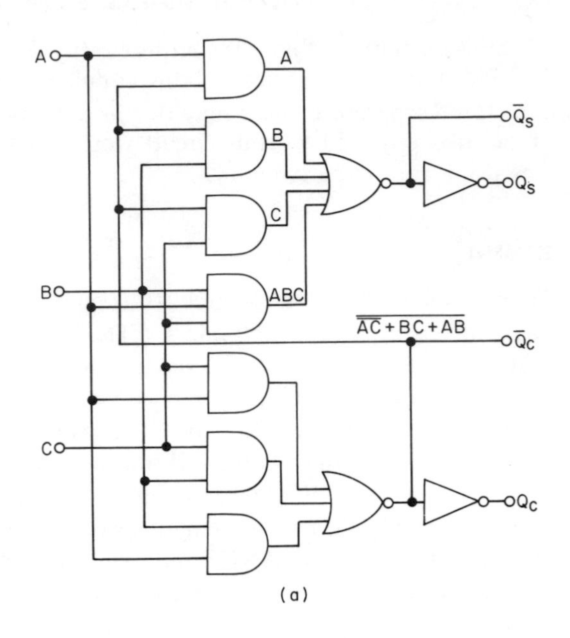

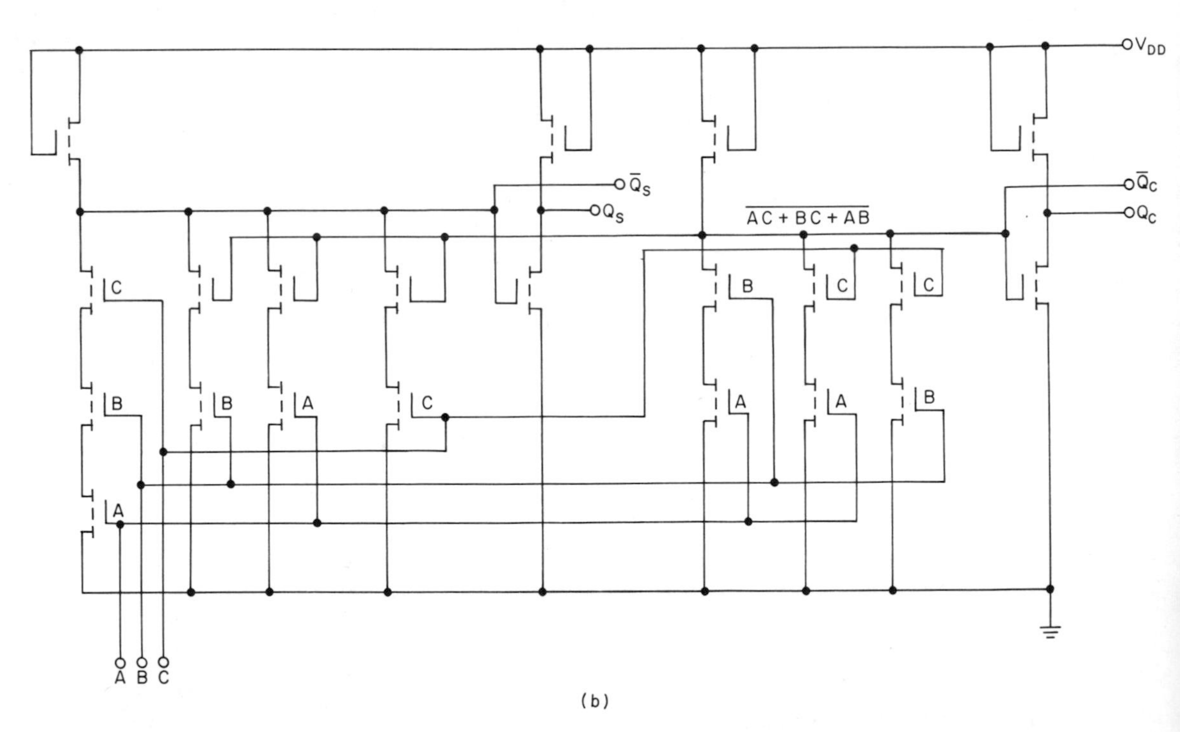

Fig. 4.17. Full-adder derived directly from $Q_C = AB + BC + AB$ and $Q_S = Q_C(A + B + C) + ABC$: (*a*) logic diagram; (*b*) circuit schematic.

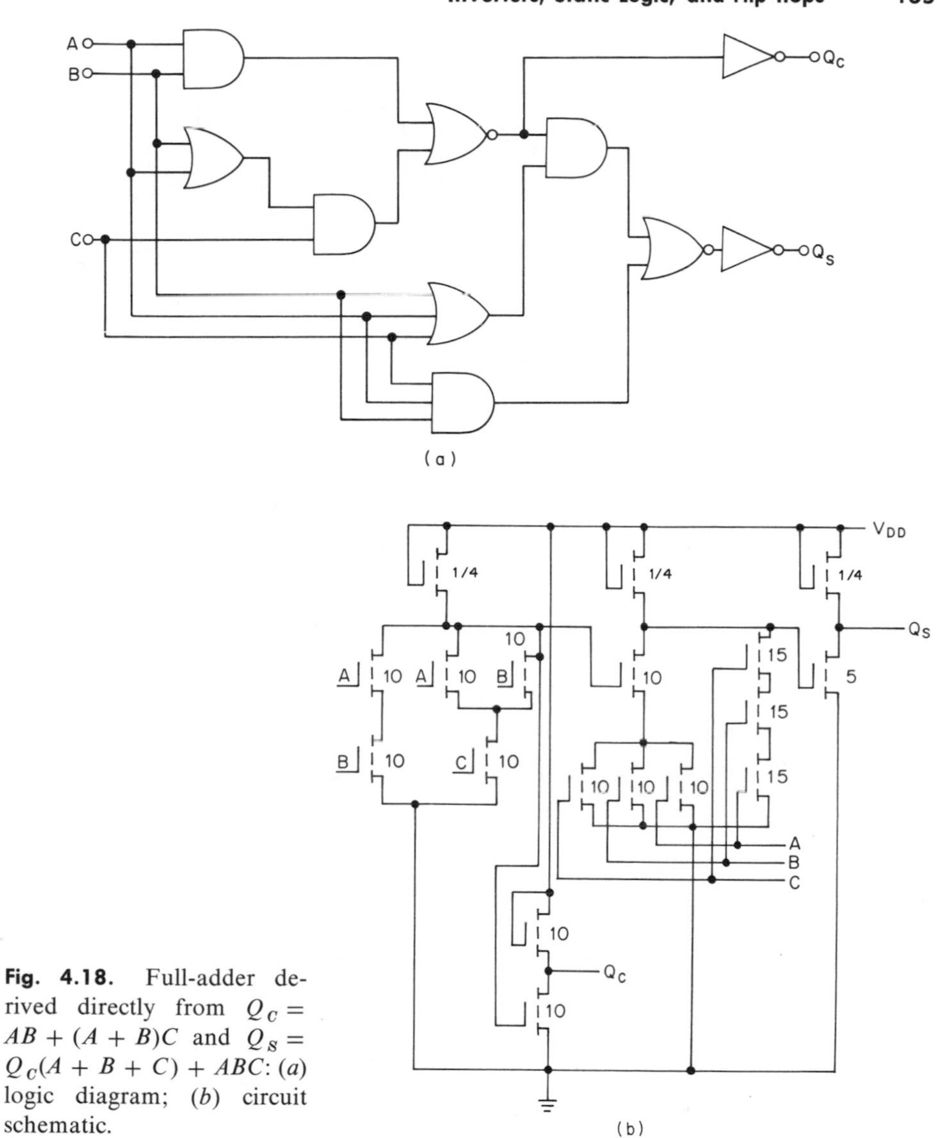

Fig. 4.18. Full-adder derived directly from $Q_C = AB + (A + B)C$ and $Q_S = Q_C(A + B + C) + ABC$: (*a*) logic diagram; (*b*) circuit schematic.

in a logical 1 or high position. When the clock goes low, the information is retained at the Q output until the clock is permitted to go high again. The latch does not have a reset input at a separate circuit node; it must be reset by clocking in data through the D node in a time-serial manner.

4.13 TOGGLE *T*-TYPE FLIP-FLOP CIRCUITRY

The toggle T-type flip-flop gets its name from its ability to toggle or change state with each input data pulse T. The truth table and transition diagram for

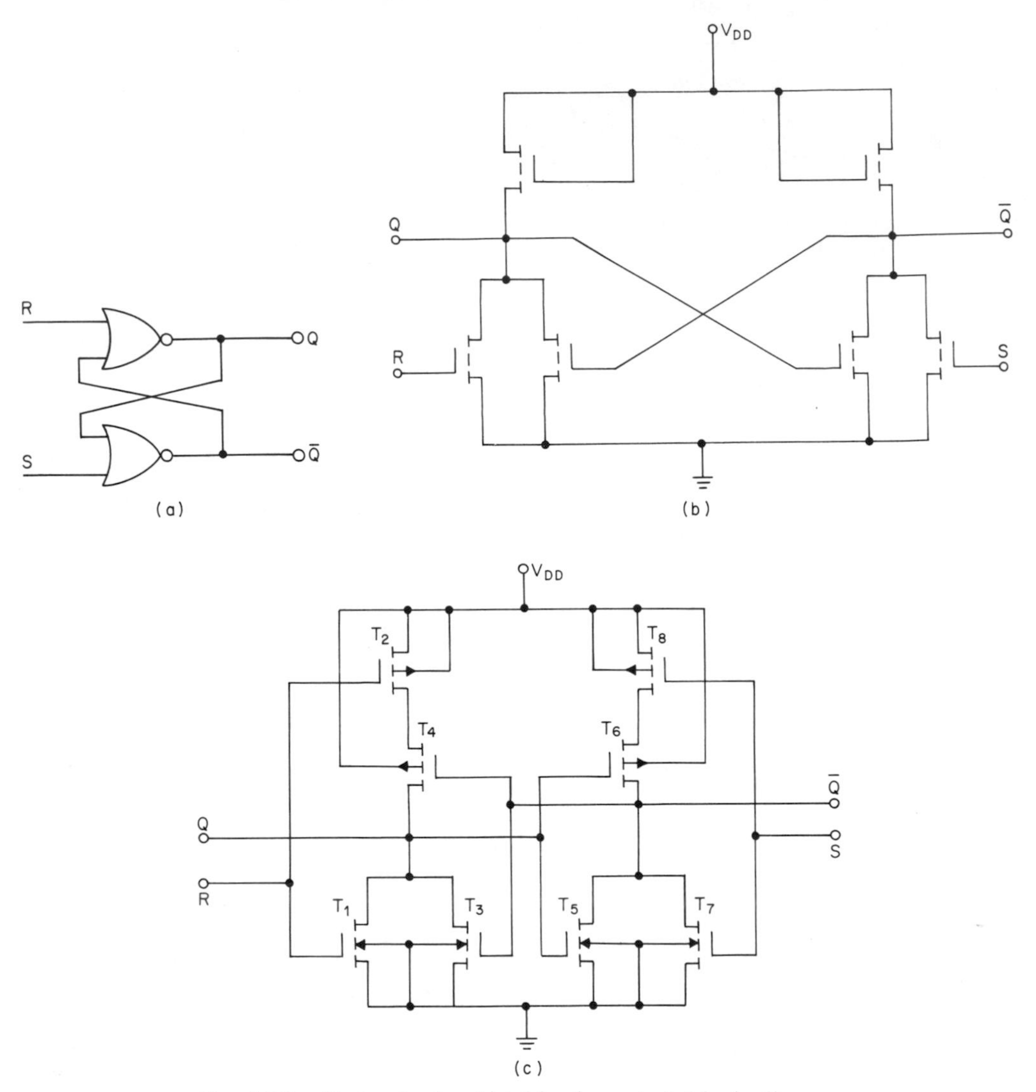

Fig. 4.19. Eccles-Jordan bistable element: (*a*) logic diagram; (*b*) PELS circuit schematic; (*c*) CMOS circuit schematic.

the *T*-type flip-flop are shown in Fig. 4.21*a* and *b*. A logic diagram for this flip-flop together with a PELS negative-logic circuit is presented in Fig. 4.21*c* and *d*. The toggle circuit consists of the basic bistable element in which the input data is gated through AND gates connected within a feedback circuit. In the circuit schematic, devices T_1 and T_2 drive the basic flip-flop from which the double-rail outputs are obtained directly. When the input pulse T is at a negative logical 1, the steering devices T_7 and T_1 are conducting, and the output of the flip-flop T_1 and T_2 is coupled into the gates of T_4 and T_6 with corresponding storage capacitance C_1 and C_2.

These storage capacitances will maintain the gates at devices T_4 and T_6 constant during the period when the input data function T is switching. If we assume that the input function T is at a logical 0, the drain of T_2 is set to a large negative magnitude (logical 1), and capacitances C_2 and C_1 will be charged and discharged, respectively. As the input level T goes to a high voltage magnitude, the potential on the gate of device T_5 completes the low resistance path from node Q through device T_6 down to ground. As a result, the output level Q moves to a logical low, device T_2 turns on, and the flip-flop toggles. As the input pulse T returns to ground potential, the flip-flop remains in the same stage and does not toggle until the input T goes to a negative voltage again. The toggling action depends upon temporary charge storage at capacitor C_1 and C_2. The total device count for the circuit in Fig. 4.21*d* is 12, counting the two devices within the inverter.

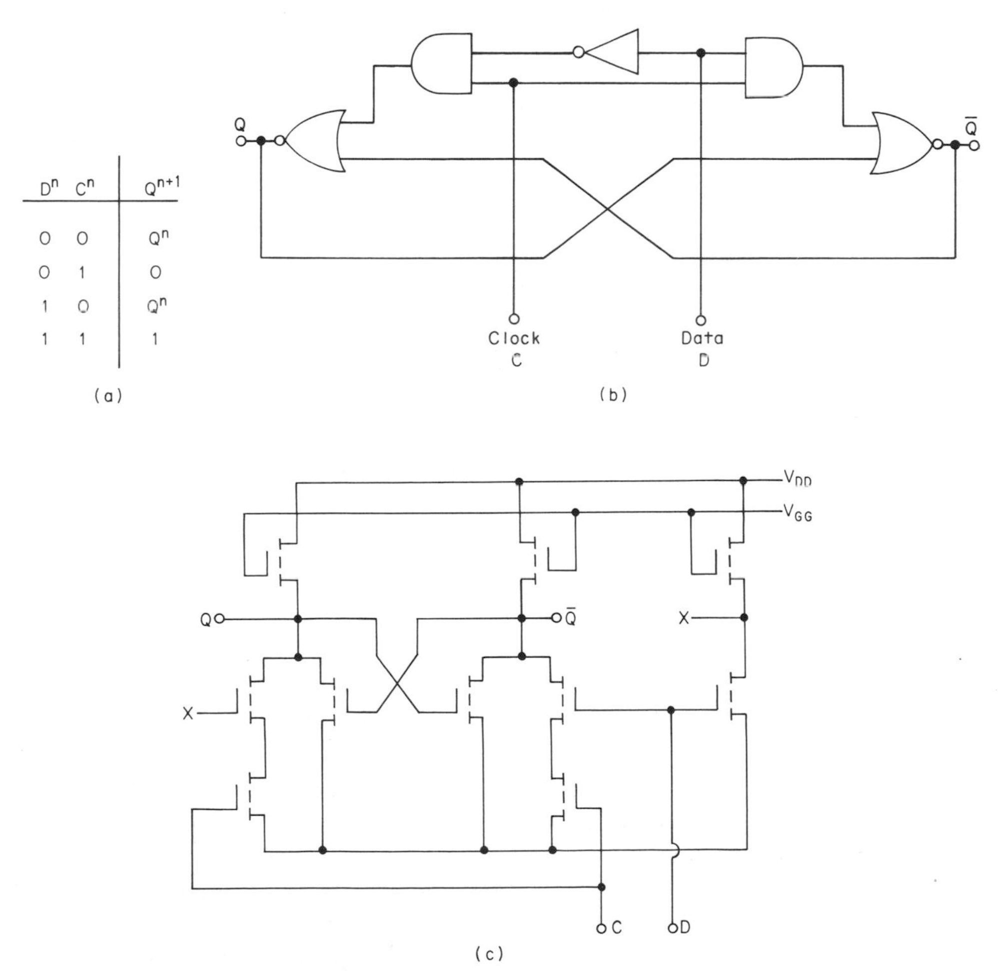

Fig. 4.20. Static logic latch: (*a*) truth table; (*b*) logic diagram; (*c*) PELS circuit schematic.

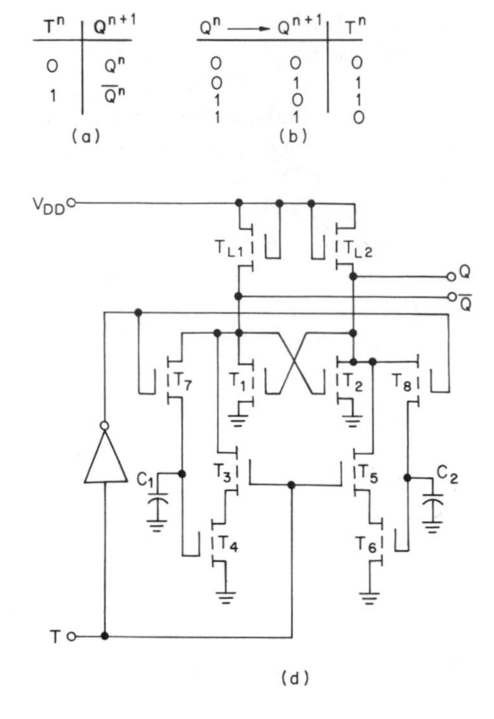

T^n	Q^{n+1}
0	Q^n
1	$\overline{Q}^n$

(a)

$Q^n \longrightarrow$	Q^{n+1}	T^n
0	0	0
0	1	1
1	0	1
1	1	0

(b)

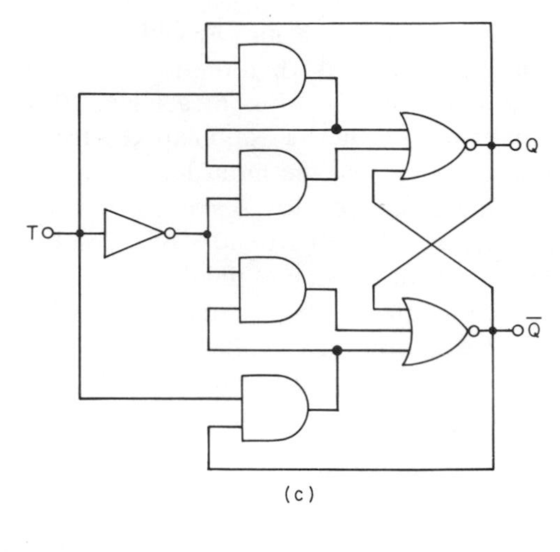

Fig. 4.21. Toggle (T-type) static flip-flop: (a) truth table; (b) excitation table; (c) logic diagram; (d) PELS circuit schematic.

An alternative circuit design is shown in Fig. 4.22. It uses only data-steering gates and does not require a temporary charge storage capacitance. In Fig. 4.22a the toggle T input is gated through an AND gate to the bistable element. The PELS equivalent circuit for this function is shown in Fig. 4.22b. A dual-level gating scheme is required in which a NOR gate is used to compare the toggle input with the bistable output function Q. The circuit of Fig. 4.22 is normally preferred for the static toggle implementation in MOS.

4.14 *R-S*-TYPE FLIP-FLOPS

The reset-set, or R-S, flip-flop is derived directly from the bistable element of Fig. 4.23. The truth table and excitation table for this flip-flop are shown in Fig. 4.23a and b. The R-S flip-flop can be programed or steered to any desired state by programing the input according to the truth table or excitation table. The function characterized here remains stable as long as the R and S inputs are in the logical 0 state. If S increases to a high level, the flip-flop assumes a 1 state in its output, whereas if R is raised to a 1, the flip-flop assumes a logical 0 on its output node Q. When S and R are both at the 1 level simultaneously, the output is intermediate—that is, unknown—and there is generally no way to predict what state the output Q will be in. Therefore, $S = R = 1$ is called a *forbidden state*, and it is a state to be avoided by appropriate programing of the inputs S and R.

The excitation table is developed from the truth table by deductive analysis. The symbol ϕ represents a "don't care" condition and does not affect the output transition. Figure 4.23*c* shows a master-slave combination for providing the *R-S* flip-flop function shown in the excitation table. Here one has direct-input S_D and R_D lines available to override the clocked inputs $\bar{S}$ and $\bar{R}$. Two separate bistable elements are presented in the master-slave combination. The master flip-flop contains set *S* and reset *R* functions that are clocked in with the clocking pulse *C*. This master-slave circuit in Fig. 4.23*d* requires a total of 24 devices.

4.15 *J-K*-TYPE FLIP-FLOPS

The *J-K*-type flip-flop can be considered a refinement of the *R-S*-type flip-flop, in which the indeterminate state is now defined. The truth table for the *J-K*-type flip-flop and its excitation table are shown in Fig. 4.24*a* and *b*. Here, if the control inputs *J* and *K* are high and the flip-flop is initially in the 0 state, it will change to a 1 state when a pulse occurs, resulting in toggling action. Conversely, if it is

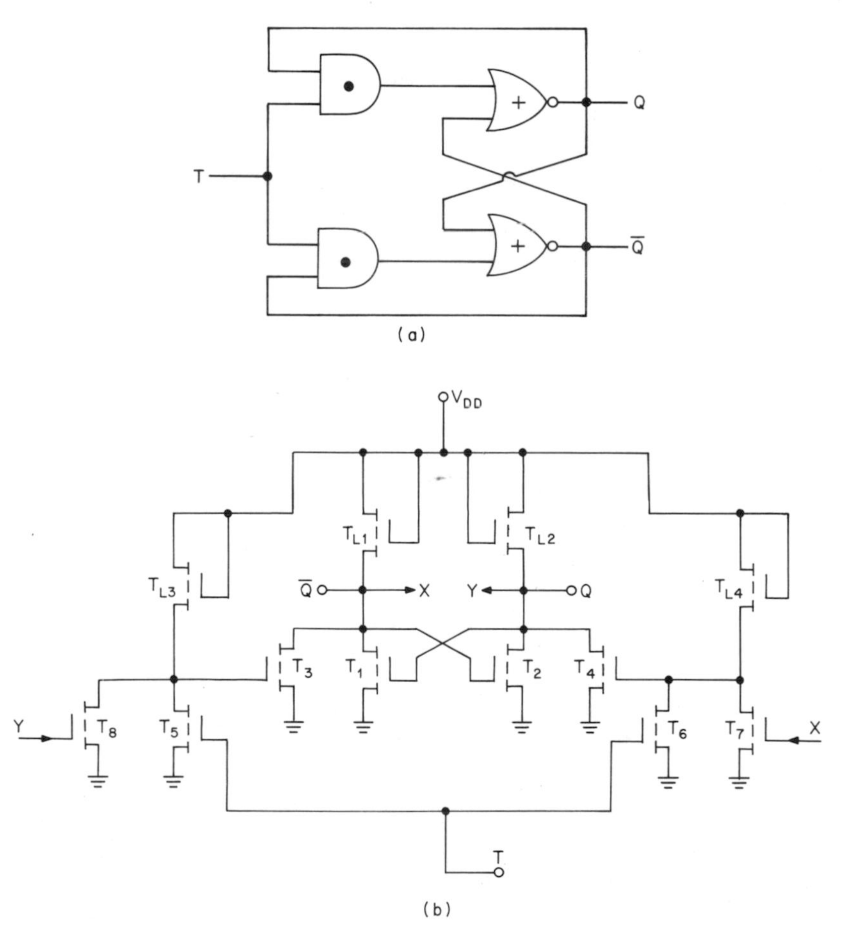

Fig. 4.22. Alternative *T*-type static flip-flop circuit: (*a*) logic diagram; (*b*) PELS circuit schematic.

in the 1 state and both *J* and *K* inputs are high when the pulse occurs, the flip-flop will toggle into the 0 state. The *J-K* flip-flop therefore has the features of both a *R-S*-type and a *T*-type flip-flop. The excitation table for the *J-K*-type flip-flop further illustrates the versatility of this device by the generous number of "don't care" terms. The "don't care" terms are illustrated by the ϕ's which are shared with the *R-S*-type flip-flop. Other "don't care" terms unique to the *J-K*-type flip-flop are denoted by the letter *J* in the excitation table.

In Fig. 4.24*c* the standard logic diagram for a *J-K*-type flip-flop is shown, and it may be compared with the equivalent logic diagram shown in Fig. 4.24*d* which permits a reduced component count using MOS gates. The direct circuit implementation using PELS inverter combinations for the circuit in Fig. 4.24*d* is shown in Fig. 4.24*e*. It contains 30 devices. The circuit implementation in Fig. 4.24*e* again uses the AND gates in the feedback and clocking portion of the flip-flop. Notice that the *J* and *K* inputs are both connected into a series string of three transistors. The series AND gate with three devices requires correspondingly larger transconductance for each device and increases the total area required for the flip-flop. It should be emphasized, however, that the use of the AND gates in Fig. 4.24*e* provides a reduction in the number of load devices and an increase in switching speed compared with a direct implementation of the circuit in Fig. 4.24*c*. The circuit in Fig. 4.24*c* uses all-NOR gates, but the increase in number of load devices increases the propagation delay per stage and decreases the overall switching speed.

S^n	R^n	Q^{n+1}
0	0	Q^n
0	1	0
1	0	1
1	1	?

(a)

Q^n	$\rightarrow Q^{n+1}$	S^n	R^n
0	0	0	ϕ
0	1	1	0
1	0	0	1
1	1	ϕ	0

(b)

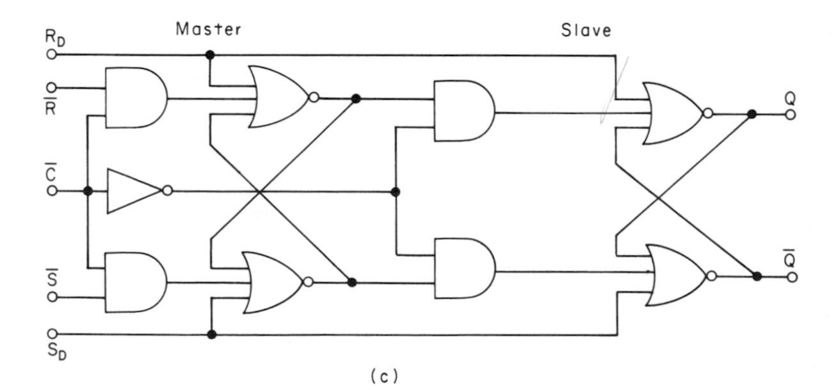

(c)

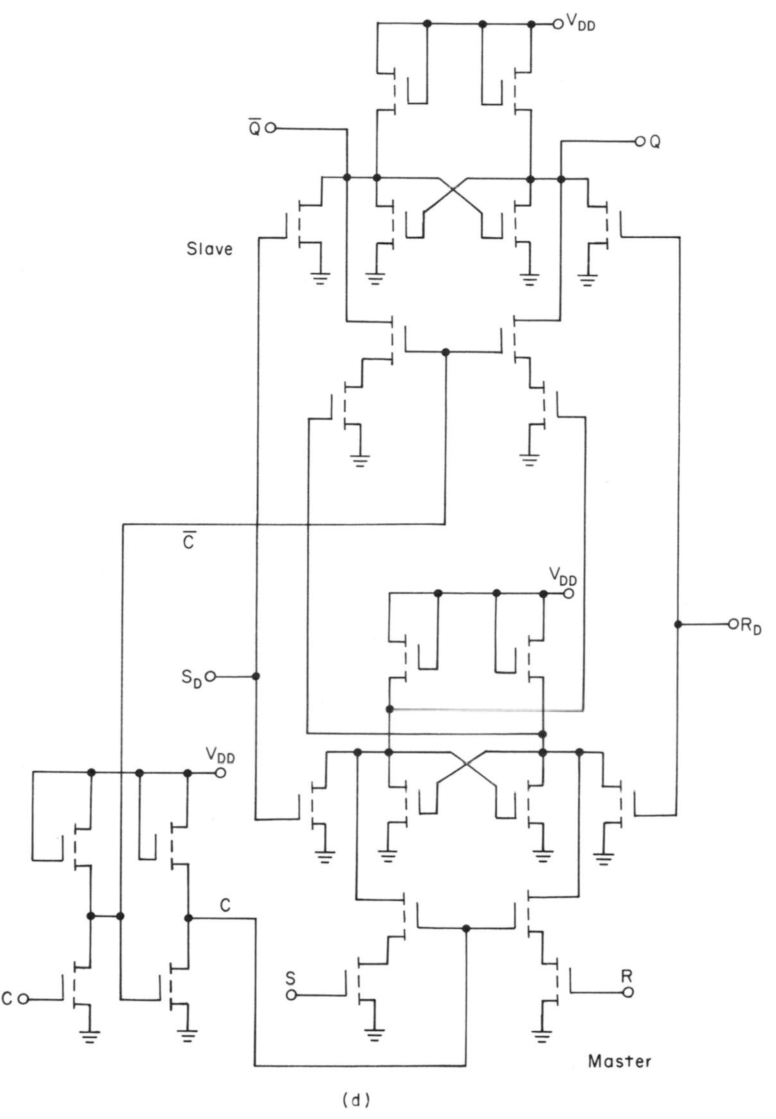

(d)

Fig. 4.23. *R-S*-type static flip-flop: (*a*) truth table; (*b*) excitation table; (*c*) logic diagram for master-slave; (*d*) PELS circuit schematic.

4.16 *D*-TYPE FLIP-FLOPS

The *D*-type flip-flop represents the simplest type of device from a control point of view. The inputs at the time the clock pulse occurs completely determine the form the flip-flop will assume. If a logical 1 is applied to the input line *D*, regardless of what state the flip-flop was in before the pulse occurs, it will assume a 1 stage on its true *Q* output when the clocking occurs. Similarly, if the input line *D* is at a logical 0 when the clocking pulse appears, it will assume a logical 0 state in

J^n	K^n	Q^{n+1}
0	0	Q^n
0	1	0
1	0	1
1	1	$\overline{Q}^n$

(a)

$Q^n \rightarrow Q^{n+1}$		J^n	K^n
0	0	0	ϕ
0	1	1	J
1	0	J	1
1	1	ϕ	0

(b)

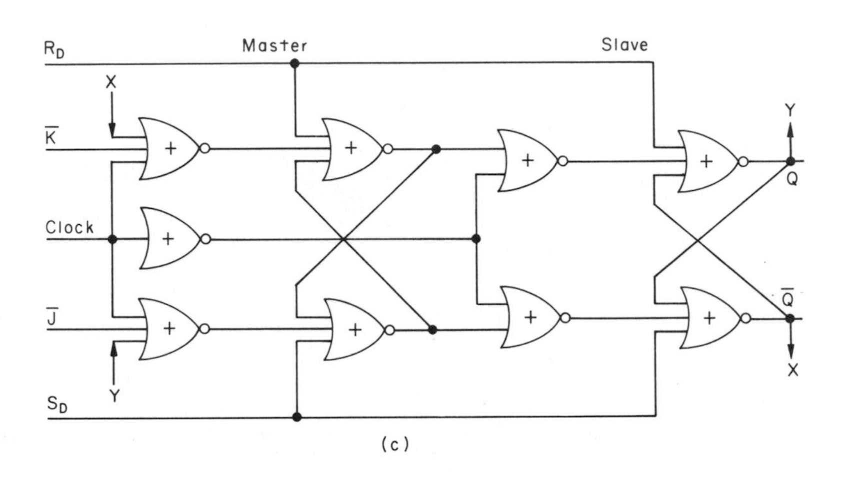

(c)

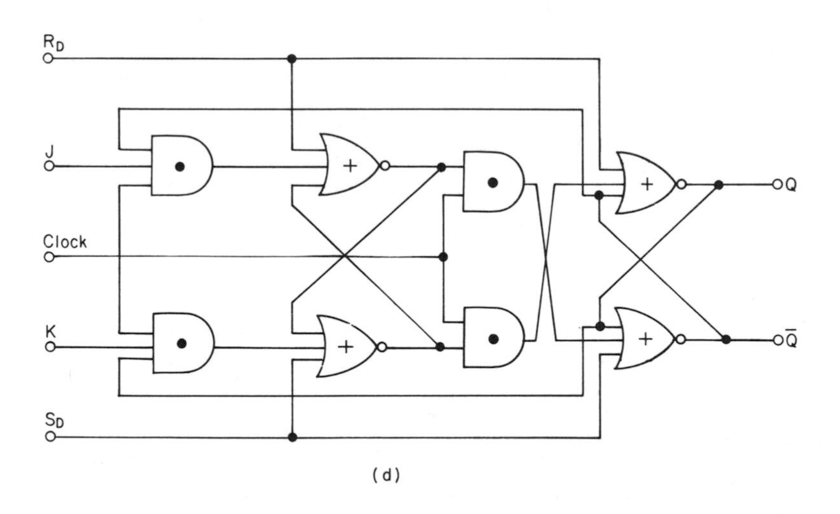

(d)

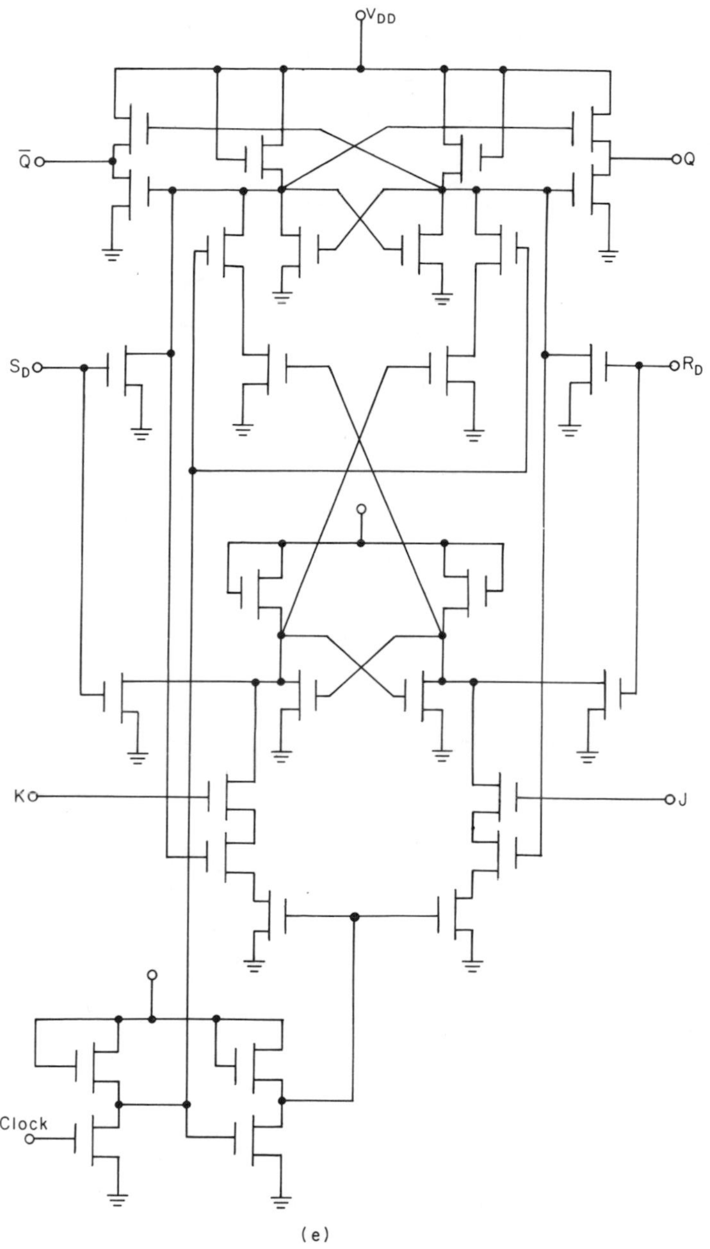

(e)

Fig. 4.24. *J-K*-type static flip-flop: (*a*) truth table; (*b*) excitation table; (*c*) logic diagram for master-slave; (*d*) logic diagram for reduced component count; (*e*) PELS circuit schematic.

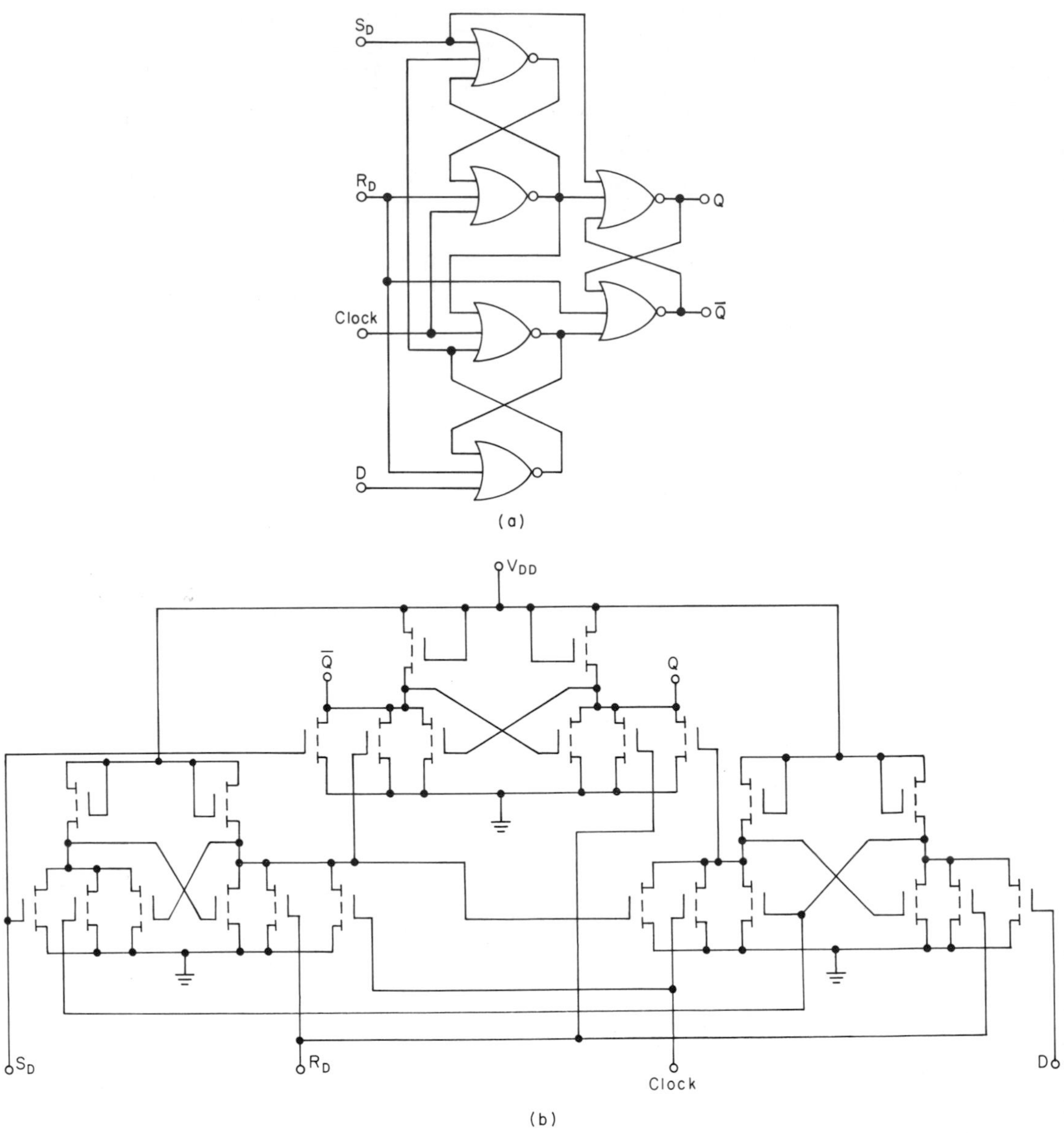

Fig. 4.25. Delay (*D*-type) static flip-flop: (*a*) logic diagram; (*b*) PELS circuit schematic.

its output *Q*. The truth table and excitation table for the delay-type function are given in Fig. 4.25*a* and *b*. This type of flip-flop is useful when one needs to delay data for a clocking interval while a separate function is simultaneously being processed in another circuit. Later the desired logic functions are available for further processing during the same clocking interval. Another use of the *D*-type flip-flop is to provide an echo signal following a data transfer. For example, suppose one register has transferred data to a separate portion of the circuit and a validating signal is required in return. The D-type flip-flop will then generate a pulse for return following the complete transfer of information.

One can derive the D-type flip-flop function from other flip-flops. For instance, the J-K-type flip-flop will perform as a D-type flip-flop if input lines $\bar{J}$ and $\bar{K}$ are tied together. In Fig. 4.25*a* the D-function is obtained using an edge-triggered circuit. Here, clock-triggering occurs at a voltage level of a clock pulse and is not directly related to the transition time of that pulse. After the clock input threshold voltage has been passed, the data input D is locked out of the circuit and is ineffective. The bistable element can be set and reset directly through lines S_D and R_D. The master-slave combination circuit actually contains three bistable elements. The slave portion contains the standard bistable circuit which contains, in this case, eight devices. The direct set and reset portion of the circuit also contains eight devices in the master section. The data input together with its clocked gating is contained in a third bistable element which also contains eight devices. Thus, the circuit shown in Fig. 4.25*b* contains a total of 24 devices. This includes 6 load devices and 18 driver devices each, of minimum transconductance values.

The various bistable circuit configurations in this chapter all provide a static (dc-stable) memory function. Design of circuitry using the static functions follows rather directly from standard bipolar transistor designs. One should realize, however, that in most cases it is a mistake to attempt a direct implementation in MOS based upon a previous static logic diagram for bipolar functions. The primary reason is that matching with static elements using MOS circuits are not the optimum design in a large fraction of design cases. MOS design generally provides for a greater flexibility in circuit clocking and function. In the next chapter we will consider another type of MOS circuitry, dynamic MOS, in which the temporary charge storage at capacitive nodes is exploited as a design aid. MOS dynamic circuitry generally has no direct bipolar equivalent.

REFERENCES

1. M. H. White and J. R. Gricchi, Complementary MOS Transistors, *Solid-State Electronics,* **9:** 991–1008 (1966).
2. J. Seely, Advances in the State-of-the-Art of MOS Device Technology, *SCP and Solid-State Technology,* **8:** 59–62 (1967).
3. R. H. Crawford, "MOSFET in Circuit Design," McGraw-Hill Book Company, New York, 1967.

5

Shift Registers for Data Delay, Logic, and Memory

5.1 INTRODUCTION

One of the first MOS/LSI circuit functions to find extensive acceptance has been the digital delay line or, as it is more commonly termed, the shift register. Several attributes of MOS are uniquely valuable in shift register design: (1) the high impedance of an MOS device gate permits temporary data storage in the form of charge in parasitic capacitance; (2) MOS technology permits realization of bidirectional transmission with a zero voltage offset across the device; and (3) the load devices may be turned off as desired by multiphase clocks to reduce power dissipation.

The MOS shift register has layout advantages as well, since most of the delay cells require little area on the chip. Each of these advantages is exploited in the various shift register circuits in this chapter.

MOS shift registers have found their most extensive application in computer display terminals, where the data to be displayed are circulated through the delay register in synchronization with the CRT raster. Also, shift registers are often used in electronic calculators and computer peripherals as memory elements. By providing address counters and a comparator, the shift register can be used as a random-access (sequential-address) memory for applications where circulation time is not a restriction.

Two circuit families of MOS registers—static and dynamic—are used in digital delay lines. The simplest circuit type is the dynamic shift register, which may contain as few as six MOS devices per master-slave delay cell. Dynamic registers depend upon charge storage for data retention. If the clock runs too slowly—or stops—in a dynamic register, the information stored is lost. The data volatility problem, however, is offset by the economy and speed available with the dynamic circuitry.

Static registers, in contrast, contain additional devices that form a bistable (static) latch within each register cell or delay element. The static register will not lose data when the clock is stopped. Static cells are generally larger and slower than dynamic cells.

5.2 TWO-PHASE DYNAMIC RATIO SHIFT REGISTER

The dynamic shift register of the ratio type uses a series of inverter stages whose outputs are clocked between successive inputs with a two-phase power supply. The ratio-type delay stages are simply serially coupled inverters, each with greater than unity gain. They effectively amplify and reshape the digital pulse amplitude with each clocking interval. Figure 5.1*a* illustrates a typical two-phase ratio-type shift register delay section within a cascade of shift register elements. Specifically, transistor pairs T_2-T_3, T_5-T_6, and T_8-T_9 constitute three separate inverters where the data are amplified and coupled between temporary storage capacitances at C_1, C_2, and C_3, respectively.

The shift register action in which data are shifted toward the right with successive nonoverlapping clock pulses is illustrated in the pulse timing diagram of Fig. 5.1*b*. When ϕ_1 is at its most negative value, T_1 is turned on, allowing node A and capacitance C_1 to either charge or discharge depending upon the data input V_{in}. The ratio-type inverter places the logical complement of the level at A on node B only when the load device T_3 is energized by clock pulse ϕ_2. With ϕ_2 at its most negative level, the transistor combination T_2-T_3 functions as an inverter and power is dissipated in the T_2-T_3 combination if transistor T_2 is conducting. Also with ϕ_2 high, node C is now connected to node B. During this timing period, gate node C is matched to that voltage from the preceding half-bit stage. When ϕ_2 returns to zero, the logical level at node C remains, and we are ready for clock pulse ϕ_1 to go high (return to its most negative state). As ϕ_1 goes high, we find that the node V_o level becomes the inversion of the logic level from node C, and the output voltage V_o now represents a full bit-time of delay for the input signal V_{in}.

The 1-bit delay section for the two-phase ratio-type shift register contains six transistors. These transistors are not minimum-area devices because the inverter combinations T_2-T_3, T_5-T_6 require a reasonably large β_R, and minimum geometries cannot be used. Figure 5.1*b* shows the timing sequence for voltage levels at the various nodes as the positive- and negative-going magnitudes are delayed through the shift register stage. Note that the ratio-type shift register inverter dissipates a considerable amount of power during the period in which driver devices T_2, T_5, and T_8 are conducting. The ratio-type shift register stages such as Fig. 5.1*a* dissipate more power than those circuits that employ ratioless-type circuitry.

The geometrical layout for a single cell of the ratio-type two-phase register is shown in Fig. 5.1*c*. The power V_{DD} and clock ϕ lines require as much area within each cell as the devices themselves. The cell shown requires approximately 40 sq mils. With self-aligning gates the area may be reduced.

The maximum clocking rate at which the ratio-type shift register operates depends primarily upon the time to move charge to and from capacitances at nodes A, B, C, and D. The load device T_3 must be a high-resistance device to provide a sufficiently large beta ratio in the inverter. Since T_3 is a high resistance in its *on* state, it does not permit charging of capacitances at nodes B and C with the speed comparable to other transients existing in the circuit. For instance, when transistor T_2 is turned on and a negative-going voltage must be discharged from nodes B and C, we find a very fast discharge transient. The corresponding charging of nodes B and C through device T_3 is slower. One can increase the speed of the ratio-type

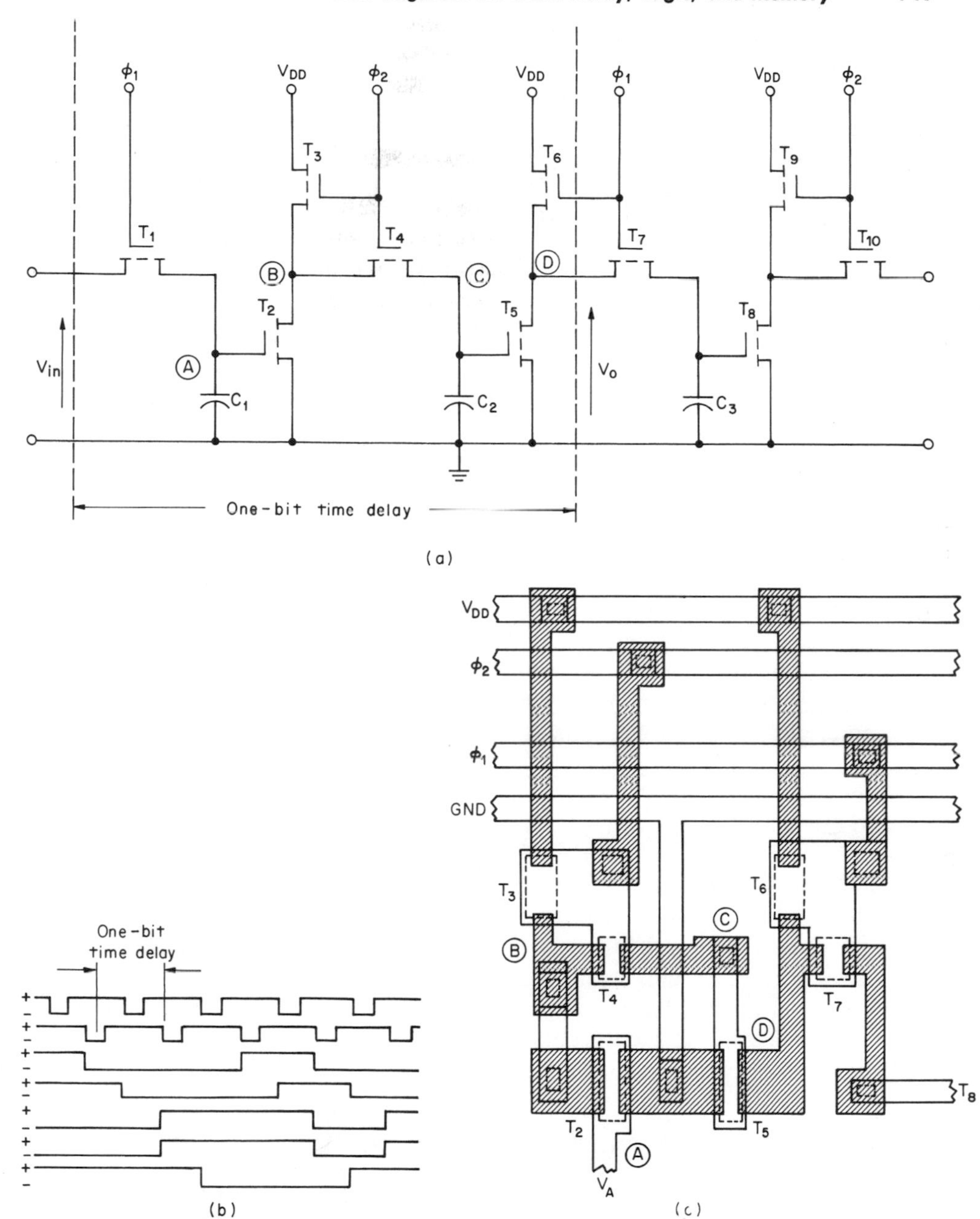

Fig. 5.1. Two-phase ratio shift register: (*a*) circuit schematic; (*b*) timing sequence; (*c*) layout geometry for 1 bit.

shift register by bringing out the gate on transistors T_3, T_6, and T_9 to a large external dc supply voltage V_{GG}. The inverters, with the dc V_{GG} supply, result in an increased dissipation, determined only by the presence or absence of a negative-going voltage at the input node to the driver portion of the inverter. The price we pay for this increase in speed with the V_{GG} connection is, of course, increased power dissipation.

The minimum clocking rate for most shift registers is determined by the time during which we can successfully maintain charge at nodes *A*, *B*, *C*, and *D* without refresh from clocking pulses. The leakage current discharge from each of these nodes is determined primarily by the reverse leakage of the p-n junctions. We find that the minimum clocking rate for shift registers generally falls between 10 Hz and 1 KHz, depending upon the temperature ambient, the particular technology used in circuit fabrication, and the capacitance at nodes *A*, *B*, *C*, and *D*. One can anticipate a doubling of the minimum operating frequency for each 10°C of circuit temperature rise.

One can "ripple through" data, by maintaining both clock pulses at most negative values simultaneously. Data will be shifted through the entire length of the register at maximum speed under these conditions. This technique can be used to set all data bits either high or low for logic level initialization.

If the input V_{in} is connected to positive voltage magnitudes with respect to the substrate, the p-n junction on device T_1 can be forward biased with an accompanying injection of minority carriers into the substrate. Heavy loading and charge-storage time-delay generally exists under these conditions. At no time should the inputs be biased more positive than 0.3 V with respect to the substrate for the p-channel enhancement-mode shift register circuitry. Similarly, the inputs to n-channel devices diffused into p-type substrates should not swing more negative than 0.3 V. On the other hand, the problem of forward-biased substrate junctions does not exist in circuitry using insulating substrates such as spinel and sapphire.

Another important consideration for the design and application of the MOS shift register in systems is the capacitive loading presented by the shift register to the clocking circuitry. The clock driver generally must drive a substantial capacitive load and be able to swing voltages over a rather wide range, sometimes up to 28 V. Depending upon the particular technology used, one finds varying degrees of loading presented by the MOS shift register to the clock drivers. The capacitive loading presented to the clock supply by the two-phase ratio-type shift register in Fig. 5.1*a* is moderate, compared with other circuits that we will consider later. Power is drawn from the clock drivers only during the transient.

Next we will consider the operation of an MOS shift register (ratioless) in which no dc power dissipation occurs, but which does demand increased transient power from the clock drivers.

5.3 TWO-PHASE DYNAMIC RATIOLESS SHIFT REGISTERS

It is possible to use the same clocking pulse waveforms used in the ratio-type register to implement a ratioless design. A ratioless design permits no dc paths to ground. The ratioless-type shift register has a power dissipation which is proportional to the clocking frequency. Lower power operation is obtained with lower clock frequency, and again, the low-frequency minimum is determined by the leakage current associated with those reverse-biased p-n junctions shunting charge-storage capacitances. Figure 5.2*a* shows a circuit schematic for a single bit of signal delay using a ratioless charge-transfer scheme. This circuit we will designate as type I

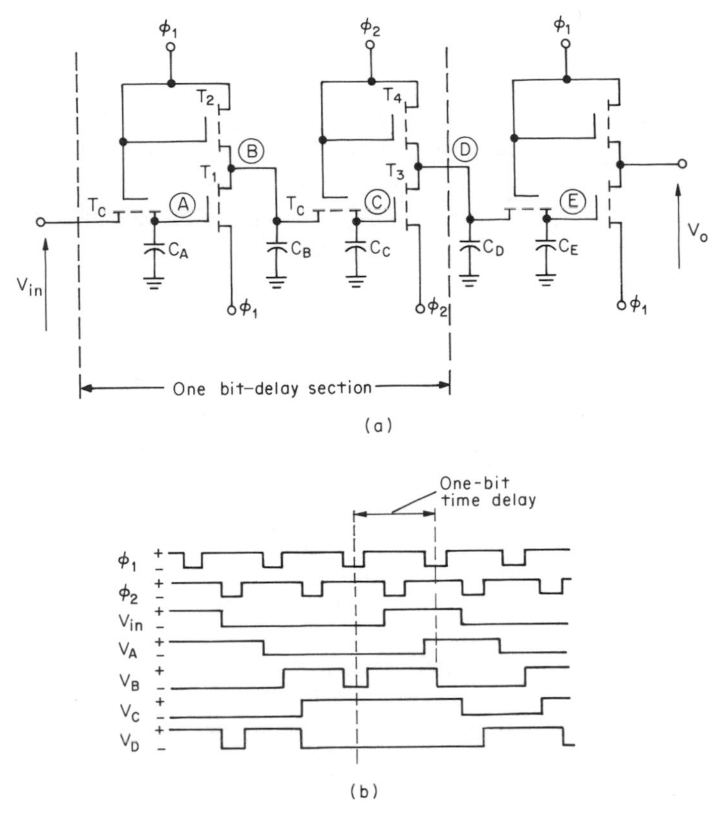

Fig. 5.2. Two-phase ratioless type I shift register: (*a*) circuit schematic; (*b*) timing sequence.

ratioless. When clock pulse ϕ_1 goes to its negative value, we find that the input voltage V_{in} is gated onto node A with shunt capacitance C_A. In addition, if node A goes negative, corresponding to a negative-going V_{in} value, then transistor T_1 is turned on and node B is charged to a negative voltage also through the clock supply ϕ_1 (paths T_1 and T_2). On the other hand, if the input voltage had been at its zero voltage level, then transistor T_1 would not turn on, and node B would remain *precharged* with the negative voltage supplied through clock pulse ϕ_1. As clock pulse ϕ_1 goes to its zero level, node A retains its initial charge and is electrically isolated by T_c from the input voltage V_{in}. If transistor T_1 is conducting, we find that node B discharges to the $\phi_1 = 0$ voltage level through T_1. Thus, we have achieved a logic inversion of input voltage V_{in} onto node B.

As nonoverlapping clock pulse ϕ_2 goes to its most negative value, the charge from capacitance C_B is now shared with node C and its corresponding shunt capacitance C_C. If capacitance C_B is large enough, then the total voltage remaining at node C following the charge-sharing time corresponding to device T_c conducting will be sufficient to turn on transistor T_3 if node B had been at a negative value. On the other hand, if the voltage at node B had been at a zero voltage value during clock pulse ϕ_2, we would find that node C is reduced by charge sharing to a similar low voltage level.

Transistors T_3 and T_4 constitute an inverter which provides the desired logical inversion from node C to node D during the period while clock pulse ϕ_2 is at its most negative value, and immediately following. The logical inversion occurring with transistors T_3 and T_4 is the same as that which occurred during the preceding time interval at transistors T_1 and T_2. The total combination of transistors T_1 through T_4 (together with their coupling device T_c) constitutes a single bit of signal delay and requires six devices.

A detailed timing diagram for transferring both logic high and logic low signals through the ratioless inverter is shown in Fig. 5.2*b*. Notice that node B is always precharged to the most negative value during clock pulse ϕ_1. Similarly node D is always charged to its most negative value during clock pulse ϕ_2. This characteristic distinguishes the ratioless shift register delay section from the ratio-type delay section of Fig. 5.1. Notice that the total capacitive loading on the clock drivers is greater for the ratioless circuit. We do not need a V_{DD} supply line for the ratioless circuit.

The price we have paid for eliminating one of the power supply lines is increased capacitive loading on the clock drivers and possible noise-margin problems from the charge sharing. The circuit can be used with minimum-geometry devices. We no longer depend upon the β_R for the proper inverter action. The use of minimum-geometry devices is offset by the fact that we must add area to the chip in the form of increased capacitance at node B for C_B and node D for C_D. Capacitances C_B and C_D must be considerably larger than their corresponding charge-share capacitances C_C and C_E if we are to maintain useful noise margins in the two-phase ratioless circuits.

We can reduce the loading on the clock drivers by adding another transistor to each inverter stage and designing with bit-delay sections of eight devices each. We will designate the second two-phase ratioless circuit as type II. Figure 5.3*a* shows the higher performance circuit in which clock loading is reduced and area of the chip is increased slightly as compared with the type I ratioless circuit. Figure 5.3*b* shows the pulse timing of this circuit. In this circuit we note that when clock ϕ_1 goes to its most negative state, the input voltage V_{in} charges node A with shunt capacitance C_A. Simultaneously, transistor T_4 is conducting and node B is charged to a most negative value with its corresponding shunt capacitance C_B. We may note that capacitance C_B is precharged during the clock ϕ_1 most negative value. When clock ϕ_2 goes to its most negative value, we *sample* with transistor T_2 to determine if we are to discharge node B. Specifically, if the voltage at node A is at its most negative value during the clock ϕ_2 sampling period, we will find that any voltage existing at node B is discharged to ground through conducting transistors T_3 and T_2. Further, if node A had been at a zero voltage value, then transistor T_2 would not be conducting and the negative voltage existing at node B would not be discharged to ground during sampling period ϕ_2. Therefore, an effective inversion for an input V_{in} now appears at node B and across capacitance C_B. Clock pulse ϕ_2 simultaneously, through T_5, has permitted charge sharing between C_B and C_C and has precharged node D to a negative voltage.

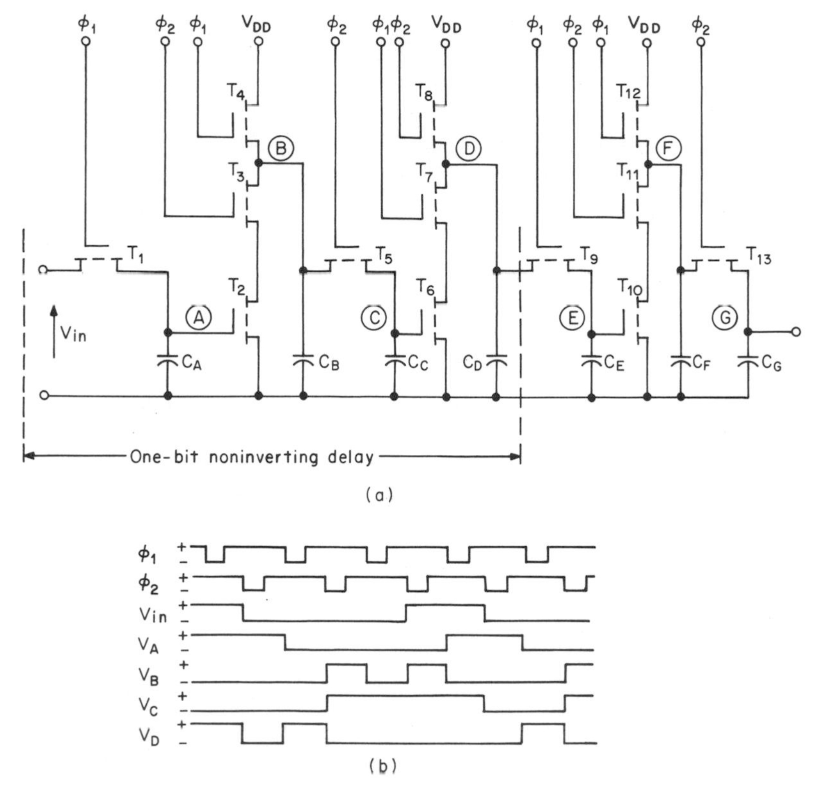

Fig. 5.3. Two-phase ratioless type II shift register: (*a*) circuit schematic; (*b*) timing diagram.

When clock pulse ϕ_1 goes to its most negative value, we again sample to determine if that voltage at node D is to be discharged to ground through T_6. In order to transfer the charge effectively from capacitance C_B to C_C, capacitance C_B must be much larger than C_C to preserve the logic levels during the charge sharing. Notice that the inversion of the data does not depend upon a β_R for a driver and inverter stage, but instead depends upon the precharge and successive sampling of nodes B and D. The gating of data through the first and second inverters may be considered as a master-slave combination. We may consider the first four devices to be the master and the second four devices to be the slave. This master-slave equivalency exists in all MOS shift registers and may be found in each delay cell.

Clocks ϕ_1 and ϕ_2 are not capacitively loaded to the extent that they were in the circuit in Fig. 5.2. Note that the clocks are loaded in the type II circuit only by gates of individual transistors in the delay section, and that all source and drain connections are either to the common ground or to a power supply V_{DD} static line. The transistors for the circuit of Fig. 5.3 may all be of uniform size and minimum geometry. The maximum frequency of operation is limited primarily by the large capacitances at C_B, C_D, and C_F, which must be intentionally designed into the circuitry. We should expect that higher clocking rates could be achieved if we go to a ratioless-type circuit that does not depend upon a gated charge sharing between such capacitors as C_B and C_C. It is possible to achieve such performance at the expense of an increased number of clocks.

Since minimum geometry design does not provide this large capacitance at node C_B, one must intentionally add area at node B for this capacitance, and as a result we have offset those gains obtained by designing with minimum-geometry transistors. One finds, in practice, that the two-phase ratioless-type shift register is seldom used, and other designs are generally preferred.

5.4 FOUR-PHASE DYNAMIC RATIOLESS SHIFT REGISTER

If we are willing to add two additional clocks, we can design a shift register element that can operate at higher speed. Figure 5.4*a* shows a 1-bit delay section coupled into a shift register circuit which contains six devices per delay section. In this circuit neither the transistors of pair T_2-T_3 nor those of the pair T_5-T_6 are conducting simultaneously. Thus, there is never a dc current path through any series transistor combination from power supply V_{DD} to circuit ground at any time.

If we clock the register of Fig. 5.4*a* according to the clocking sequence shown in Fig. 5.4*c* with nonoverlapping clock pulses, we find that initially node B is precharged to its most negative value as clock pulse ϕ_1 goes to its corresponding negative potential. During clock pulse ϕ_2, if transistor T_1 is conducting, any voltage appearing at node B will be discharged to ground through T_1 and T_2. We depend upon capacitance C_B to store charge and maintain constant voltage between clocking pulses at node B. Thus, while clock ϕ_1 is at a negative level, node B is unconditionally precharged to a voltage V_{DD} through conducting transistor T_3. After being

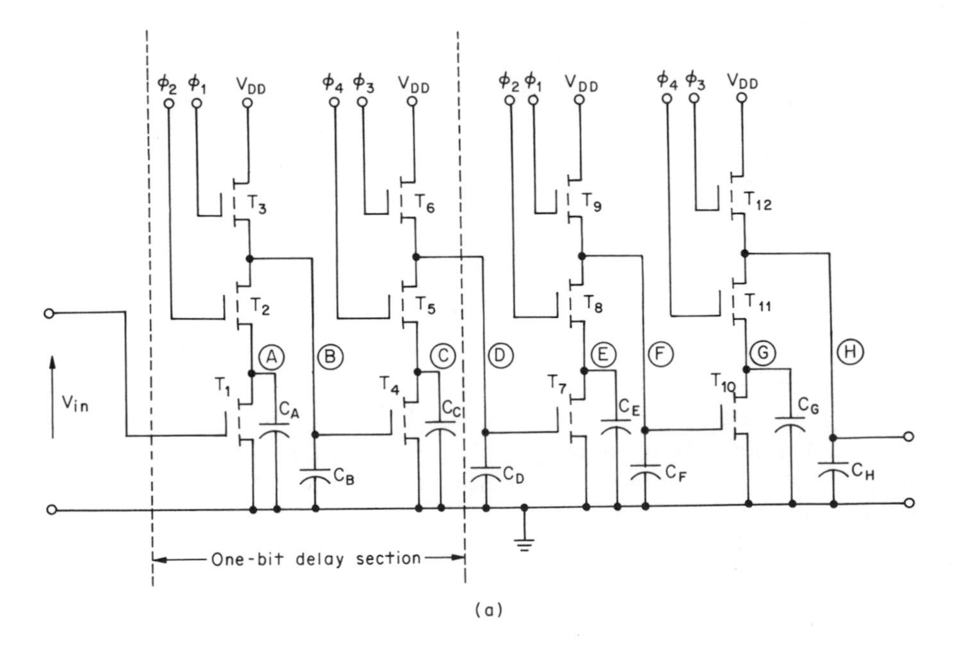

(a)

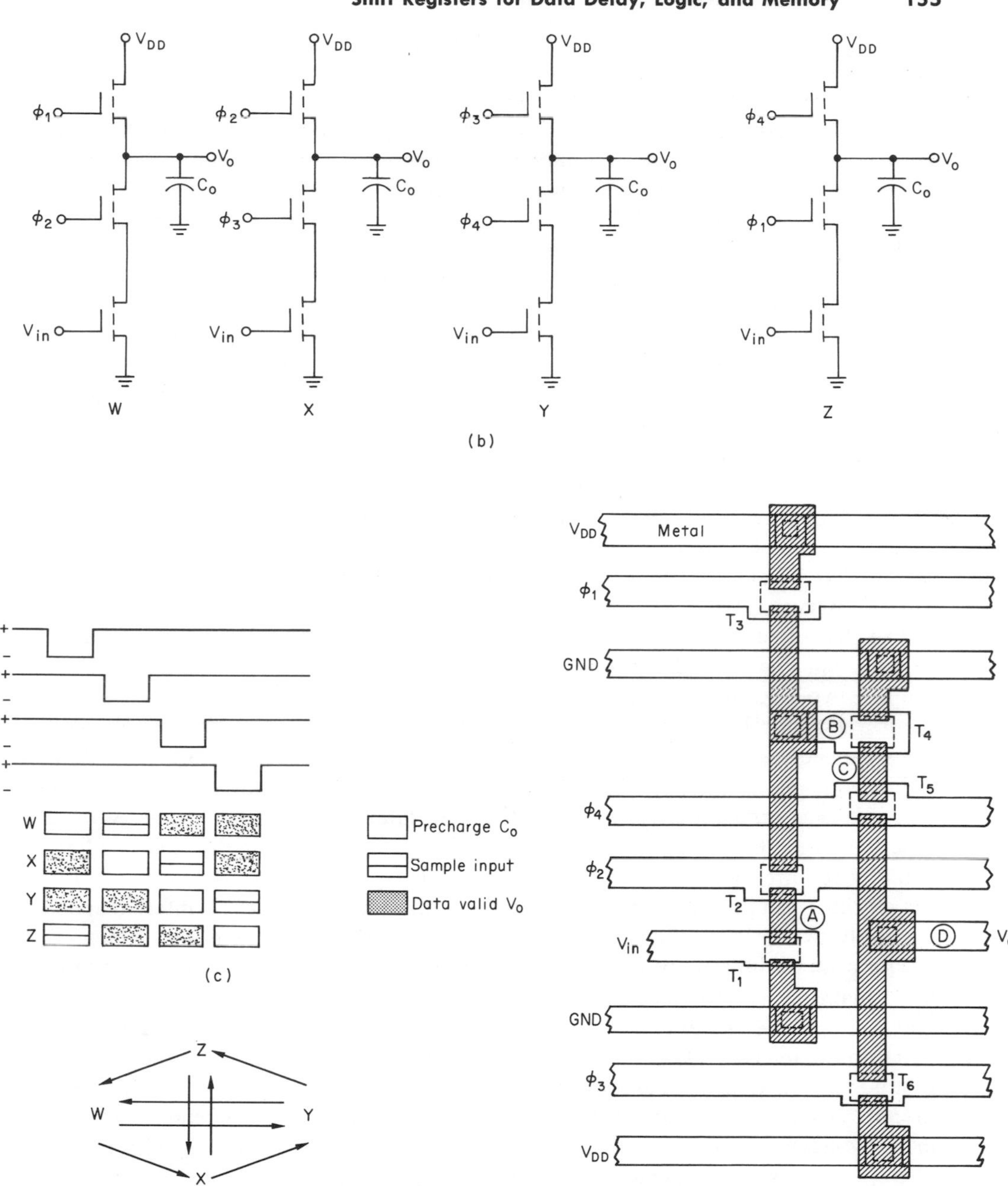

Fig. 5.4. Four-phase ratioless type I shift register: (*a*) circuit schematic; (*b*) phasing circuit combinations; (*c*) timing of pulses for (*b*); (*d*) permitted fan-outs; (*e*) layout geometry for 1 *W*-bit.

precharged, this node is selectively discharged during clock time ϕ_2 through transistors T_1 and T_2, only if the V_{in} input causes T_1 to conduct. Otherwise, node B remains charged to the most negative voltage level.

After ϕ_2 returns to its zero level, ϕ_3 goes to a negative level and precharges node D just as ϕ_1 precharged node B earlier. Finally, transistors T_4 and T_5 selectively discharge node D during ϕ_4 time only if the signal V_B at node B was active, causing transistor T_4 to conduct. Thus, in four clock pulses the input data are inverted twice and transferred to the output with one net clocking period of delay. Capacitances C_B, C_D, C_F, and C_H appear at the input of each master and slave and must maintain their charge levels during that short interval between clock pulses.

We do not depend, in this circuit, upon the sharing of charge between two capacitors. This type of shift register cell is dynamic because we do depend upon multiphase clocking and the capacitance associated with nodes such as B, D, F, and H to maintain proper operation. The lower limit of the clock frequency is determined primarily by those leakage currents that discharge the storage capacitances.

Various circuit combinations of the type I four-phase shift register may be used as shown in Fig. 5.4*b*. These four combinations labeled W, X, Y, and Z provide precharge, sample input, and data valid V_o levels according to the timing sequence of Fig. 5.4*c*. For instance, a W-gate samples the input data during the second ϕ_2 timing interval.

The permitted fan-outs can be understood quickly from the clocking sequence summary in Fig. 5.4*c*. For instance, when a W-gate has its data valid at V_o during the last two clocking intervals, we find that we can sample that level by either an X- or a Y-gate. The W-gate cannot be used to drive the Z-gate. Similar observations can be made to complete the fan-out possibilities summarized in Fig. 5.4*d*.

The four-phase ratioless shift register in Fig. 5.4 we will term type I. This will distinguish it from other types to be discussed later. The layout geometry for a type I shift register delay bit is shown in Fig. 5.4*e*. Notice that the clocking lines actually require more area on the chip per cell than do the active devices. But the clocking can permit operation of the dynamic shift register at an internal speed generally two to three times faster than that obtainable using two-phase clocking.

The primary limitation of the type I four-phase ratioless shift register is one of noise margin. The worst case for noise margin occurs when the voltage differential between nodes A and B is maximum and transistor T_1 does not conduct during the ϕ_2 sampling interval. For this case the capacitance C_A (between node A and circuit ground) is shunting node B and charge is shared at a time when we do not wish to share charge. If the capacitance C_A is too large, then we find a substantial change in the voltage existing at node B at the termination of the negative-going period of pulse ϕ_2. It is this capacitive loading by C_A which limits the noise margin of the type I four-phase ratioless shift register.

There is another four-phase circuit which eliminates the V_{DD} power line and, in turn, reduces the chip area required per bit of delay by using the circuit of Fig. 5.5*a*. This circuit uses six devices per delay element. When ϕ_1 goes to its negative value, the capacitance C_B is charged through device T_5. When ϕ_1 returns to a zero level, grounding the source of T_3, we observe that C_B is discharged if T_3 is conducting and as ϕ_2 goes to a negative level. This circuit requires overlapping clock pulses

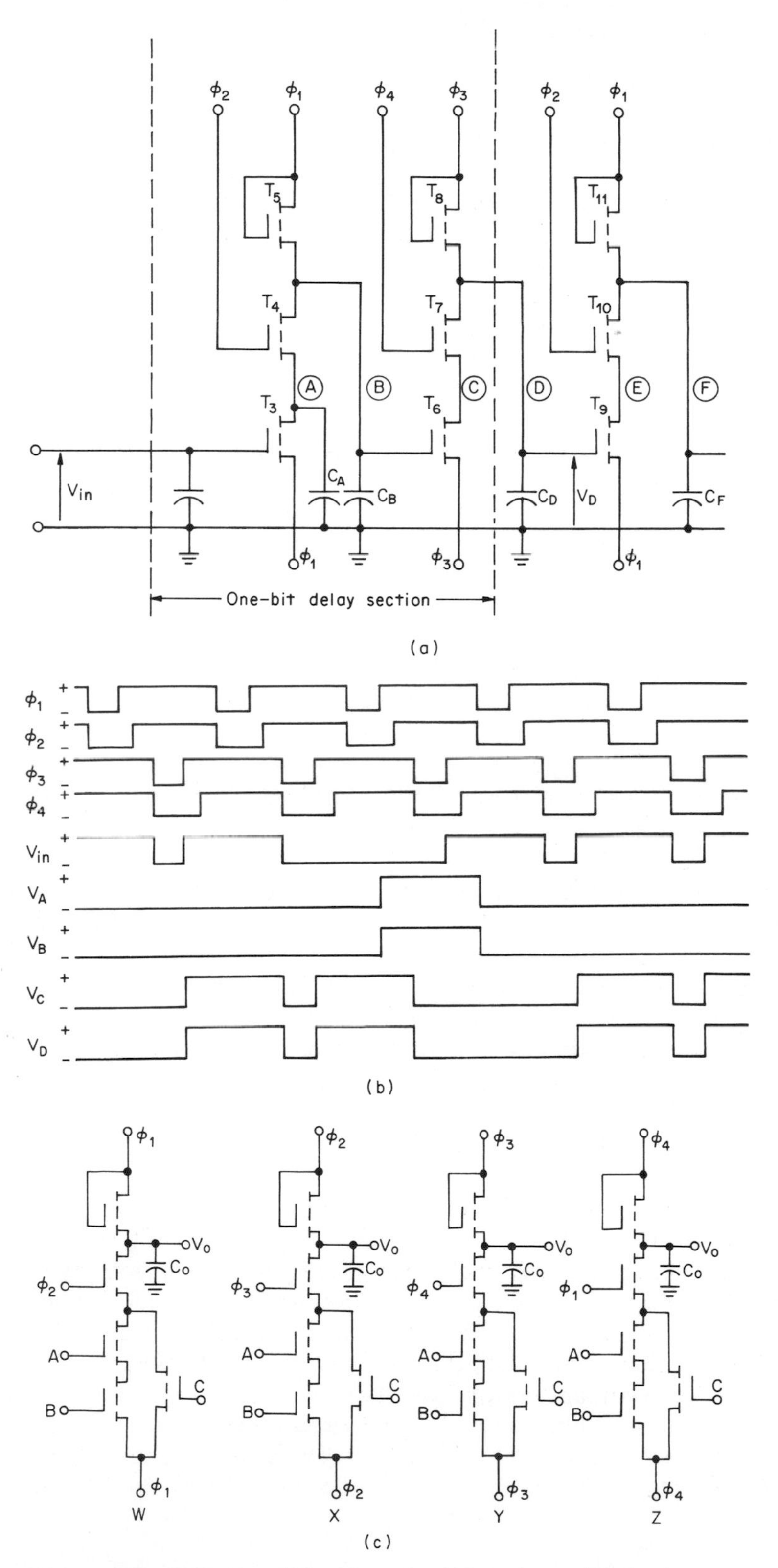

Fig. 5.5. Four-phase ratioless type II shift register: (*a*) circuit schematic; (*b*) timing of pulses for (*a*); (*c*) circuit combinations; (*d*), (*e*) follow.

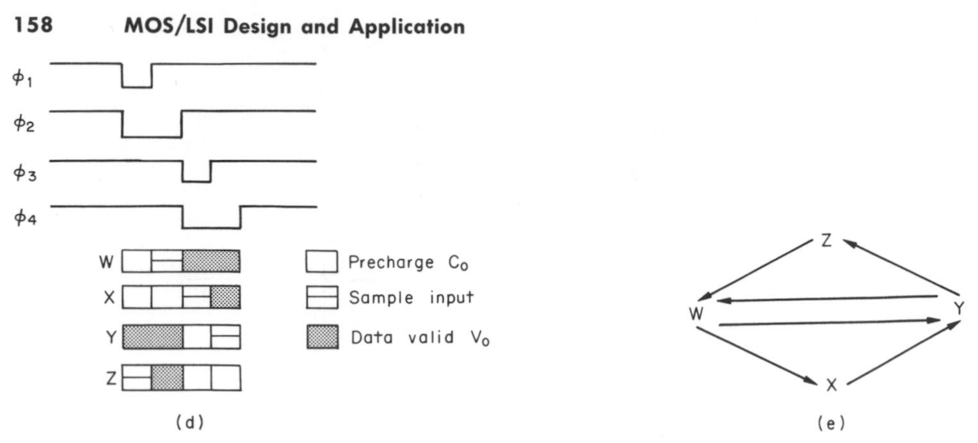

Fig. 5.5 Four-phase ratioless type II shift register (*continued*): (*d*) timing sequence; (*e*) permitted fan-outs.

in ϕ_1 and ϕ_2. If transistor T_3 is cut off, then capacitance C_B retains its charge after clock ϕ_2 returns to a zero level. The type II circuit is identical with that of the type I circuit except that the loading now has been increased on the clock drivers and the V_{DD} power supply line has been eliminated. Also the overlapping clock pulse requirement (see Fig. 5.5*b*) distinguishes the operation of the type II four-phase circuit from the type I.

The timing of clocks and data is shown in Fig. 5.5*b*. Note that the output voltage V_D or V_o always goes to its negative value while clock pulse ϕ_2 is simultaneously at a negative value. Figure 5.5*c* illustrates the four clocking combinations *WXYZ* available for the type II four-phase circuit. The clock lines ϕ_1, ϕ_2, ϕ_3, or ϕ_4 are used for both precharge and the data return to ground in *WXYZ*, respectively. In the timing diagram, Fig. 5.5*d*, the *W* and *Y* circuits provide valid data at V_o during ϕ_2 and ϕ_4 negative, respectively. Circuits *X* and *Z* correspondingly precharge the output capacitance at V_o during the same time intervals ϕ_2 and ϕ_4 negative.

The clock pulses of the type II register are partially overlapping during successive half-periods. The overlapping pulses are advantageous because ϕ_2 and ϕ_4 can easily be generated on the chip from ϕ_1 and ϕ_3 inputs, respectively. A simple MOS inverter which has a certain amount of delay can easily be used to generate ϕ_2 and ϕ_4 on the MOS chip. Figure 5.5*e* indicates the fan-out possibilities using the type II register elements.

The inputs *A*, *B*, and *C* of Fig. 5.5*c* are examples of how the shift register may also be used to implement random logic within each stage. The technique illustrated in Fig. 5.5*c* is not limited to this particular type II circuit. In fact, any shift register stage, static or dynamic, can be tailored to include the implementation of random logic. The added nodal capacitance that results from the increased number of devices with random logic must be carefully considered because it effects (1) noise margin and (2) maximum clocking speed.

A major problem with the type II circuit is again the loading of the signal voltage due to the capacitance C_A of Fig. 5.5*a*. If one can precharge the capacitance at node *A*, then the loading problem will no longer exist as charge is shared with node B. We can introduce an additional transistor into each inverter of the shift register delay to precharge node A if we desire. Such a precharging transistor has been added in the circuit example of Fig. 5.6 for the four-phase type III circuit. In this circuit a single bit of delay for a shift register element with a single data input contains eight separate transistors. If one desires to implement random logic in addition to a delay function, larger numbers of transistors per master-slave section would, of course, be required, depending upon the complexity of the logic implemented. The circuit of Fig. 5.6 will precharge both nodes A and B to a negative voltage while clock ϕ_1 is at its negative value. The timing for this shift resistor may use either the overlapping or nonoverlapping clock pulses of Figs. 5.5d and 5.4c, respectively. During the latter portion of clock pulse ϕ_2 negative, the clock ϕ_1 must be zero to permit sampling of the data input. With $\phi_1 = 0$ and $\phi_2 = -V_{\phi 2}$, the data at A, B, and C are sampled. A discharge path exists from V_o to ground if the input $AB + C$ is true during the sampling interval.

The clocking of type III logic gates with either overlapping or nonoverlapping four-phase clock pulses depends upon several factors. One important parameter is the loading which is presented to the clocks by the circuit. Given relatively light loading on the MOS chip, both ϕ_2 and ϕ_4 may be generated on the chip with MOS circuitry while only ϕ_1 and ϕ_3 are generated external to the chip. On the other hand, if the circuit presents large capacitive loads to all four phases, then it may be preferable to generate all four clock pulses with bipolar or JFET circuitry. Other factors are, of course, the maximum chip size permissible and availability of clock drivers.

In general, the MOS clock generators on the chip will drive only relatively small capacitances. The type III circuit presents a maximum capacitive load to the clocks as compared with other four-phase circuit types. In four-phase circuits one finds that for sufficiently complex functions the designer must give very careful consideration to loading capacitances on the clock lines if error-free logic at high clocking rates is to be implemented.

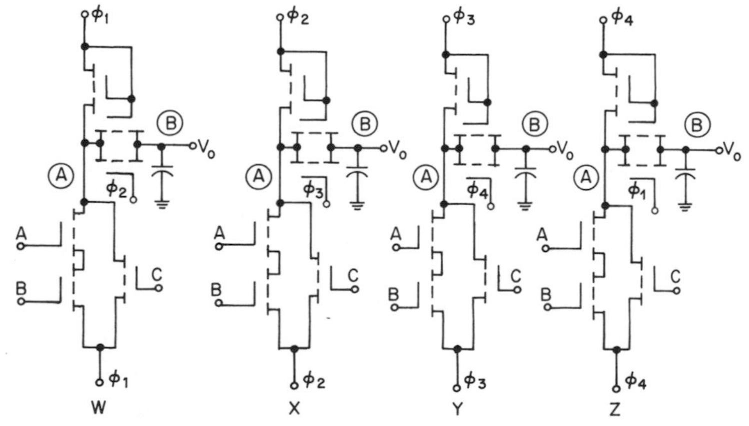

Fig. 5.6. Four-phase ratioless type III shift register circuit combinations.

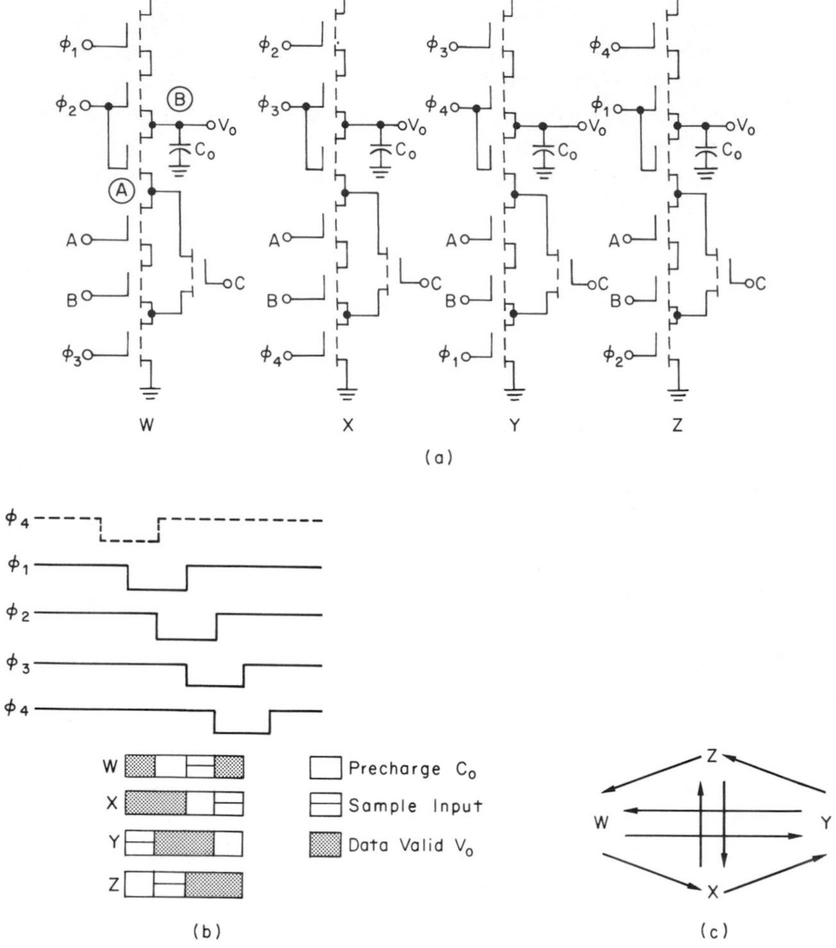

Fig. 5.7. Four-phase ratioless type IV shift register: (*a*) circuit combinations; (*b*) timing sequence; (*c*) permitted fan-outs.

In Fig. 5.7*a* is shown another four-phase circuit, termed type IV, which provides the reduced clock driver loading feature and also precharges node *A* to improve the noise margin. This is a ratioless circuit where no dc path exists between V_{DD} and circuit ground at any time for the clock pulse scheme of Fig. 5.7*b*. The power transferred to and from the static supply and clock drivers is only that required for charging and discharging capacitors.

The example illustrated in Fig. 5.7*a* shows a random logic function being implemented in four different clocking arrangements, *W*, *X*, *Y*, and *Z*, with the type IV circuit scheme. In this case overlapping clock pulses are shown for illustration in Fig. 5.7*b* and the resulting fan-out possibilities are summarized in Fig. 5.7*c*. For a circuit such as this we would normally find that, because of the reduced loading on the clock drivers, up to three of the clock phases may be generated on the chip.

The type IV four-phase circuit phasing diagram shown has a partial overlapping of successive clock pulses ϕ_1, ϕ_2, ϕ_3, and ϕ_4 to obtain the desired precharging at the lower node *A* in Fig. 5.7*a*. In addition, a third nonoverlapping phase is added into each string of series devices to eliminate the dc path between V_{DD} and common ground. Thus, we need a minimum of ten transistors for each 1-bit, noninverting, delay section of the type IV four-phase ratioless shift register.

A fifth circuit possibility exists for the four-phase delay cell, which we will term type V. It uses nonoverlapping clock pulses and a complete precharge of all internal nodes for improved noise margin. Figure 5.8*a* shows four classes (*W*, *X*, *Y*, and *Z*) of the type V four-phase circuit. Eight transistors are needed to implement a single bit of noninverting shift register delay. Figure 5.8*b* shows the clocking sequence using the nonoverlapping pulses. From each stage data are valid at V_o 50 percent of the time and are sampled readily using this clocking scheme. The capacitive loading on the clock drivers is reduced to a minimum as compared with

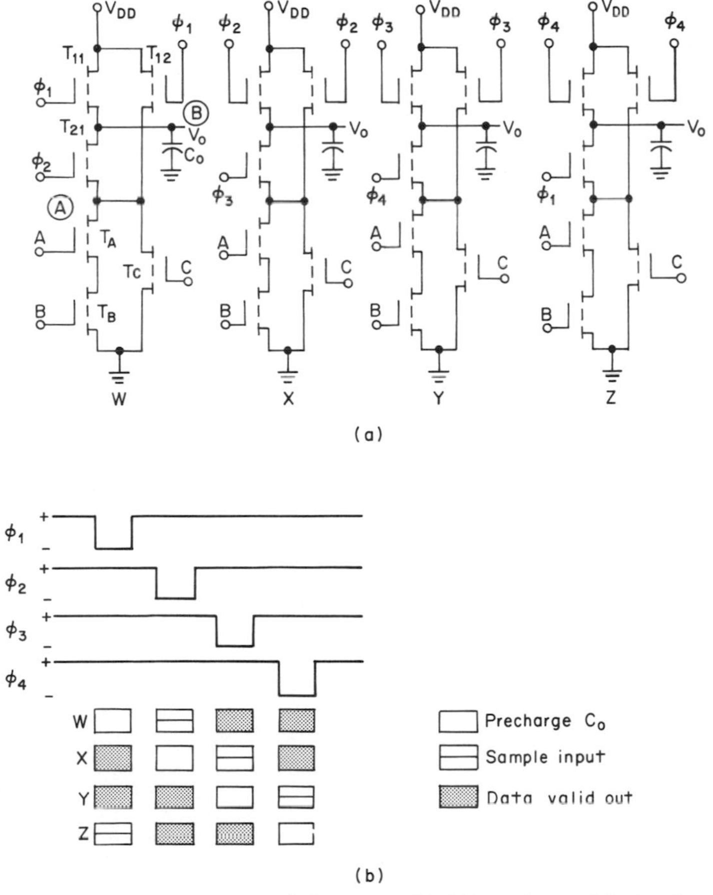

Fig. 5.8. Four-phase ratioless type V shift register: (*a*) circuit combinations; (*b*) timing sequence.

other four-phase clocking possibilities and precharging of the charge-sharing node *A* is used. In this circuit power dissipation increases and noise margin problems appear if the clock pulses overlap.

5.5 SINGLE-PHASE STATIC RATIO-TYPE SHIFT REGISTER

Early MOS designs often served to duplicate the Boolean functions used in bipolar circuitry. An example of this type of design might use a direct MOS implementation of Fig. 5.9*a* for the *R-S* flip-flop function. This flip-flop might be used as a single delay section in a single-phase shift register with static holding capability. The static circuit for such a delay element using MOS enhancement-mode devices is shown in Fig. 5.9*b* using negative-logic. This circuit contains 18 devices, but the total reduces to 15 if we use only a single data input to each static shift register element, as would be the case in a typical dynamic MOS design. The clamping feedback

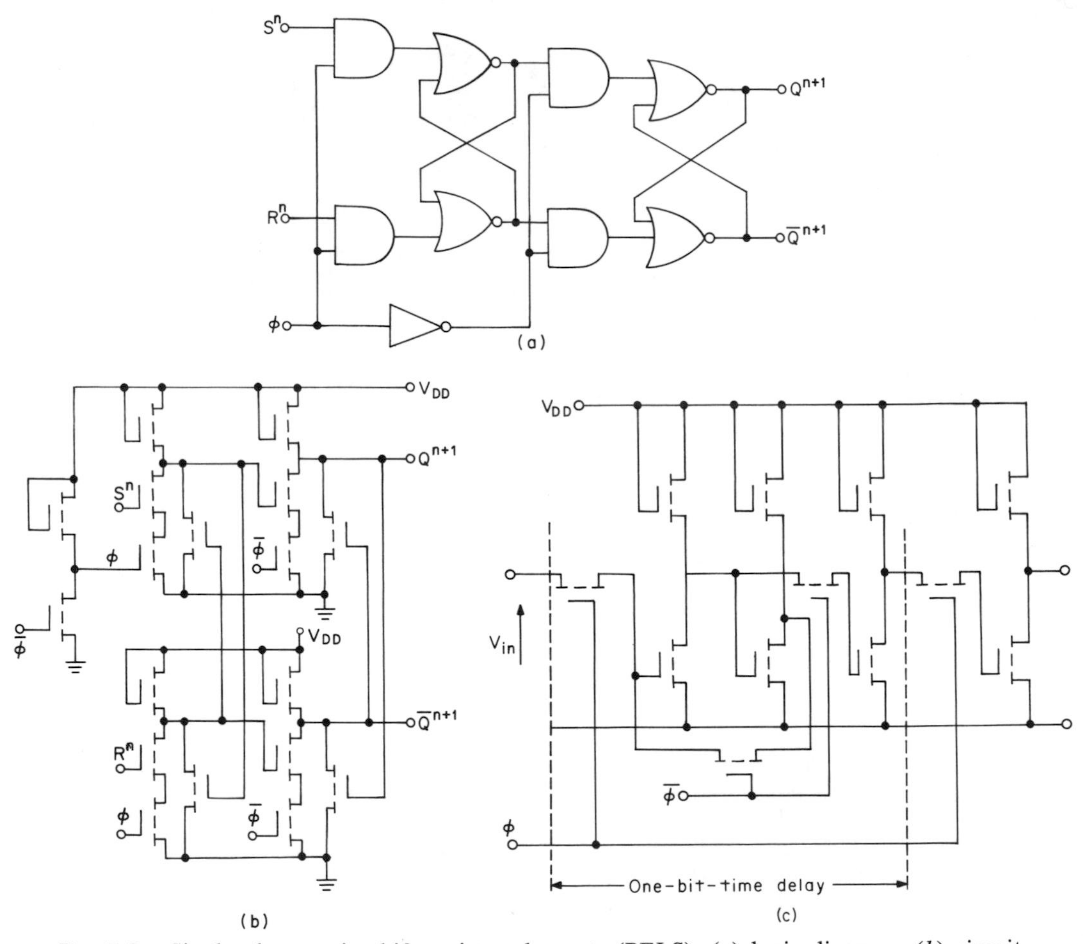

Fig. 5.9. Single-phase ratio shift register elements (PELS): (*a*) logic diagram; (*b*) circuit schematic; (*c*) improved circuit schematic.

is needed in only one direction for the static delay element. The device count in the circuit of Fig. 5.9*b* can be reduced still further.

The necessary elements for a feedback latch implementing an *R-S* flip-flop function are included in the more optimum design of Fig. 5.9*c* for the single-phase MOS static shift register element. In this circuit each cell contains nine transistors (including three load devices). Data are held indefinitely within each cell for $\phi = 0$. Data are shifted on the negative-going clock pulse transient through the master-slave circuit combination.

The implementation of a static delay element is also straightforward using CMOS technology. Among the several possibilities, the logic net of Fig. 5.10*a* has been selected. The feedback latch is for the circuit implementation in Fig. 5.10*b* with the complementary transistors. The latch will hold data indefinitely for either $\phi = 0$ or $\phi = 1$. New data are shifted into the latch of Fig. 5.10*b* on the negative-going edge of ϕ. Data are shifted out of a cell during $\phi = 0$ when the output coupling

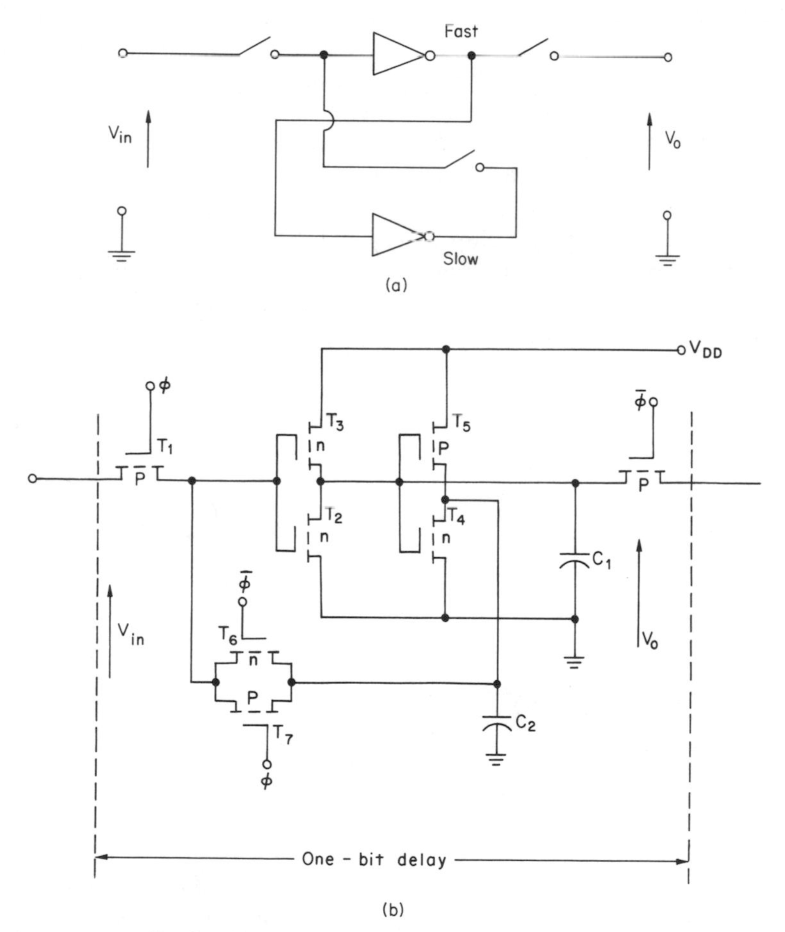

Fig. 5.10. Single-phase CMOS shift register: (*a*) logic net; (*b*) circuit schematic.

transistor is conducting. The substrate connections of each transistor are omitted for the sake of schematic clarity. The n and P denote n- and p-channel devices, respectively, in Fig. 5.10*b*. The capacitor C_2 is considerably larger than C_1 and permits the shift of level V_{in} to node *A* very quickly for $\phi = 1$, and prior to the setting of the feedback clamping voltage at node *B*. The inverter with node 1 at the output must be faster than the inverter driving node 2. Thus, the notation "Fast" and "Slow" in Fig. 5.10*a*. The CMOS static delay stage requires eight devices, compared with the all p-channel device count of nine in Fig. 5.9*c*. Two additional devices must be added to the CMOS circuit to obtain noninverting action.

5.6 THREE-PHASE STATIC/DYNAMIC RATIO SHIFT REGISTER

The bistable latch cannot be used directly to advantage with two-phase clocking. One can, however, take advantage of the dc storage properties of the gate capacitors in a bistable element by adding a third holding phase in the cell as shown in Fig. 5.11*a*. The bistable element will hold data indefinitely if $\phi_3 = 1$. The shift between stages occurs when ϕ_1 is negative. The shift within each bistable element occurs during the trailing rise time in which ϕ_2 is faster than ϕ_3, permitting the desired double inversion of the level between node *A* onto the output node V_o. As shown in Fig. 5.11*b*, clock ϕ_2 is designed with a faster trailing edge than clock ϕ_3, causing information always to shift toward the right. Clock ϕ_3 is the clamping clock which holds data in the static mode. In a practical shift register, phases ϕ_2 and ϕ_3 may be generated internally on the chip at an inverter driven by a master clock ϕ_1. When $\phi_1 = 0$ and $\phi_3 = 1$, the shift register goes into a static holding mode. In Fig. 5.11*a* each delay cell of master-slave shifting requires seven devices. It offers little advantage over the circuit cell of Fig. 5.11*a* converted to static operation (eight devices).

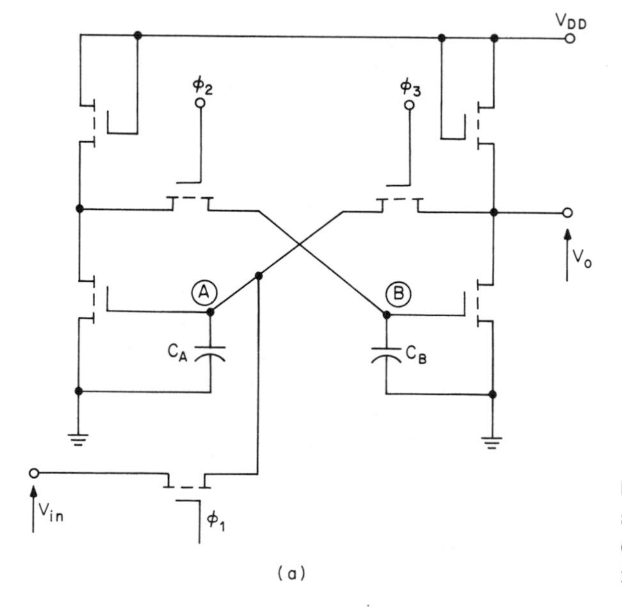

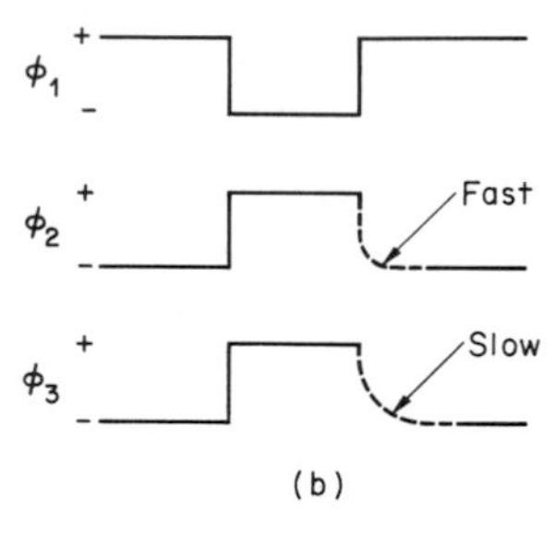

Fig. 5.11. Two-phase ratio static shift register cell: (*a*) circuit schematic; (*b*) pulse clocking.

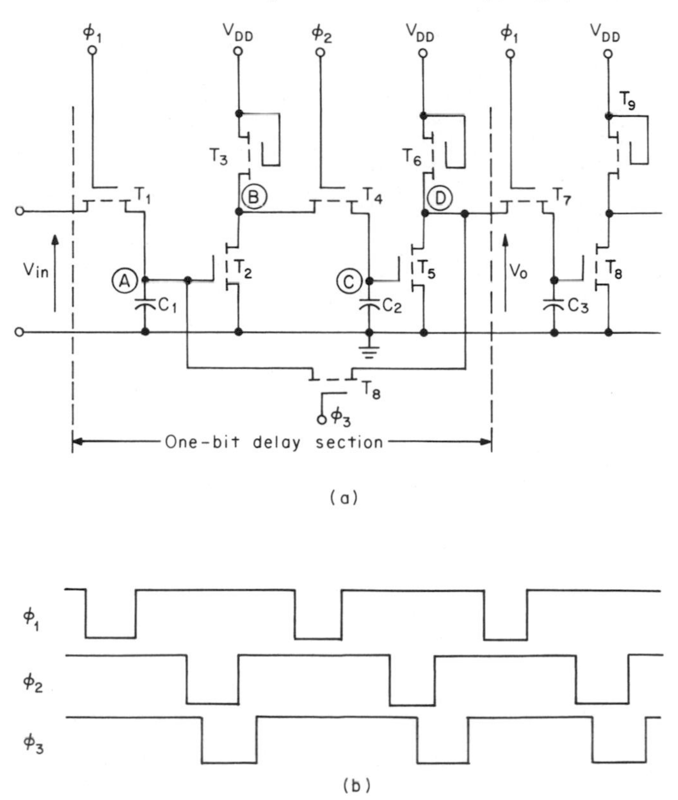

Fig. 5.12. Three-phase ratio shift register with static capability: (*a*) circuit schematic; (*b*) clocking sequence.

One can add static holding capability to any shift register element by adding clocked feedback devices. This technique applies to each of the various dynamic shift register delay sections already discussed. For instance, the standard two-phase ratio-type shift register delay section can be converted into a static register simply by adding a third phase and a feedback transistor as shown in Fig. 5.12*a*, with the clocking sequence for holding data in Fig. 5.12*b*. Phase ϕ_3 gates the output of an element back to the input, and therefore ϕ_3 must extend for a slightly longer time generation than ϕ_2 during the negative-going portion of the pulse. If the phase ϕ_3 does not go into its negative state, we find that normal shift register action with data shifting toward the right exists, and the circuit performs as a standard two-phase ratio-type shift register. If one wishes to stop the clocks without losing data, the circuit requires that the clock stop with phase ϕ_2 in its negative-going state and $\phi_1 = 0$. With ϕ_2 negative, the level at node *C* will track that at node *B* and will remain valid as long as *B* is valid. The level on node *B* is simply the data at node *A* inverted. Similarly the level at node *D* is simply the inverted data from node *C*.

If clock ϕ_3 remains high, then the level at node *A* will remain, whereas otherwise it would have been lost due to self-discharge through circuit leakage. Thus the ϕ_3 phase is used as an input which locks data into a feedback loop and implements flip-flop action. To hold data within a given bit while the clocks are running, we may use the clocking scheme shown in Fig. 5.12*b*. Here the clock ϕ_3 does provide the desired feedback loop, and static operation results with no net transfer of data out of the shift register bit. Notice that data is lost if the clocks are stopped during the wrong portion of phasing with ϕ_1 negative. It is possible to add additional feedback transistors that will permit one to latch into a static mode for stopping on either ϕ_1 or ϕ_2 negative-going pulses. The total system considerations of permitting the clock to stop in either state normally do not justify the added expense of incorporating the extra transistor per bit required for such operations. If one desires to hold data in a static mode indefinitely with reduced power dissipation, one should turn clock ϕ_1 off. Notice that in the static holding mode with ϕ_1 in its positive-going state, the circuit, for a single bit, is very similar to that of Fig. 5.9*c*.

5.7 FOUR-PHASE STATIC/DYNAMIC RATIOLESS SHIFT REGISTER

A four-phase static holding register can be obtained by using the circuit of Fig. 5.13 shown in the logic diagram of Fig. 5.13*a* plus the Eccles-Jordan equivalent of Fig. 5.13*b*. In this circuit a ratioless type I static element is implemented. The advantage here is that no dc power dissipation exists while a static holding capability is maintained.

A novel shift-right/shift-left register using four-phase clocking is illustrated in Fig. 5.14*a* with the timing sequence shown in Fig. 5.14*b*. If ϕ_3 and ϕ_4 are suppressed and maintained constant in their positive level, then the register can function as a standard two-phase ratio shift register. If one desires to operate in the static holding mode, he can use ϕ_3 or ϕ_4 or both to stop the shift register in either clocking period. Furthermore, the register can operate as a shift-left register by suppressing ϕ_1 and ϕ_2 in their positive state and clocking with ϕ_3 and ϕ_4. Notice that ϕ_3 and ϕ_4 do not overlap each other and permit the register to shift data toward the left in a manner similar to that normal mode of shift-right action obtained by utilizing ϕ_1 and ϕ_2 alone. The shift-right/shift-left register of Fig. 5.14 is a very versatile register but requires four clocks and special gating to implement the desired shifting action.

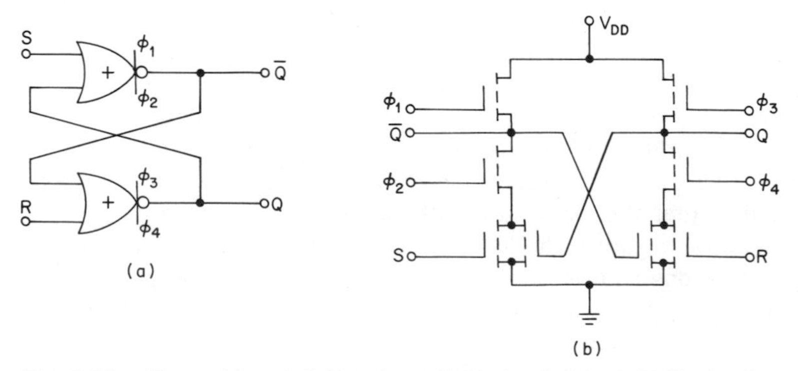

Fig. 5.13. Four-phase ratioless (type I) static element: (*a*) logic diagram; (*b*) circuit schematic.

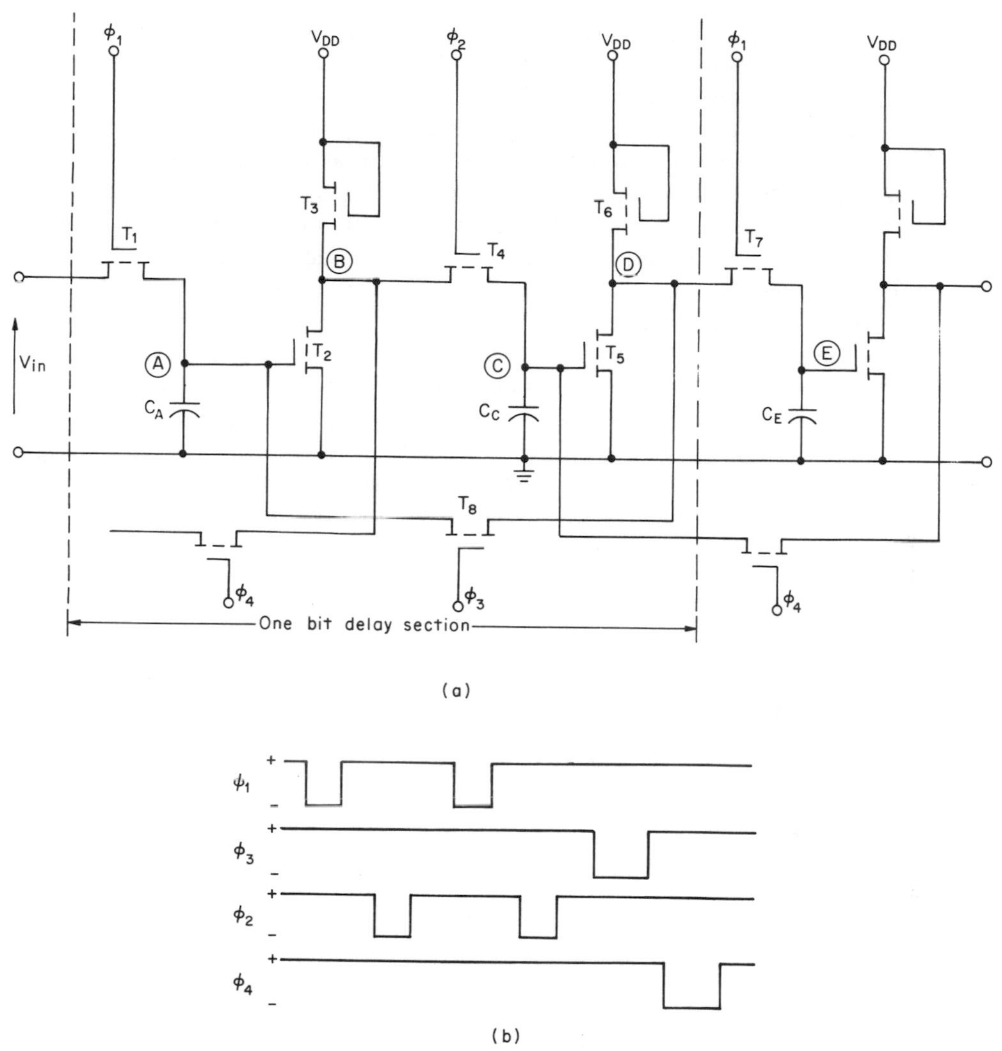

Fig. 5.14. Four-phase ratio shift-right/shift-left register with static capability: (*a*) circuit schematic; (*b*) clock pulse requirement.

5.8 PERFORMANCE COMPARISONS

The selection of clock phasing and circuit type depends on many factors for specific systems application. One finds that the layout of dynamic systems with four-phase circuitry is more complex, and many designers prefer to remain with two-phase circuitry for design convenience. The single-phase designs with MOS are generally ineffective except when static capability is required. For dynamic circuitry using mono-channel enhancement-type devices, one finds that the four-phase circuit generally can provide the maximum speed when proper clock drive speed and power are available. Also, CMOS has proven to be very fast with good design. The particular type of delay element within the range of one-, two-, three-, and four-phase systems depends on several factors. One finds that the clock loading varies within wide limits. There are cases where the multiphase clock drivers simply will not drive the large capacitive loads at the speeds desired.

If we desire a minimum power dissipation for various reasons, then we find that the ratio-type circuits are not permissible and a ratioless design is required. If we are ultimately concerned with production yield, then minimum size devices are generally useful, and we find again that the ratioless designs are desirable. For maximum frequency of operation, the various four-phase circuits or CMOS may prove to be best. For maximum noise immunity, the considerations can be complex; generally, the ratioless circuits with pullup transistors on both critical nodes will provide the maximum noise margin. Generally, one finds that the noise immunity provided by each of the circuits presented here (with the possible exception of the type II ratioless without pullup transistors) can be made adequate with careful design.

When static operation is required, we find that virtually all of the shift register elements can be operated in static modes by the addition of feedback transistors and the additional static holding phase. Other important considerations include the available power supply, the number of power supplies involved internal and external to the chip, control of the threshold voltages within desired specification ranges, circuit immunity to shifts in power supply, and threshold voltages variation on the chip. Some of these important performance comparisons are shown in summary in Table 5.1. Each dynamic circuit in Table 5.1 has been discussed separately.

5.9 MULTIPLEXING FOR INCREASED SPEED

If registers are operated in parallel, the input-output data rates of individual shift registers can, with suitable multiplexing, be multiplied by the number of delay lines in parallel. The general configuration for a high-speed multiplexed configuration

Table 5.1. Shift Register Bit Component Count Comparison (does not include CMOS)

Circuit type	Static	Dynamic	Power supply lines
1ϕ ratio	9	....	3
2ϕ ratio	7	6	4
2ϕ ratioless type I	....	6	3
2ϕ ratioless type II	8	8	4
3ϕ ratio	7	....	5
4ϕ ratioless type I	8	6	6
4ϕ ratioless type II	8	6	5
4ϕ ratioless type III	10	8	5
4ϕ ratioless type IV	12	10	6
4ϕ ratioless type V	12	10	6
4ϕ right/left	8	....	6

Note: The 4ϕ static operation requires continued clocking for "static" storage of a bit in any given cell.

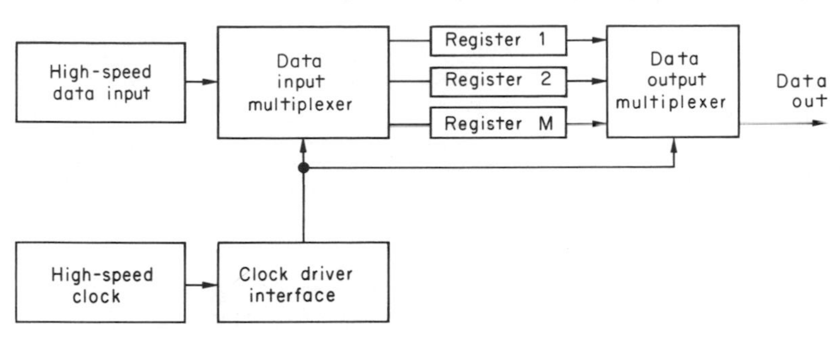

Fig. 5.15. High-speed data system. M separate shift registers, each shifting F_0 bits per second, are used in the multiplexing circuit to obtain an overall data transfer rate of MF_0 bits per second.

is shown in Fig. 5.15. Here a parallel combination of M shift registers, each shifting at a clock frequency of F_0, is used to multiply the overall data transfer rate by a factor of M. Here the data input multiplexer distributes the input data stream so that register 1 gets bits 1, 5, 9, etc.; register 2 gets bits 2, 6, 10, etc. The data modulator at the output reconstructs the bit stream and restores the original input data rate. With M registers in parallel, the delay line can accept data at M times the shift rate of the individual registers. This type of circuitry can include MOS/bipolar interface techniques and can provide data at rates limited primarily by the speed at which bipolar circuits can separate and reassemble the outputs of desired signals.

5.10 IMPLEMENTATION OF DYNAMIC LOGIC

5.10.1 Random Logic

It was noted earlier that one can substitute random logic implementation into each of the basic shift register elements. The shift register based on ratio inverters is limited in its capability for random logic in that the noise margin is degraded if one adds more than three series devices in place of the driver transistor. Power dissipation in a ratio circuit increases correspondingly as one attempts to compensate for degraded noise margin by increasing the number of transconductance devices. Random logic with ratioless shift registers is easier to implement, especially in the various four-phase shift register sections, where one has the advantage of the pullup transistor on critical capacitive nodes. As the complexity of random logic is increased, we find that an additional gating transistor is desirable in the lower end of the series chain of the inverter circuit. This additional transistor has not been discussed in the previous sections. Let it suffice now to say that useful logic implementations, such as we see in this section, can be adapted to most of the shift register schemes.

A random logic function $\overline{Q} = A\overline{B} + \overline{C}$ is implemented using a four-phase half-bit delay cell in Fig. 5.16*a*. A compact geometrical layout is shown for the half-bit in Fig. 5.16*b*. Each of the ratioless cells permits random logic to be included in the position normally used for the single data inverting device.

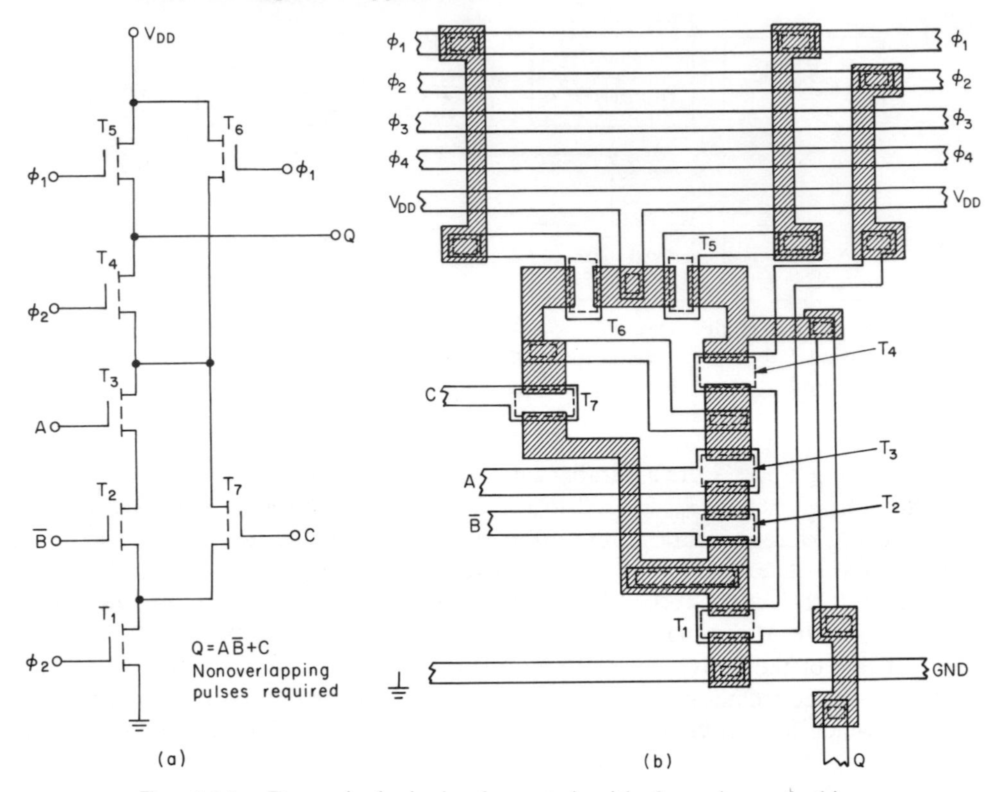

Fig. 5.16. Dynamic logic implemented with four-phase clocking; $\bar{Q} = A\bar{B} + C$ for half-bit: (*a*) circuit schematic; (*b*) geometrical layout example.

5.10.2 Flip-flop and Latch Functions

Flip-flop action can be obtained using 1-bit delay elements (*D*-type flip-flops) as basic constituents. For instance, the Boolean equation for the *R*-*S*-type flip-flop obtained directly from the transition table is

$$Q^{n+1} = S^n + \bar{R}^n Q^n \tag{5-1}$$

This states that the *R*-*S*-type flip-flop can be forced or set to a logical 1, either by raising the *S* level to high, or by holding the *R* input at zero when the previous state Q^n was 1. The circuit using a delay element *D* may be implemented readily as in Fig. 5.17.

The standard delay element would be used in conjunction with these additional four transistors to implement the *R*-*S*-type flip-flop action. Here, the random logic has been designed into a ratio-type series combination of the PELS type. One can use any of the standard delay configurations for the *D*-function. The Boolean expression for an *R*-*S*-*T* flip-flop is generally taken to be

$$Q^{n+1} = S^n\bar{T}^n + \bar{R}^n\bar{T}^nQ^n + T^n\bar{Q}^n \tag{5-2}$$

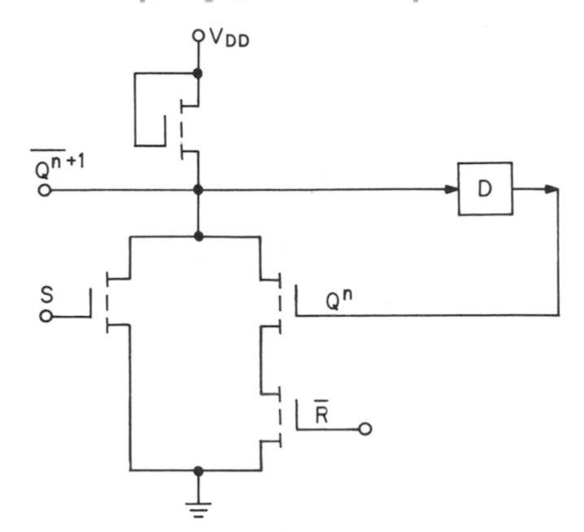

Fig. 5.17. *R-S* flip-flop function $Q^{n+1} = S + \bar{R}Q^n$ circuit schematic.

This implementation in Fig. 5.18 requires seven transistors plus the delay element. One should note that the Boolean expression above for the *R-S-T* flip-flop is somewhat arbitrary and depends upon specific application. In Fig. 5.19*a* the standard *J-K*-type flip-flop function is implemented in a very simple manner directly from the *J-K* function with the delay element indicated:

$$Q^{n+1} = J^n\bar{Q}^n + \bar{K}^nQ^n \tag{5-3}$$

The *J-K*-type flip-flop action here is implemented with an inverting delay section *D*. If one used instead the static *J-K* flip-flop with standard Eccles-Jordan equivalent circuitry, a total of 30 devices (as was shown in Chap. 4) would be needed, if direct set (S_D) and direct reset (R_D) together with clocking were provided.

To illustrate a specific design using two-phase delay elements, we can use the logic net of Fig. 5.19*b* to implement the circuit according to Fig. 5.19*c*. The two-phase

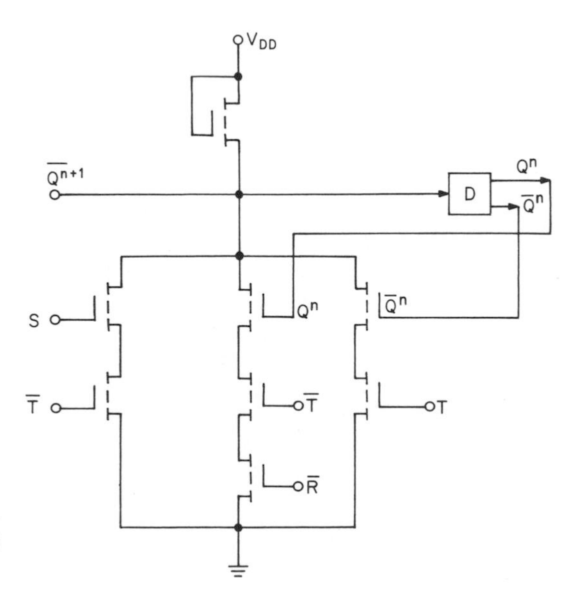

Fig. 5.18. *R-S-T* flip-flop function $Q^{n+1} = S\bar{T} + \overline{RTQ^nET\bar{Q}^n}$ circuit schematic.

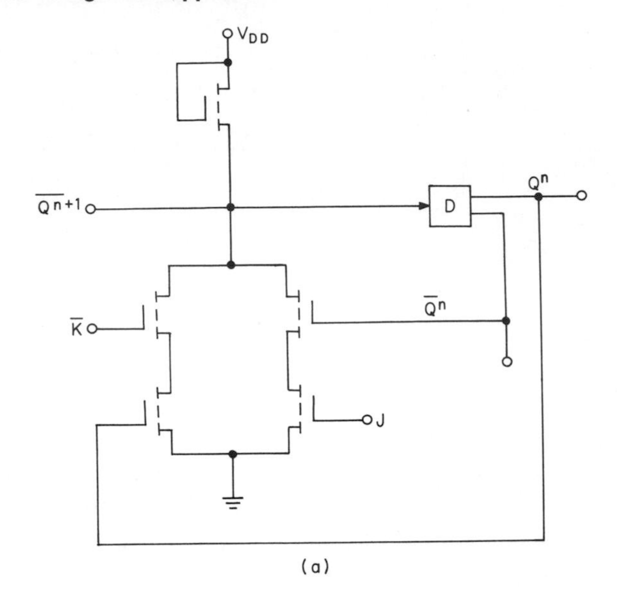

V_{DD}
$\overline{Q}^{n+1}$
D
Q^n
$\overline{K}$
$\overline{Q}^n$
J

(a)

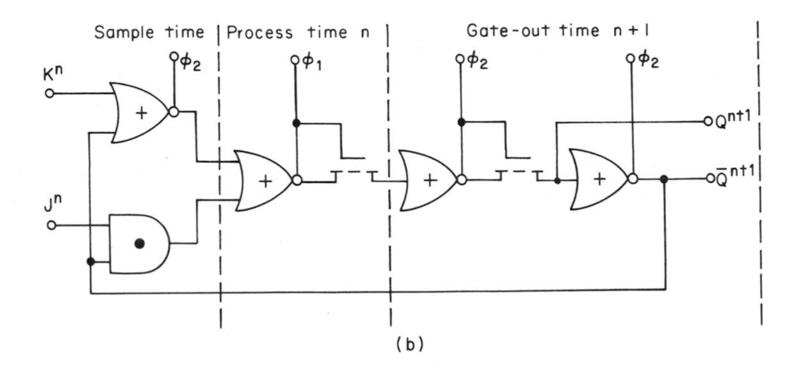

Sample time
Process time n
Gate-out time n + 1
K^n
J^n
ϕ_2
ϕ_1
ϕ_2
ϕ_2
Q^{n+1}
$\overline{Q}^{n+1}$

(b)

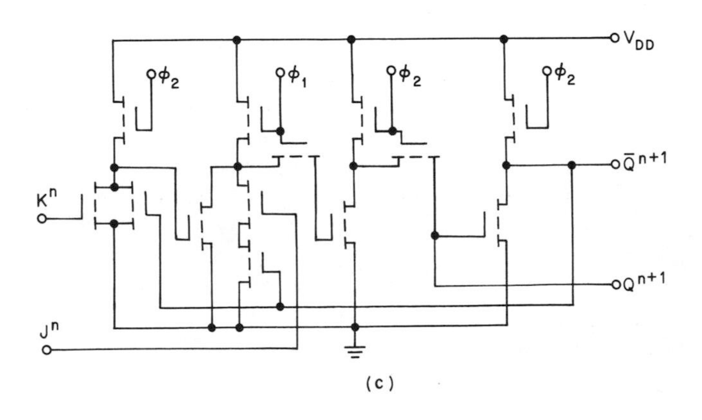

V_{DD}
ϕ_2
ϕ_1
ϕ_2
ϕ_2
K^n
J^n
$\overline{Q}^{n+1}$
Q^{n+1}

(c)

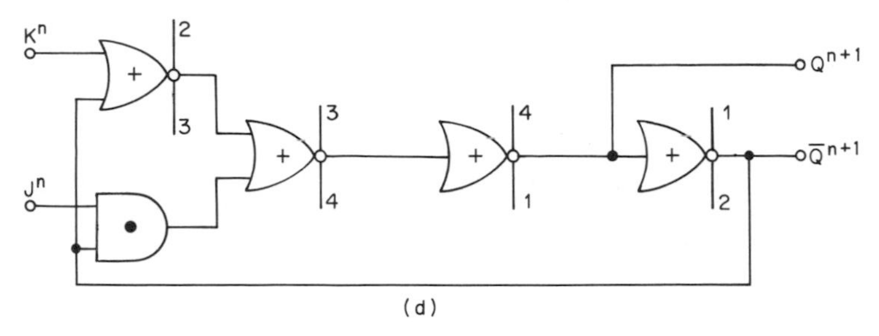

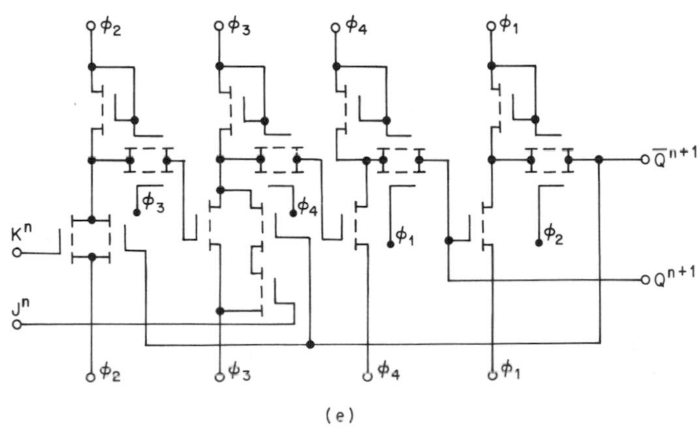

Fig. 5.19. *J-K* flip-flop; $Q^{n+1} = J\overline{Q}^n + \overline{K}Q^n$: (*a*) general circuit schematic; (*b*) logic implementation with 2ϕ ratio elements; (*c*) two-phase circuit schematic; (*d*) logic implementation with 4ϕ ratioless elements; (*e*) four-phase type circuit schematic.

delay element contains the *J-K* random logic section within the master section of the 1-bit delay. This particular design for a *J-K*-type flip-flop without direct set and reset requires a total of 13 separate devices. These devices are not minimum geometry, because the ratio-type circuitry is used here for the example. A four-phase type III ratioless circuit is shown in Fig. 5.19*e* following the logic design of Fig. 5.19*d*. The logic design is separated into four separate clocking intervals. Four separate inverters for the cases *X*, *Y*, *Z*, and *W* are used for successive time intervals. A simple logically true *J-K* input is used; this circuit can be implemented with a total device count of 19 minimum-geometry devices.

The sample-and-hold function in Fig. 5.20 may be implemented according to the logic expression

$$Q^{n+1} = D^nS^n + \overline{S}^nQ^n \tag{5-4}$$

using only four devices in the random-logic design section. The toggle function

$$Q^{n+1} = Q^n \oplus T^n \tag{5-5}$$

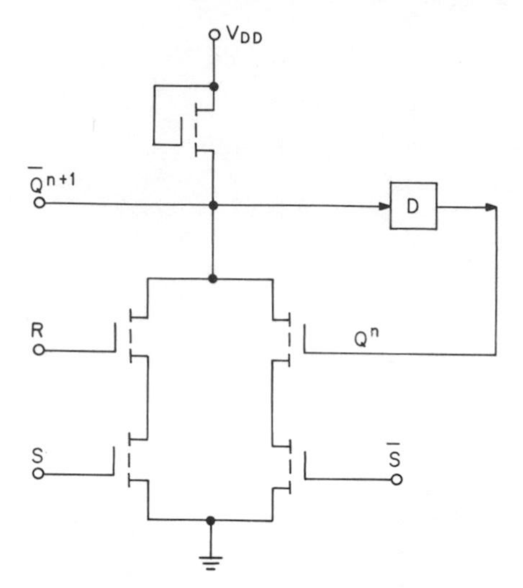

Fig. 5.20. Sample and hold function: $Q^{n+1} = DS + \bar{S}Q^n$.

may be implemented according to the logic diagram of Fig. 5.21*a*. In a particular chip we might gain some geometrical layout advantage by using the equivalent layout of Fig. 5.21*b*. The circuits of Fig. 5.21*a* and *b* are logically identical.

The latching function is another circuit with a simple MOS configuration. The latching function using D as the data input and R as the read input (Fig. 5.22):

$$Q^{n+1} = D^n R^n + \bar{R}^n Q^n \tag{5-6}$$

can be implemented with four devices in the random logic section. With both the data D and the read R input at negative levels, we read in the data. On the other

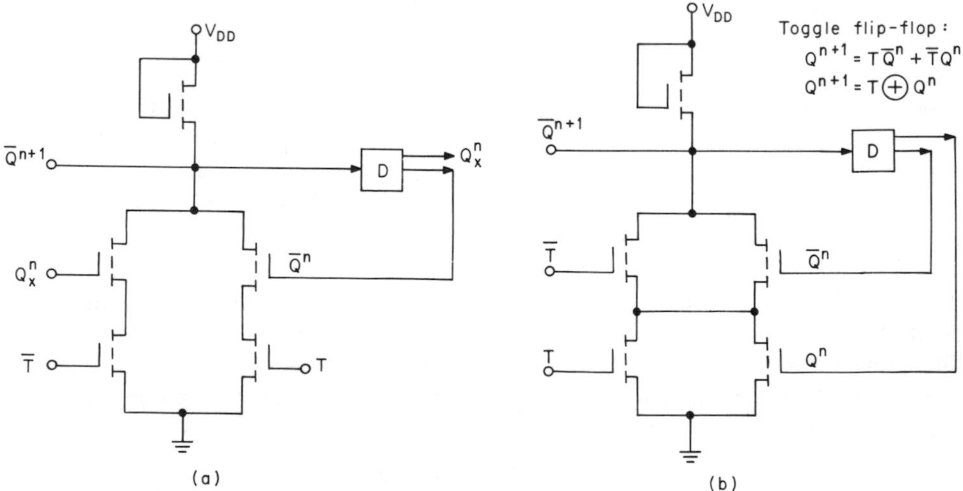

Fig. 5.21. Toggle flip-flop; $Q^{n+1} = \bar{T}Q^n + T\bar{Q}^n = T \oplus Q^n$: (*a*) equivalent I; (*b*) equivalent II.

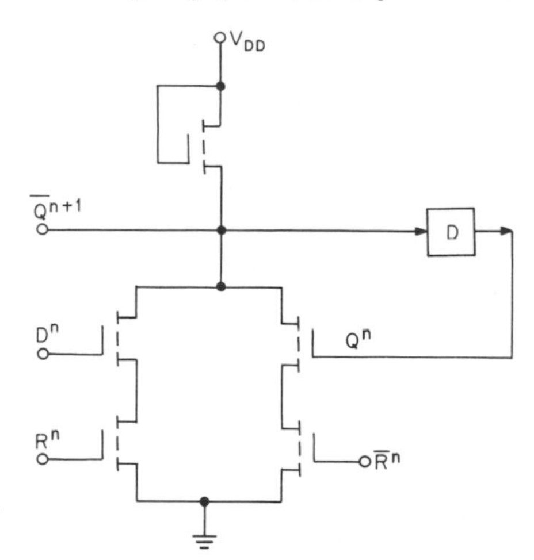

Fig. 5.22. Latching function: $Q^{n+1} = DR + \overline{R}Q^n$: circuit implementation.

hand, when the read input is at the positive-going level, the latch simply retains existing data and recirculates them through the delay section. Thus the latching element functions as a static holding register element. One can design a static holding register element using the circuit of Fig. 5.22 as an alternative to the static register elements already discussed.

5.10.3 Full-adder

A more complex logic function may be implemented as an instructive example. The full-adder circuit of Fig. 5.23*a* represents an increase in circuit complexity. The Boolean expressions for the full-adder reduce to the following and match Fig. 5.23*a* and *b*.

$$Q_s = ABC + (A + B + C)\overline{Q}_c \tag{5-7}$$
$$Q_c = AB + BC + AC \tag{5-8}$$

The building blocks of Fig. 5.23*a* require a total of 12 transistors. To complete the full-adder function, we would have to add series loads for the inverters and match the clocking to whatever logic scheme has been chosen—one-phase, two-phase, three-phase, or four-phase. We can use the building blocks of Fig. 5.23*a* to implement the full-adder of Fig. 5.23*b* with the two-phase clocked structure. The master section, clocked with ϕ_1, followed by the slave section, clocked with ϕ_2, provide the outputs Q_s and Q_c simultaneously. If we wish, on the other hand, to clock with four-phase circuitry, then we may divide a clocking interval into the four-interval sequence shown in Fig. 5.23*c*. The carry Q_c function is developed during the initial clocking interval ϕ_2-ϕ_3. This is followed by the sum Q_s function during the clocking interval ϕ_3-ϕ_4. The data-valid output is obtained during the clocking interval ϕ_1-ϕ_2. Capacitor storage is used during the period between pulses ϕ_4 and ϕ_1 in this clocking sequence. Other full-adder implementations may be specified in a similar manner.

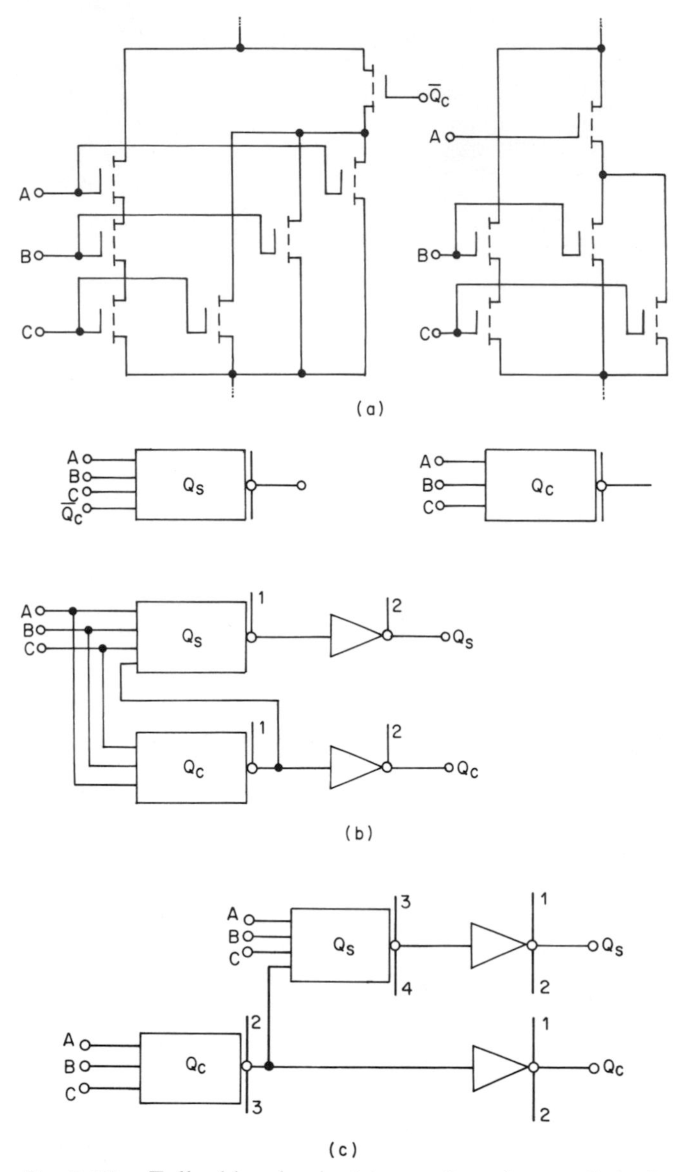

Fig. 5.23. Full-adder circuit: (*a*) sum Q_s and carry Q_c logic net for multiphase circuit implementation; (*b*) typical two-phase type II logic net; (*c*) typical four-phase logic net.

5.11 CIRCULATING MEMORY

The serial shift register may be operated in a recirculate mode by gating the shift register output back into the input. This mode of operation, illustrated in Fig. 5.24, permits one to refresh a CRT display or provide data that are used in a temporary memory, such as a scratch-pad memory. If the external gating circuitry is MOS,

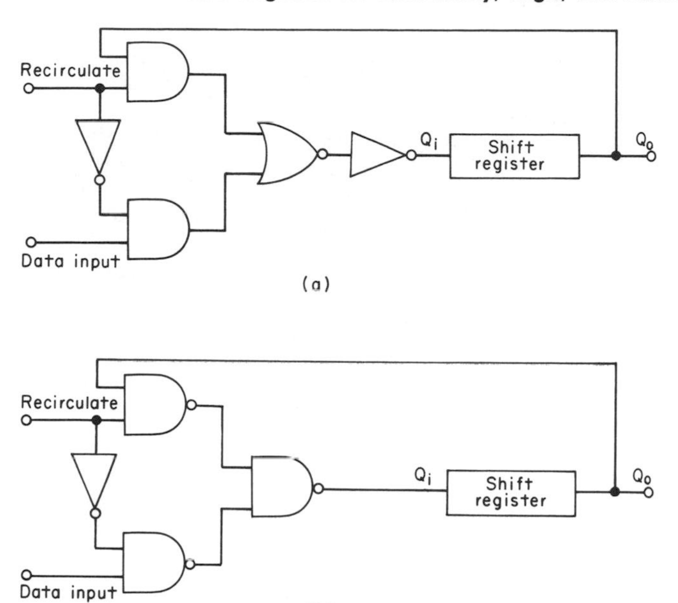

Fig. 5.24. Gated serial shift register with recirculate control: (*a*) MOS logic; (*b*) TTL (bipolar) logic.

one may implement the recirculate/data input modes using the external circuit shown in Fig. 5.24*a*, together with the standard shift register. More often, we will use bipolar circuitry in place of the MOS external gates, and therefore NAND circuits of Fig. 5.24*b* will be useful.

A general representation of a recirculating register which supplies several bits simultaneously for operation in modes such as character generation is shown in Fig. 5.25. A more specific circuit for scratch-pad application is shown in Fig. 5.26. Here

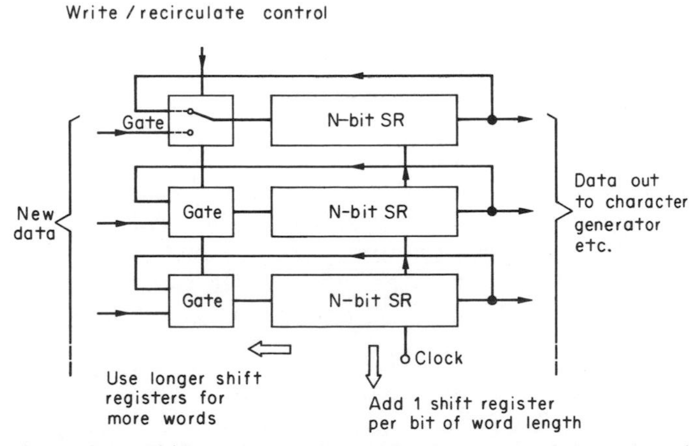

Fig. 5.25. Shift register with recirculate control for general applications.

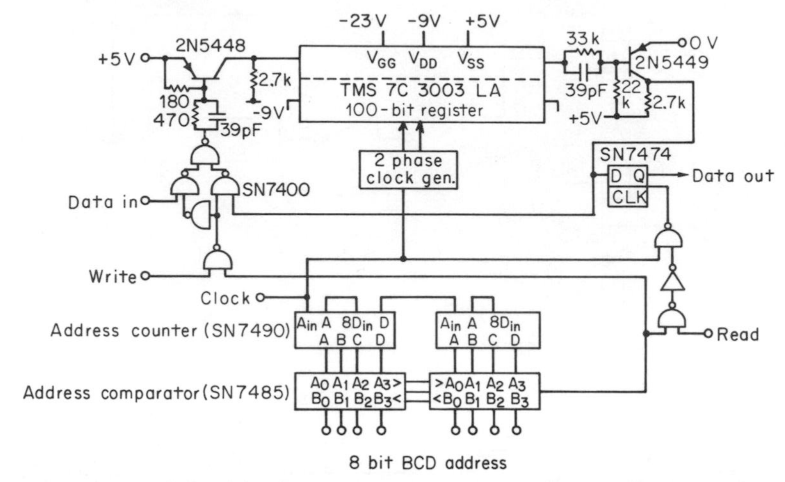

Fig. 5.26. Scratch-pad type memory with bipolar control circuitry.

a serial shift register is used in conjunction with the recirculate/data input external circuitry. In addition, an address counter from the SN7490 series of bipolar circuits has been added. A comparator such as the SN7485 is used to address specific locations along the serial time-pulse train. Separate gates for writing data into the shift register and reading data out of the register are provided as shown. When the contents of the counter corresponds to the specific address, a logical 1 will appear at the output terminal of the comparator. This state of all 1's is detected by a multiple-input NAND gate whose output is used to disable the recirculating gate.

REFERENCES

1. Petritz, R. L.: Current Status of Large-scale Integration Technology, *IEEE JSSC,* **SC-2**(4): 130–147 (1967).
2. Yen, Y. T.: Transient Analysis of Four-phase MOS Switching Circuits, *IEEE JSSC,* **SC-3**(1): 1–5 (1968).
3. Yen, Y. T.: A Mathematical Model Characterizing Four-phase MOS Circuits for Logic Simulation, *IEEE Trans. Comp.,* **C-17**(9): 822–826 (1968).
4. Hoffman, G. B.: MOS Static Shift Registers and TTL/DTL Systems, *Texas Instruments Application Report CA-114,* November 1968.
5. Lohman, R. D.: Some Applications of Metal-oxide Semiconductors to Switching Circuits, *Semiconductor Products and Solid State Technology,* May 1964.
6. Ahrons, R. W.: MOS Complementary Shift Registers and Counters, *Digest of Technical Papers,* Government Microcircuits Applications Conference, October 1968.
7. Mann, R.: Circulating Memories, *Texas Instruments Application Report.*

6

The MOS/Bipolar Interface

6.1 INTRODUCTION

The basic problem encountered in interfacing MOS and bipolar integrated circuits involves the translation of voltage levels—generally, voltage levels of opposing polarities. Typical bipolar integrated circuits, such as the TTL type, use a 5-V power supply. The typical p-channel enhancement-type MOS integrated circuit uses a negative-going power supply varying from -5 V to -30 V. MOS circuits are not standardized throughout the industry with respect to supply voltages. They operate with a variety of power-supply voltages and a wide range of logic swings. In this chapter we are concerned with translating logic levels between the bipolar and the MOS circuit for data transfer and clocking. We can use external discrete devices between the bipolar and MOS integrated circuits, but we prefer to interface directly for a cost savings. We will review those areas where direct interfacing has been possible, as well as those areas where additional discrete devices are required. Since TTL circuitry is the most common form in the bipolar world, we will be concerned primarily with interfacing between TTL and MOS.

The problems of interfacing from MOS to TTL are quite different than the TTL-to-MOS data transfer. The bipolar TTL totem pole output circuits have ample drive capability to provide the speed for driving most MOS capacitive loads, even under high fan-out conditions. Many MOS/LSI circuits are designed to operate with 5-V power supplies and thus are directly compatible with the power supply and data transfer. When the MOS power supply is larger than 5 V, a problem of voltage translation and logic swing occurs.

The MOS transistor, being a low transconductance, high impedance device, is not well suited for driving directly into a TTL load. The MOS transistor which will sink the 1.6 mA from each standard TTL load must have a large transconductance. In fact, the MOS transistors which drive external circuits from the chip are those of maximum transconductance. In cases where the logic swing for MOS is significantly larger than 5 V, the interface to TTL must contain discrete components or special bipolar interface integration.

The drive requirements are considerably reduced when MOS is used to drive low-power TTL loads. The low-power TTL gates supply a maximum of 0.18 mA to the MOS driver. The equivalent resistance of the MOS driver for the positive-logic zero level is 1,700 Ohms. These values compare with 1.6 mA and 250 Ohms for driving standard TTL gates.

6.2 DIRECT COUPLING OF BIPOLAR TO MOS WITH $V_{SS} = V_{CC}$

The task of translating logic levels from bipolar to MOS circuits is much simpler if one of the power-supply potential levels can be shared. The most positive TTL supply level (5 V) is generally connected to the most positive MOS (substrate V_{SS}) level. In fact, we can attempt to direct-couple both power supply levels as shown in Fig. 6.1*a* to discover the problems involved. Here the substrate potential V_{SS}

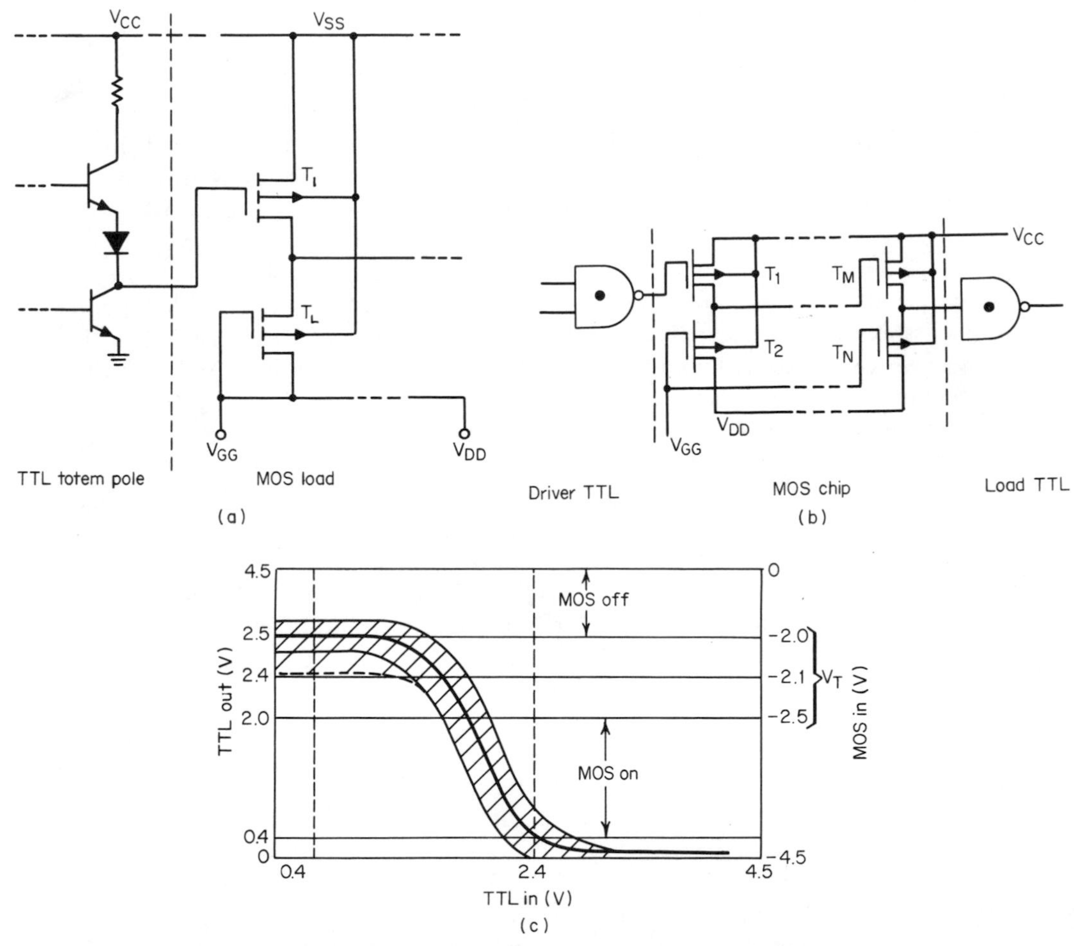

Fig. 6.1. Typical bipolar TTL circuits direct-coupled: (*a*) bipolar TTL driving an MOS PELT inverter; (*b*) MOS chip showing external TTL circuits connected at input and output; (*c*) interface transfer characteristics indicating noise-margin problem.

of the MOS chip is maintained at the positive bipolar power-supply level (V_{CC}). The MOS inverter must have a low threshold voltage for device T_1. A second circuit schematic illustrating a TTL driver and a TTL load both connected to an MOS chip is shown in Fig. 6.1*b*. The problem of voltage translation for these circuits is illustrated in Fig. 6.1*c*, which provides us with the TTL transfer characteristics. Here the vertical scale has been calibrated on the left side for voltage levels referenced to system ground, and on the right to MOS substrate. Notice that the bipolar voltage of 4.5 V corresponds to a zero voltage level in the MOS circuit. The worst-case transfer characteristics for the TTL provide a positive logic low for voltages of zero to 0.4 V. Similarly, a positive logic 1 is obtained in the TTL circuitry for levels of 2.4 V and higher. As a result, worst-case design points on the transfer characteristic of Fig. 6.1*c* are 2.4 and 0.4 V.

We wish to reference these voltages to the MOS by translating from left to right from the TTL voltage levels into the MOS voltage levels. The special voltage indicated here as being typical of our MOS circuit ranges from −2.0 to −2.5 V. The corresponding positive voltage levels for the TTL circuit are 2.5 and 2.0 V. Notice that the MOS voltage of −2.5 V corresponding to a TTL output voltage of 2.0 V may cause an MOS inverter to turn on when we do not wish it to do so. In fact, the noise margin in this worst case is negative, and therefore unacceptable. We conclude that we cannot drive the MOS circuit directly from the TTL where worst-case MOS thresholds are in excess of −2.1 V. A simple remedy is available for maintaining the voltage output from the TTL at a higher positive level. One can add a resistor R_1 between the TTL output and the positive power supply, as is shown in Fig. 6.2*a*. When this is done, the transfer characteristics of Fig. 6.2*b* can be obtained for a sufficiently small value of R_1. Notice that now the minimum voltage level from the TTL for a positive logic 1 is approximately 3.6 V, corresponding to a voltage on the MOS chip of −0.9 V. The higher voltage output obtained from the TTL gate or other bipolar driver circuit in this case, is now more acceptable and can provide noise margins that are quite adequate. If the unity gain point for thresholds of the MOS inverter ranges from −2.0 to −2.5 V, we have, in this case, a worst-case noise margin of 1.0 V, as shown in Fig. 6.2*b*. For the bipolar output voltage of a positive logic 0, we have no problem in maintaining good noise margin. For instance, one would expect the output voltage from the bipolar driver to increase toward 2 V before a false data transfer results in switching action in the MOS inverter. The two noise margins shown as derived from the transfer characteristics in Fig. 6.2*b* are based upon an idealized MOS transfer function. Specifically, these noise margins assume that the unity gain points for the MOS inverter are occurring at a voltage equal in magnitude to V_T. This assumption is of course not precisely accurate for an actual inverter, as we have seen in Chap. 4, but serves, in this instance, to indicate clearly the problem in interfacing the data from a bipolar source into an MOS load.

The pullup resistor in Fig. 6.2*a* will generally be in the range of 1,000 to 2,000 Ohms. A diffused resistor of this value would normally require a large area on the chip. A much better solution, illustrated in Fig. 6.2*c*, substitutes an MOS device for resistor R_1. This device is designed with a resistance equivalent to the value desired.

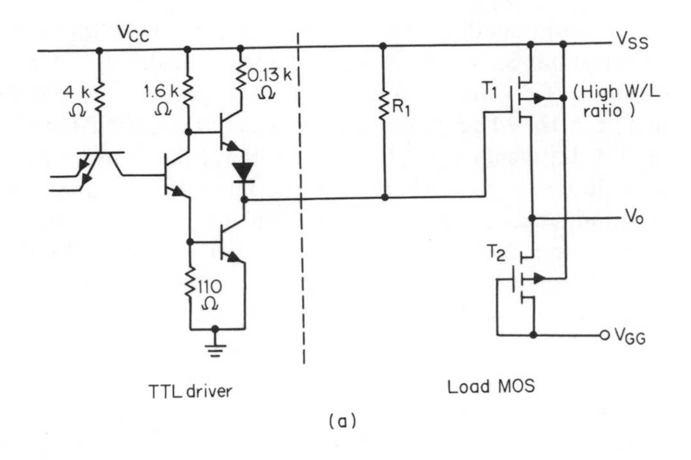

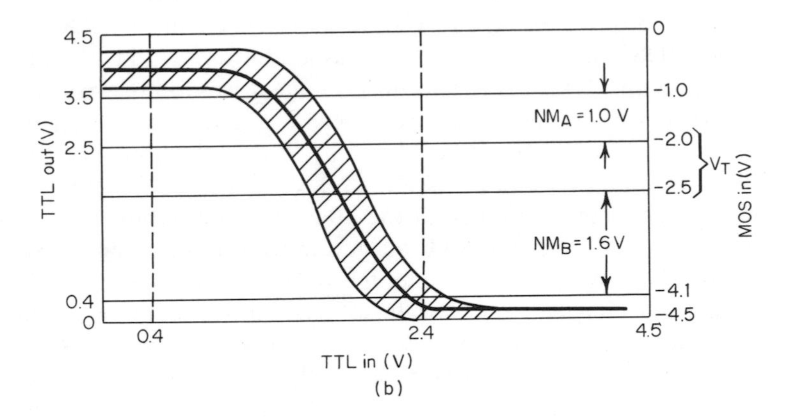

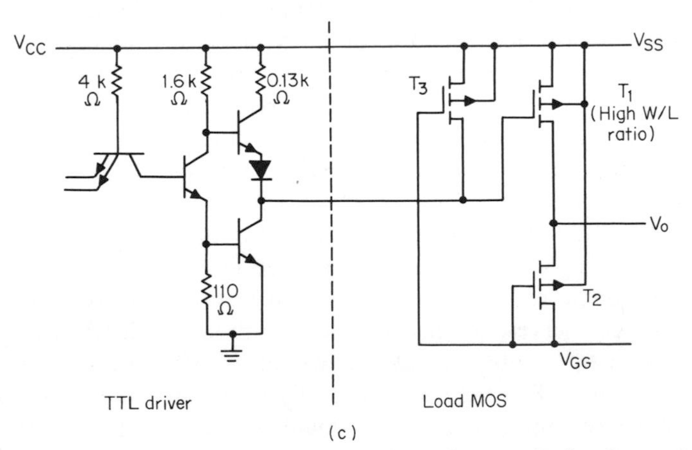

Fig. 6.2. Pullup resistance improves the noise margin for data entering the MOS chip: (*a*) diffused resistor pullup; (*b*) transfer characteristics; (*c*) improved circuit uses device for pullup.

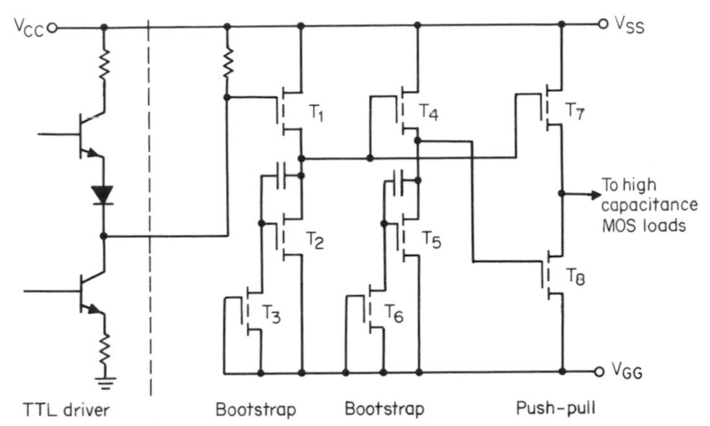

Fig. 6.3. Increased drive capability near the MOS chip data input, with bootstrap pullup inverters switching a low-impedance push-pull stage.

6.3 INCREASED DRIVE CAPABILITY ON THE MOS CHIP

The input gates on an MOS/LSI chip may have a high fan-out to other MOS. In cases where clocking is generated on the chip, or many logic gates fan out from a single input inverter, we must include a design for the interfacing inverter circuitry, which has a higher than average drive capability. Figure 6.3 provides an increased drive capability for capacitive loads branching out from this input inverter on the chip. This type of circuitry is required when the logic level or intermediate cascade gates will not permit one to drive these high capacitance loads directly from the bipolar external driving circuitry. In Fig. 6.3 a push-pull inverter consisting of transistors T_7 and T_8 is driven from two separate inverters, each having a bootstrap-pullup load. Driving the push-pull devices provides a maximum drive capability for capacitive loads on the MOS chip. This circuit has the advantage of low power dissipation but it does require dynamic clocking. An alternative circuit which also provides drive capability for highly capacitive loads is shown in Fig. 6.4*a* and uses a depletion-type load device. You will recognize this as the PDLT-type inverter. Some charging transient time values for the PDLT-type inverter driving a relatively small 0.5-pF load are shown in Fig. 6.4*b*. These values refer to the driver device with gate W/L ratios of 1, 2, and 4 and a constant β ratio of 20. The threshold voltage for the driver device is held constant at -1.6 V, and the threshold V_T for the depletion load is varied from $+6$ to $+24$ V. The charging transient time t_I is the time lapse between the driver turning off and the output rising to 90 percent of its full value. The PDLT load provides a t_I value as small as 9.7 ns in the case of the maximum V_T and W/L ratio. For loads of 10 pF, the values of t_I increase by a factor of approximately 10.

In those cases where we have a higher threshold voltage on the MOS chip and require logic voltage swings in excess of that obtainable directly from the TTL or other integrated-circuit driver, we must go to either a discrete device or an open-ended collector driver. The bipolar driver shown in Fig. 6.5*a* provides the necessary

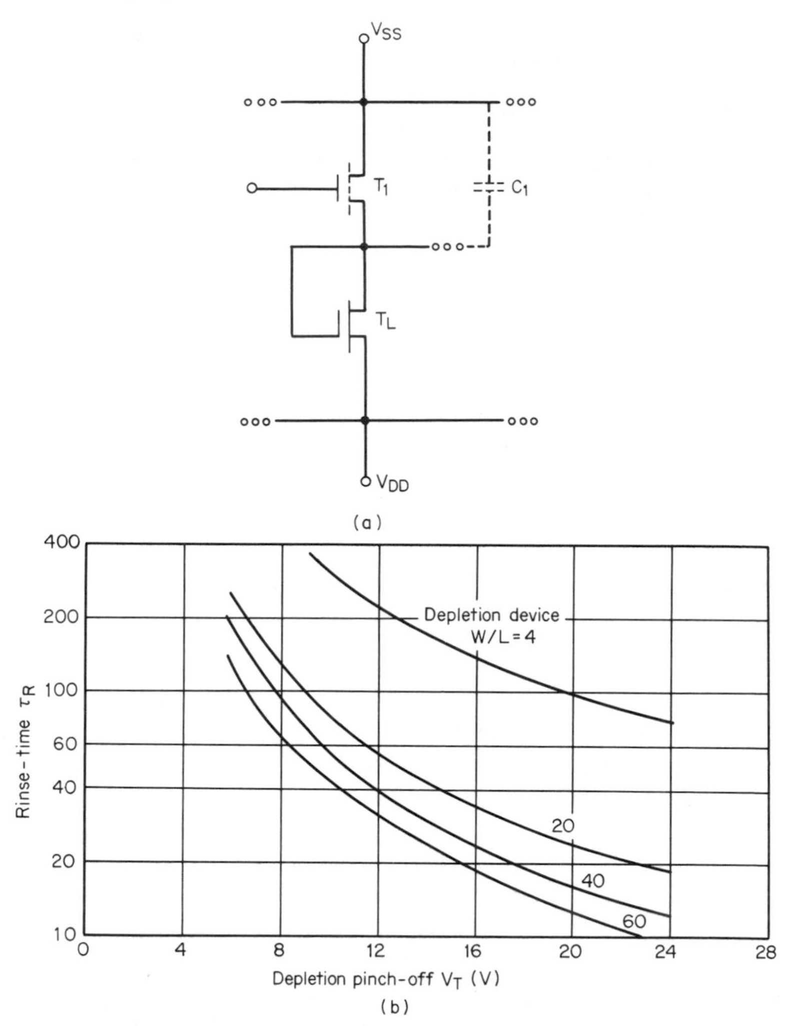

Fig. 6.4. Increased drive capability with low-impedance PDLT inverter: (*a*) circuit schematic; (*b*) rise-time t_R *vs.* threshold (pinch-off) voltage of the load device with W/L ratio of the depletion device as a parameter and $C_1 = 0.5$ pF.

increased logic voltage swing required for higher-threshold MOS devices. In Fig. 6.5*b* the V_{SS} voltage of 8.0 V is matched to the bipolar gate transfer curve. The load resistance T_L provides the necessary pullup action and results in a logic swing of approximately 7 V. The resulting noise margins are 2.4 and 4.1 V. A similar large noise margin is obtained when one uses a substrate maintained at +14 V and threshold values for the MOS circuit ranging from −3 to −5 V. Figure 6.5*c* illustrates the transfer characteristics obtained under these conditions. The 14-V power supply results in a logic swing from the bipolar driver of approximately 13.5 V. Corresponding noise margins refers to the MOS inverter of 2.9 and 8.6 V.

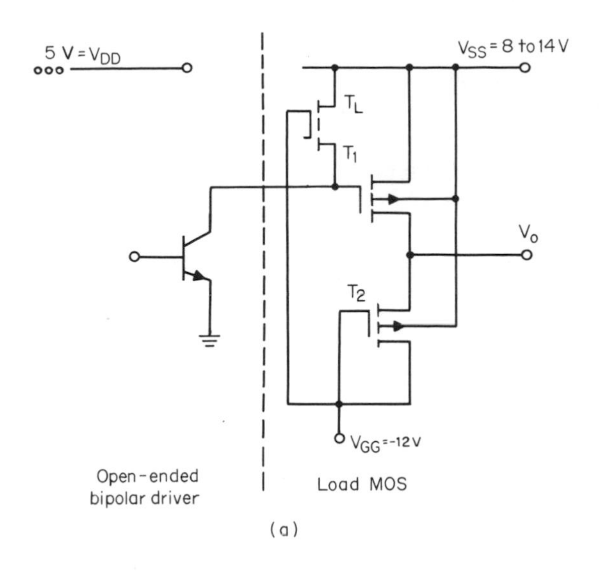

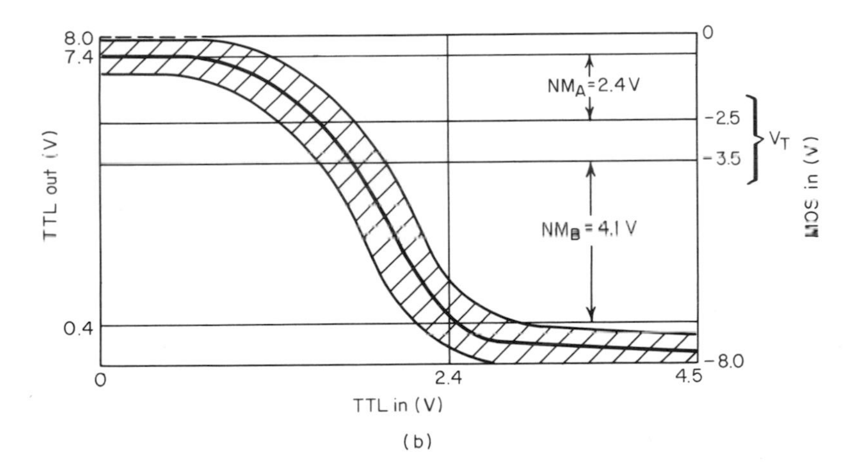

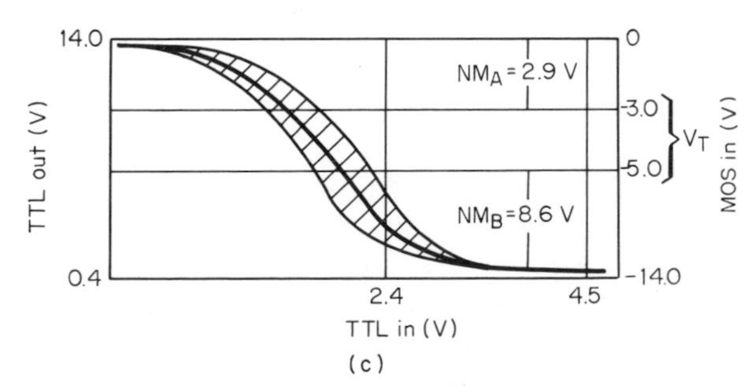

Fig. 6.5. Increased logic swing obtained for MOS from a TTL data source using proper selection of V_{SS} and V_{GG}: (*a*) circuit schematic; (*b*) noise-margin relationship for V_{SS} = 8 V; (*c*) V_{SS} = 14 V.

It is relatively easy to transfer data from a bipolar driver circuit into an MOS integrated circuit when the substrate of the MOS circuit is maintained at the most positive system voltage level. Figure 6.6*a* and *b* illustrates the relative ease of this data transfer. In Fig. 6.6*a* the single bipolar interface device operating with the pullup resistor provides an excellent noise margin for the MOS circuit. In Fig. 6.6*b* a Zener voltage translation diode is added to permit an increase in speed for the driving circuitry.

In those cases where the substrate of the MOS load is not maintained at the V_{CC} positive power-supply potential, the interface circuitry becomes more difficult. In Fig. 6.7*a* a TTL driver is interfaced into an MOS load where the substrate of the MOS circuit is maintained at system ground. Here device T_1 operates as a switched current source and provides current drive for T_2. The necessary voltage translation

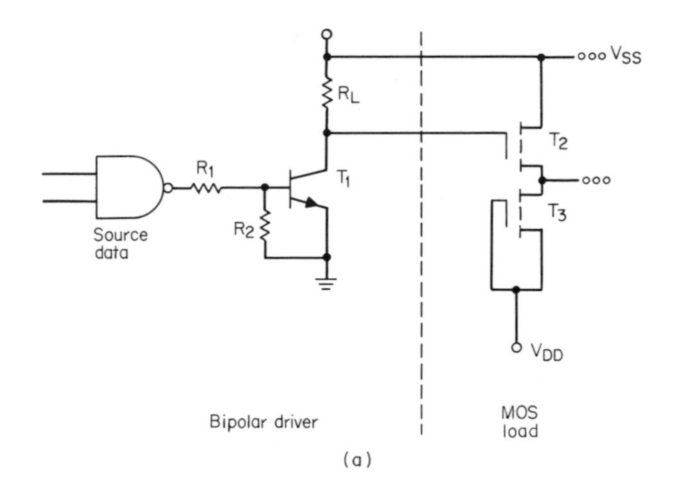

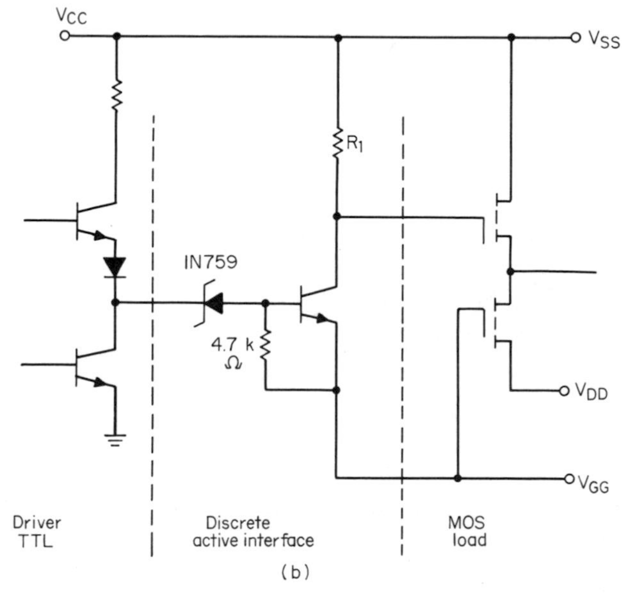

Fig. 6.6. Interface circuitry with increased logic swing for driving MOS circuitry from a bipolar driver. MOS substrate is at a voltage V_{CC} level: (*a*) inverter with base-limiting resistor; (*b*) inverter with Zener voltage translation.

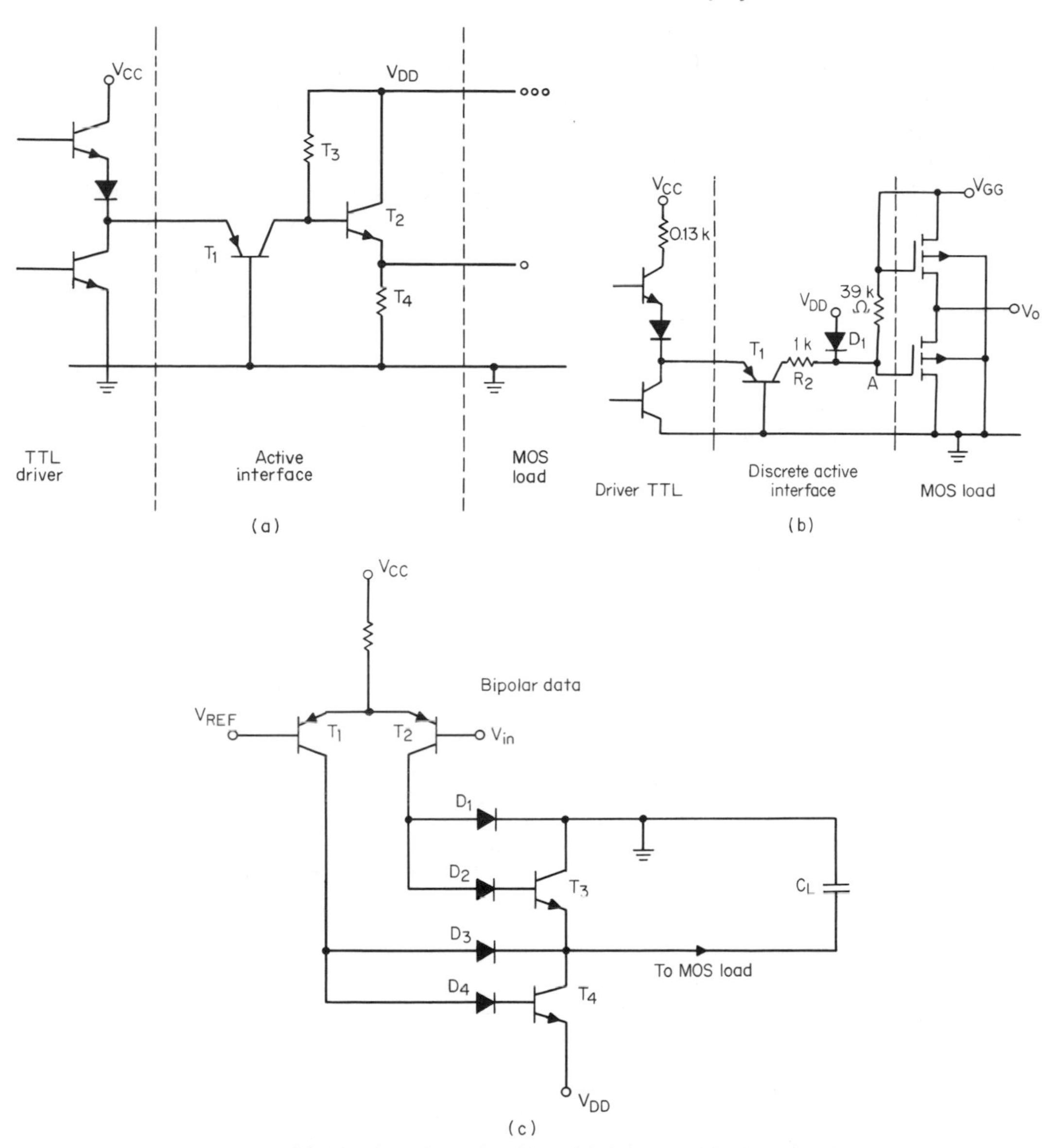

Fig. 6.7. Bipolar interface circuitry with increased logic swing for MOS load. MOS substrate is at system ground potential. (*a*) two-transistor translator; (*b*) single-transistor interface with diode pull-to-V_{DD}; (*c*) complementary drivers with increased drive capability.

occurs through device T_1, and transistor T_4 is switched in and out of saturation conditions. The PNP device contained in the active interface circuit is generally not found on bipolar integrated-circuit structures such as the TTL type, and represents an external discrete device. The PNP transistor T_1 is again used for the circuit of Fig. 6.7*b* for obtaining the desired voltage translation. Here, device T_1 is operating more as a saturating switch. Node A is reduced to ground potential as T_1 saturates. When device T_1 is not conducting, the pullup diode D_1 connected to the V_{DD} power supply clamps this diode at a high negative voltage level. These two interface circuits in a and b are not particularly fast and, therefore, may on occasion prove to be the limiting factor for system switching speed.

For those cases where the large voltage swing must be maintained and the bipolar driver circuitry requires high speed, then a circuit such as that shown in Fig. 6.7c using PNP transistors at T_1 and T_2, may be used. The PNP transistors of this circuit are no longer as critical as in the previous case, and may be of the lateral type. The circuit may be integrated on a single chip to improve the system economics.

This nonsaturating circuit includes the differential amplifier pair T_1 and T_2 which isolates the input signal V_{in}, shifts its level, and provides a convenient means of setting threshold voltage. The differential pair drives a quasi-complementary output state T_3 and T_4. Each output transistor may be thought of as an operational amplifier with a diode feedback that reduces the gain to unity when the base drive is large enough to produce saturation. The diodes D_1 through D_4 are clamping devices which will prevent these transistors from saturating. By proper selection of the reference voltage in this circuit, the threshold of the logic level delivered to the MOS load can be adjusted to particular values as required. This circuit can drive a capacitive load of 100 pF through a 20-V logic swing with rise and fall times of 4.0 ns. Thus, the circuit of Fig. 6.7c can be readily integrated as an interface level shifter.

6.4 MOS CIRCUITS FOR DRIVING A BIPOLAR LOAD

The problem of interfacing from MOS into a bipolar load is more difficult than interfacing from bipolar into a MOS load. The TTL gate, in addition to placing a 4- to 8-pF load on an MOS driver, is an active current source. In Fig. 6.8a the

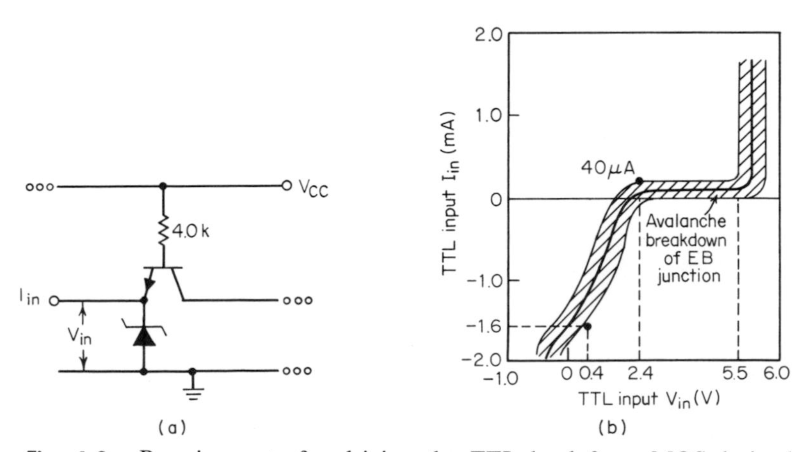

Fig. 6.8. Requirements for driving the TTL load from MOS-derived levels: (*a*) TTL gate input; (*b*) input I-V characteristics.

input device, together with its Zener protective diode, is shown for a standard TTL gate. Here, when the input is at a positive logic 0, a worst-case 1.6 mA flows from the TTL load into the driver circuit. Major consideration must be given to the fact that the TTL load is an active load and differs from the passive load that one encounters in many types of circuit situations. Another problem is the avalanching voltage associated with the Zener protective diode. The worst-case avalanche voltage is 5.5 V, a level which should not be exceeded by the voltage delivered from an MOS driver. These operating regions are shown on the I-V curve of Fig. 6.8*b* representing the input function for the TTL gate. The effective resistance that a driver device can present to the TTL load and ensure a logic 0 level is 0.4 V/1.6 mA $\cong$ 250 Ohms.

A circuit for interfacing from an MOS driver into a TTL load is shown in Fig. 6.9*a* which includes an intermediate interfacing resistance R_1. For a fan-out of unity

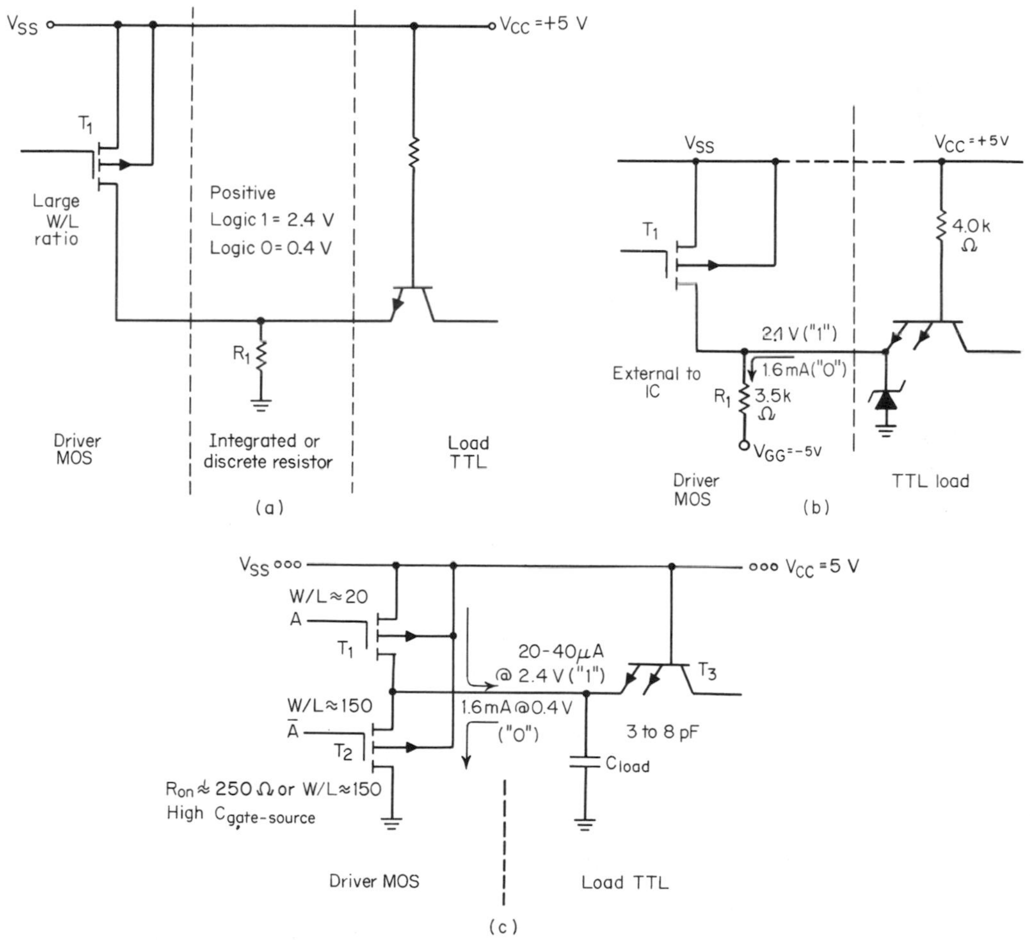

Fig. 6.9. Circuit schematics for MOS-to-TTL interface: (*a*) 250-Ohm current-sinking resistor R_1; (*b*) with additional power supply V_{GG}; (*c*) push-pull MOS without external resistors.

the interface resistance must have a maximum value of 250 Ohms. The driver MOS device T_1 must be a high-transconductance device if the TTL input logic level can swing to the positive-high state. The output voltage rises to 2.4 V when the effective resistance r_{DS} of device T_1 has approximately the same value as R_1. The W/L ratio required in device T_1 to provide an r_{DS} of 250 Ohms ranges from 150 to 200. Furthermore, the high power dissipation (50 mW) in device T_1 and resistance R_1 makes this particular interface scheme undesirable. The power dissipation in device T_1 may be reduced by using a separate power-supply voltage V_{GG} and a larger resistance R_1, as shown in Fig. 6.9*b*.

Another circuit scheme permitting the use of the single power supply utilizes a push-pull output inverter, as shown in Fig. 6.9*c*. This circuit does not require external resistors or additional power supplies, and it can be integrated completely on the MOS chip. A large transconductance device T_2 sinks the required 1.6 mA of current for the positive 0 state. If devices T_1 and T_2 are driven out of phase, the upper device, T_1, does not require a high transconductance, since this is a ratioless inverter. Typical W/L ratios for devices T_1 and T_2 are 20 and 150, respectively. The real difficulty in designing circuits of the type shown in Fig. 6.9*c* appears when one realizes that the large-area MOS devices constitute a highly capacitive load to the driver stage on the MOS chip. The problem can be solved by utilizing a cascade of inverters, each successive stage of which has higher transconductance than the preceding stage to provide the desired drive capability for these large-area push-pull output stages, as shown in Fig. 6.10*a*. The proper tapering-up for the driving inverters requires optimization of the circuit parameters. The drive capability of the tapered inverter can also be improved by using the bootstrap-pullup technique shown in Fig. 6.10*b*. This tapered inverter cascade is used to drive the push-pull MOS inverter of Fig. 6.9*c* for an optimum configuration. This total circuit is also illustrated by Fig. 6.10*a*.

Representative calculations have been made using the SCEPTRE CAD program for the switching transient of the MOS push-pull output pair loaded by a single TTL gate and driven by separate inverters each without the bootstrap-pullup feature. The simulation includes eight MOS transistors. The load TTL is a 6-pF capacitance. The dc levels are all within the 0.4-V tolerance for the logic 0 level. The transistors T_1 and T_2 of Fig. 6.10*a* have W/L ratios of 150 and 20, respectively. Transistor T_1 is driven by an inverter using an active device with a W/L_D ratio which varies. The beta ratio for the drivers is held constant at 20. The W/L ratio for the active device driving directly to the push-pull output is varied from 20 through 80. Table 6.1 summarizes the results for the transient switching response. The parameters used are W/L_D for the driver to the output push-pull, the dc output voltage from T_1 (V_{MIN}), the delay time t_D for the cascade, the fall-time t_F for charging the capacitor, and the rise-time t_R for developing the V_{MIN} voltage out. For pulsing at 2 MHz, the total power dissipation P_T is given in milliwatts. The speed/power product is found to reach an optimum for W/L_D values of approximately 60. This particular circuit can be used with a TTL fan-out of two.

Separate calculations have been made for several MOS integrated output configurations with fan-out to a single TTL gate. One especially promising circuit uses push-pull inverters for the last two stages of the tapered cascade illustrated in Fig.

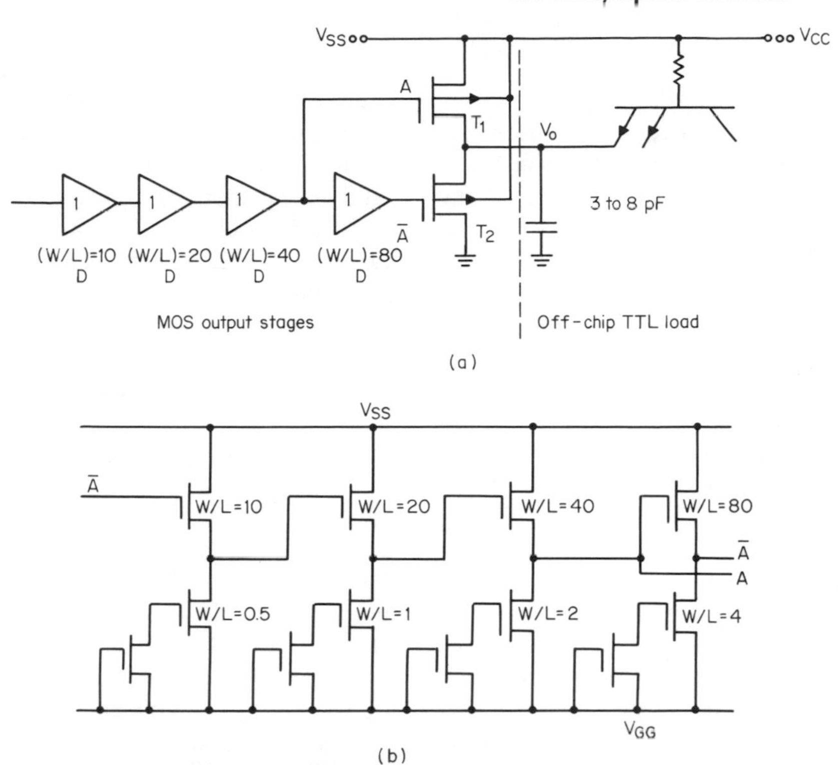

Fig. 6.10. Tapered cascade required for driving the MOS push-pull: (*a*) diagram; (*b*) circuit schematic using bootstrap-pullup inverters.

6.10*a*. Here an output push-pull pair with W/L values of 175 and 20 was used. The output pair is driven by another push-pull inverter with devices of W/L equal to 30 and 5. This double-push-pull circuit provides t_F values approximately equal to t_R to the TTL load. In this case $t_F \approx t_R = 15$ ns. The power dissipation is 3 mW, providing an overall speed/power product approximately equal to that outlined in Table 6.1.

When fan-out values greater than unity are required directly from the MOS chip, a bipolar transistor can be integrated together with the MOS. This configuration

Table 6.1. Performance Parameters Calculated for an MOS-to-TTL Push-pull Interface Circuit

W/L_D	V_{MIN}, Volts	t_D, nsec	t_F, nsec	t_R, nsec	P_T, mW
20	0.347	216	177	6.1	5.3
40	0.232	120	94	5.3	8.0
50	0.223	100	76	5.3	9.3
80	0.218	72	55	5.5	12.9

provides yet another possibility for obtaining the desired drive capability for a TTL load from an MOS chip and is shown in Fig. 6.11. Here a higher-transconductance bipolar device T_3 is integrated on the chip with the MOS circuitry. Unfortunately, this technique still requires the low resistance value of R_1 to sink the current required. The problem arises from the fact that device T_3 is integrated with a single additional diffusion, the emitter diffusion, and relies upon the substrate for the bipolar collector. This type of circuit permits us to obtain a higher-than-unity fan-out. This is one of the few ways one can obtain a higher-than-unity fan-out from the MOS chip without adding more than a single external device. The single external device in this case is the resistance R_1. For a fan-out of three, we require an R_1 value of 83 Ohms. The power dissipation in R_1 for 83 Ohms is 300 mW. This high power dissipation necessitates the placement of R_1 as an external discrete resistor.

When the low-power TTL circuits are used, the drive requirements from MOS are greatly relaxed. For low-power TTL the 250-Ohm value for the MOS output increases to 1,700 Ohms. Thus, the slower low-power TTL gates will find wide use for interfacing to the output of MOS/LSI circuits.

We have another interface problem area. The interface from high-level MOS logic to any TTL requires external discrete circuitry or special-level shifting TTL logic. The interface from a high-level MOS circuit into a TTL-type load shown in Fig. 6.12 utilizes a voltage-dividing network consisting of resistances R_1 and R_2. These values are selected according to the power-supply voltages available and the effective resistance of device T_1 in the *on* state. A fan-out higher than unity is generally not available for this configuration for standard TTL loads. In these circuits one must be careful not to exceed the 5.5-V worst-case avalanche threshold in the TTL circuit.

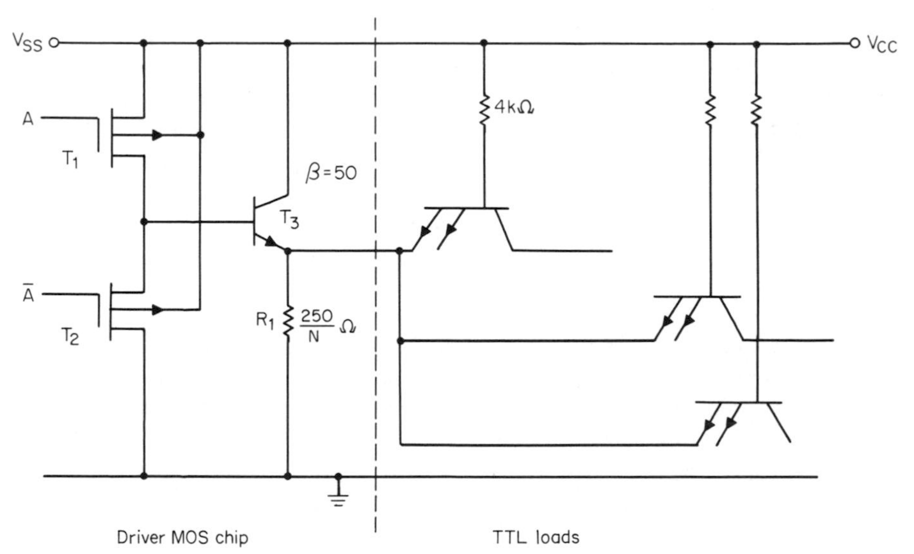

Fig. 6.11. Bipolar device integrated on the MOS substrate increased fan-out capability.

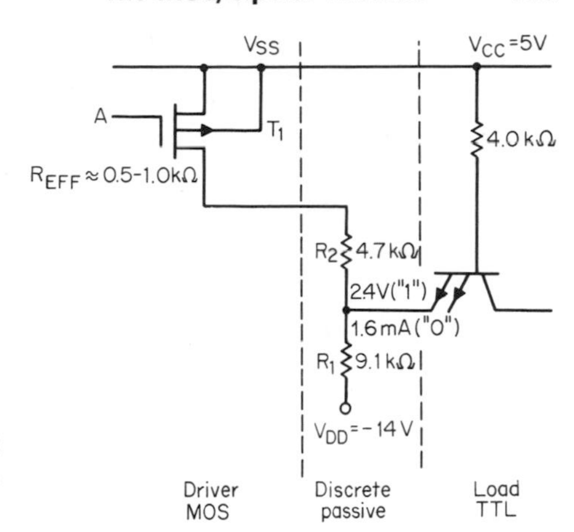

Fig. 6.12. Interface circuitry for fan-out from high-threshold MOS, using discrete resistors in voltage translator.

6.5 CLOCKS FOR SINGLE AND MULTIPHASE MOS

The design of clocking circuitry for MOS application represents a challenging problem. Clocking circuits generally drive highly capacitive loads in the order of 50 to 100 pF, and sometimes as high as 1,000 pF. This imposes a stringent requirement, and requires that we use the higher-transconductance bipolar or JFET devices. Figure 6.13 shows a circuit with complementary switching output devices T_2 and T_3 and with a control obtained from a logic gate. This clock driver must produce the desired switching speed and delay time. The switching speed is required to provide high clocking rates, and the delay is required to space the clock pulses and avoid such things as overlapping clock pulses in certain cases. The charge storage in the base of transistor T_1 during saturation introduces a delay into the generation of the clock pulse. The circuit of Fig. 6.13 has 60 ns of storage time

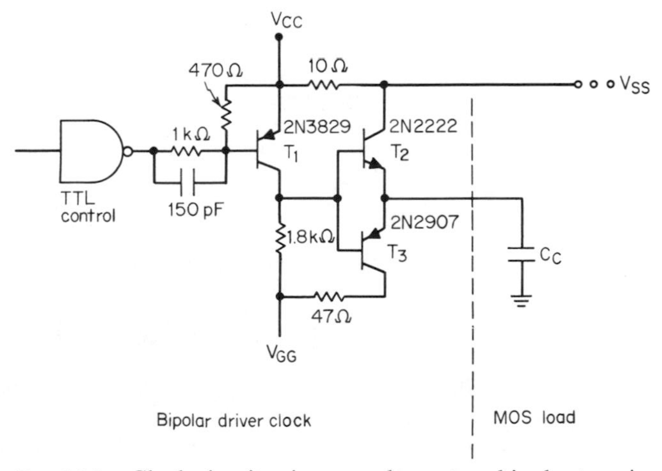

Fig. 6.13. Clock circuit using complementary bipolar transistors.

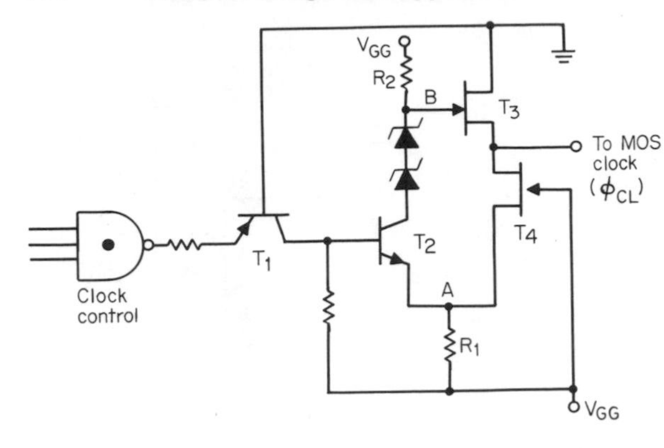

Fig. 6.14. Clock circuit using complementary JFET transistors.

caused by the saturation of transistor T_1, and will provide a rise time of less than 20 ns when driving a 1-pF load.

Another clock circuit using JFET devices T_3 and T_4 is shown in Fig. 6.14. Here another complementary clock circuit derives its data control from a logic gate. The delay of the circuit can be determined to some extent by devices T_1 and T_2. Two Zener diodes are used for voltage translation and provide the voltage drop required to ensure that transistors T_3 and T_4 are never conducting simultaneously. The JFET devices with the zero offset voltage provide a useful clock source. The logic high coming in from the TTL data source causes devices T_1 and T_2 to conduct, resulting in a voltage drop across resistance R_1 and the cut-off of JFET device T_4. At the same time, nodes *A* and *B* move toward increased positive potential, causing T_3 to conduct. Similarly, when the data input goes to a positive logic 0, devices T_1 and T_2 cease to conduct, and the voltage drop across resistance R_1 drops to zero. As a result, there is no gate-to-source bias on device T_4, and that transistor goes into strong diode conduction. Simultaneously, the supply voltage V_{GG} through pullup resistance R_2 places a large negative gate-to-source bias on device T_3, forcing that transistor into a cut-off region. As a result, devices T_3 and T_4 alternately conduct, swinging the output node between ground and supply V_{GG} voltage levels.

REFERENCES

1. D. Pippenger, "Peripheral Interface Circuits with SN75450," Texas Instruments Application Report CA-150, March 1970.
2. T. E. O'Brien, Monolithic Level Shifter Lets MOS, TTL Share Same Network, *Electronics,* **44:** 70–72, July 1971.
3. R. H. Crawford, "Current Directions in MOS/Bipolar Interfacing," IEEE International Convention Proceedings, March 1969.
4. T. Reynolds, Interfacing MOS and Bipolar Logic, *The Electronic Engineer,* **30:** 62–65, April 1970.
5. G. Hoffman, "MOS Static Shift Registers and TTL/DTL Systems," Texas Instruments Application Note CA-114, August 1969.
6. P. Hawkins, Master Thesis, Southern Methodist University, Electronics Sciences Center, 1972.

7

Memory Applications

7.1 INTRODUCTION

The earliest semiconductor memories used diode matrices. The presence or absence of a diode indicated the precoded logic level at each memory cell location. The diode memories are read only memory (ROM), and their function should be distinguished from that of read/write memory (RAM). Recently, silicon monolithic integrated-circuit technology has greatly reduced the manufacturing cost for the all-semiconductor memory. At the present time, semiconductor memory provides a very competitive alternative to magnetic core memory and other often specified memory technologies. Semiconductor memory is used for ROM, for RAM, and in addition, for a newer configuration, content-addressable memory, CAM. In every case the monolithic integration of address and decoding circuit complexity, together with special memory cells, permits unique advantages. Cost-effective applications using semiconductor MOS memories are evolving so rapidly that it is virtually impossible to predict the final systems applications.

A limitation for semiconductor RAM and CAM is the data volatility inherent in most cell technologies. Nondestructive read-out is provided by most MOS memory circuits, but the data is destroyed by a power-supply failure. Auxiliary power supplies can be used to provide the "holding" voltages necessary to retain MOS memory data intact over extended periods of time. In many applications it has been discovered that programs stored in magnetic mass disk or tape memory do not require a nonvolatile high-speed RAM in the central processor anyway. There are nonvolatile MOS memories in development that are available for systems applications. These memories are each discussed separately in this chapter.

Before MOS memory entered the scene, all computer memory had been static in nature. A static memory is one in which the logic levels are maintained indefinitely within each storage cell. Some static memories require that the power-supply levels be maintained, but special pulsing techniques are not required. With the availability of the MOSFET for monolithic memory, we can utilize charge-storage at individual device gates to specify logic levels. The MOS memory circuits, with this unique charge-storage at individual circuit nodes, permit a considerable reduction in memory cell complexity. These MOS memory cells with charge-storage at capacitors are termed dynamic memory. Dynamic memory permits smaller memory cell areas and reduced power dissipation, but must be refreshed periodically with appropriate clocking and refresh circuitry.

At the present time, the dynamic memory cell concept is applied only to read/write RAM cells. It is quite possible, however, that the dynamic charge-storage concept will be extended to specialized ROM designs.

This chapter details the various types of semiconductor MOS memories available at the present time. It is our intent to provide the reader with fundamental knowledge. A fair amount of discussion is devoted to comparisons of the various MOS memory cells and associated address/decode circuits. It would be futile, of course, to assume that these same relative merits presently existing among the various MOS memory techniques and technologies will continue indefinitely into the future. The technology is evolving quickly, and new circuit innovations come rapidly.

7.2 READ ONLY MEMORY (ROM)

7.2.1 The Basic Matrix-type ROM and Programable Variations

The simplest of the various MOS memory configurations is the read only memory (ROM). The ROM can take a digital code at its input terminals and provide a unique digital code on its output terminals. The relationship between the input and output codes is relatively fixed, usually alterable only by relatively slow techniques, and for this reason we term it "read only." The difference between a read only memory and a read/write memory is the level of difficulty of changing the stored information.

The read only memory is generally a random-access type, and permits us to access information stored within the memory with a random addressing code. The ROM is not a new circuit technique, since diode arrays have been used in this application since the earliest digital computers. The storage element in a ROM matrix may be a resistor, inductor, capacitor, light-emitting diode, optically transmitting hole in an encoded card, or other predetermined element. The metal key used to actuate a mechanical lock is actually a form of ROM in which the logic levels are encoded into the shape of the key itself. In the MOS ROM, presently available designs use active transistor elements to provide the logic level. In an MOS ROM rectangular matrix, the presence of transistor action in a memory cell is determined by programing a gate onto the storage transistor or adjusting the threshold voltage level. This provides a logic 1 or 0 in the ROM.

MOS technology is ideal for fabricating ROMs. The very dense geometrical layouts possible in MOS permit one to design MOS devices into matrix structures containing many thousands of MOS storage elements. The basic storage element of the MOS matrix type ROM using p-channel enhancement-type devices is shown in Fig. 7.1 for a rectangular array. In Fig. 7.1*a* the geometry including a single programed gate is shown; a larger memory matrix schematic is shown in Fig. 7.1*b*. To code a positive logical 1 into a particular cell, the designer specifies an active transistor for that position.

By selectively masking for a standard gate thickness to match the device location desired, one can program in the desired logical 1. To program a logic 0 into a particular cell, the gate oxide is maintained thick, and therefore no transistor action occurs at that cell. Insertion into, or deletion from, the matrix of an active transistor can be accomplished by changing a single photomask used in the MOS fabrication process.

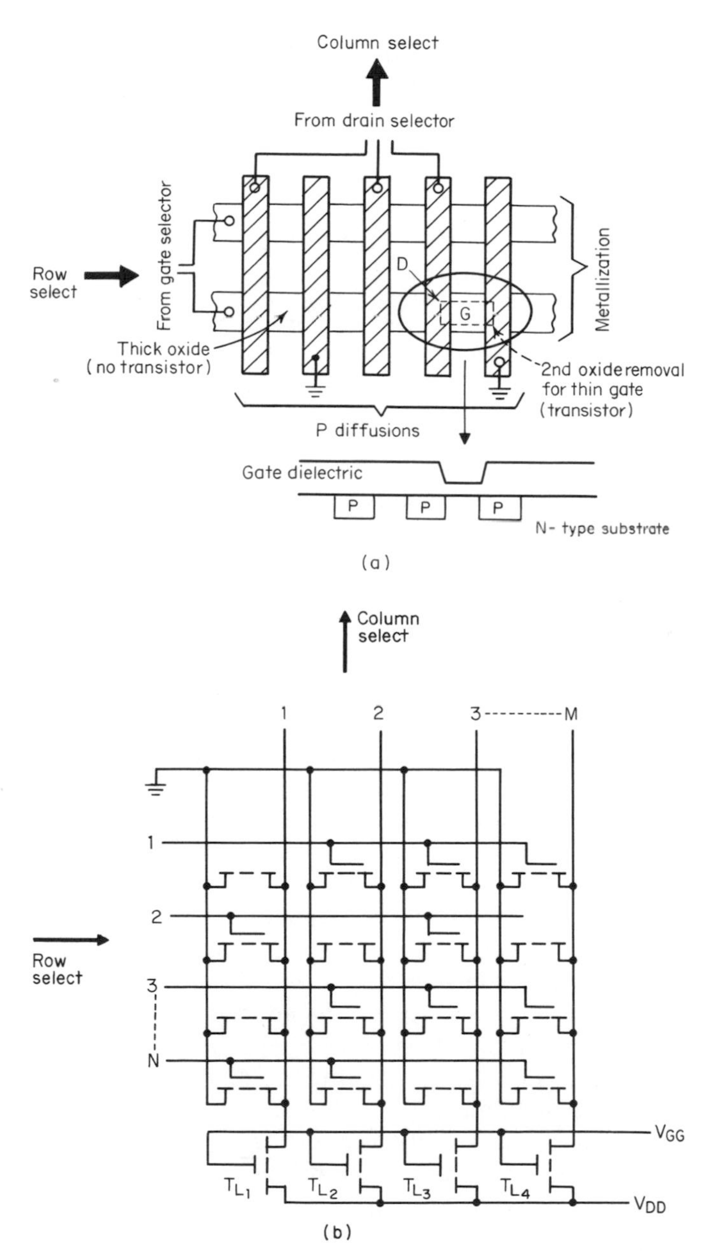

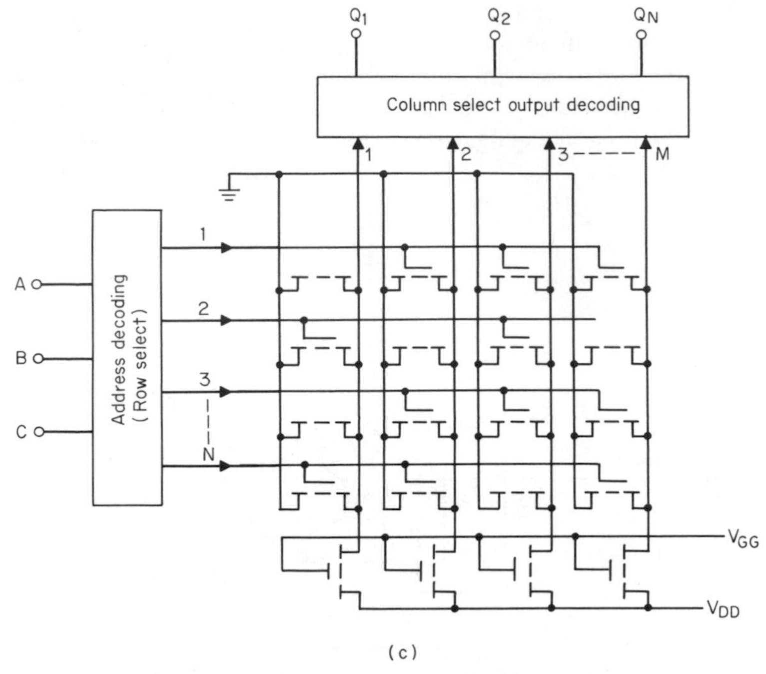

Fig. 7.1. Matrix-type ROM: (*a*) geometry layout; (*b*) circuit schematic; (*c*) input-output decoding shown.

In a few early ROM designs, the transistor cells were programed at the gates by the metalization mask only. These circuits provided very erratic operation because thin gate oxides without covering metalization or open-circuit metalization collected mobile surface charge. By collecting charge, an exposed gate sometimes resulted in transistor action where none was desired.

In Fig. 7.1*b* the circuit action can be understood easily if each column of devices within the storage matrix is considered to be nothing more than a simple multiinput inverter (negative-NOR or positive-NAND gate). For instance, if row 1 is selected with a negative-going voltage, the resulting data output that appears on column lines 1, 2, and 3 to *M* is, for positive logic, a 0, 1, 1, and 1. For negative logic the column outputs would provide the corresponding complemented number 1000.

Since the number of package connections is generally limited, it is desirable to keep the number of circuit inputs to the ROM chip to a minimum. As a result, the binary address only is used to specify the address word from external circuits. Almost all available MOS ROMs have full address decoding on the same substrate as the memory matrix. In Fig. 7.1*c* the address decoding for the row select and the corresponding output decoding with column select are shown diagramatically. The ROM control circuit is shown with more detail in Fig. 7.2. A coincidence-select technique is used to address a particular row or column line. The addressing circuitry is a random logic design.

The coding of the ROM, using gate masking as in Fig. 7.2, is the most common of several coding technologies. The ROM may also be coded by adjusting the effective threshold by control of the gate dielectric. Polarizable dielectrics, such as silicon nitride and alumina, have been used developmentally for electrically programable read only memory circuitry. These types of memories, often termed "read mostly memories" and "programable ROM memories," have exhibited slow

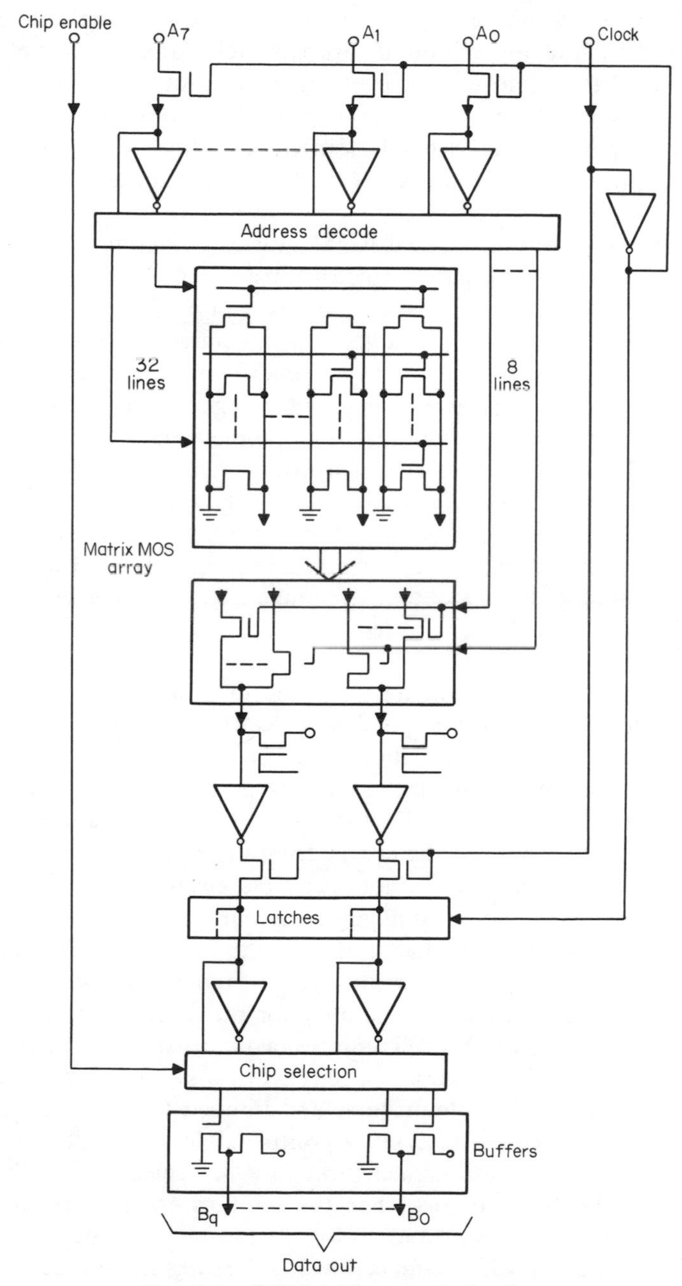

Fig. 7.2. ROM with control circuitry.

write times on the order of milliseconds to microseconds. The read time is comparable to that of standard gate mask-programed circuitry. Read times range from a few nanoseconds to one microsecond. The electrically programable read only memories using dielectrics such as silicon nitride require high-voltage pulses at the present time for gate polarization, and are therefore difficult to design with full address decoding on a monolithic substrate. The programing voltage for these polarizable dielectric memories is often approximately equal to the dielectric breakdown voltage, and it therefore introduces reliability problems.

Another read mostly type of memory utilizes thin amorphous films of bistable glass, deposited upon a diffused bipolar structure. This memory is not generally available and appears to have problems of decoding, reliability, and permanence of logical encoding.

A promising electrically programed ROM utilizes a floating gate on a transistor within each storage cell. This gate is electrically isolated, and charge is transported from the source or drain to the floating gate by a programed avalanche injection. Being isolated, the gate traps the charge, and in conduction, the electrons flow in through the oxide of negative charge to induce a p-type surface in the silicon under the gate. For p-channel devices, this negative charge inverts the silicon under the gate, and creates a permanent channel. The geometry is shown in Fig. 7.3*a*, and a circuit schematic for the device is shown in Fig. 7.3*b*. The presence or absence of gate charge (at logical 1 or 0) is sensed by measuring the source-drain conductance of the resulting transistor action. Since the floating gate upon which the charge is dropped is surrounded by silicon dioxide, a high-quality dielectric, the trapped charge remains for long periods of time. Storage time extrapolated to tens of years at room temperature has been reported for these structures fabricated with good dielectrics.

In operation, a voltage of enough magnitude to avalanche the source or drain junction of the storage element is applied, and electrons are injected into the floating silicon gate, resulting in the accumulation of a negative charge on this gate, to provide a positive 0 storage datum. The amount of charge transferred to the floating gate is a function of the amplitude and the duration of the applied avalanching voltage. When one must write 1s or clear the charge from the gates, special techniques must be used, since the gate electrode is not accessible electrically. By irradiating the gate with ultraviolet light, the resulting photocurrent will neutralize the gate charge. A similar neutralization of gate charge also results from x-ray irradiation.

The avalanche-programed ROM matrix requires an additional gating transistor to provide the desired addressing capability for programing the cell. Therefore, each avalanche-programed ROM cell requires two transistors as shown in the circuit schematic of Fig. 7.3*c*. The column select circuitry is used here for both writing and reading purposes. The column lines are dropped to a high negative voltage polarity when one desires to set a cell into a positive-logic high position in coincidence with a similar gating voltage obtained through a row-select line. The reading of information from the ROM occurs when the columns are selected and gated into level-sensing circuitry such as a random logic array of decoding gates.

For instance, if the lower left-hand cell in Fig. 7.3*c* has its storage transistor fixed in a conducting state, and row select line 3 is brought to a negative-going voltage,

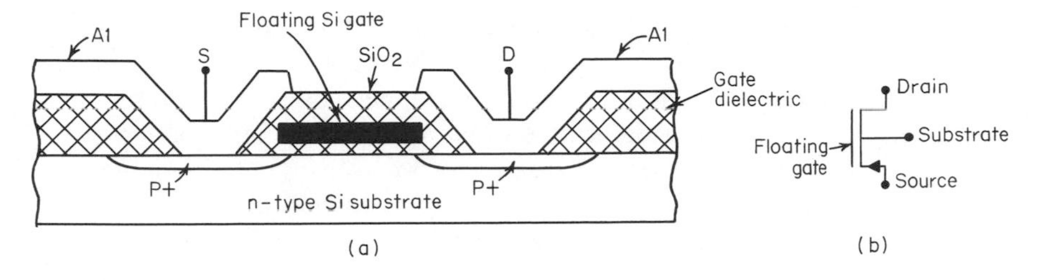

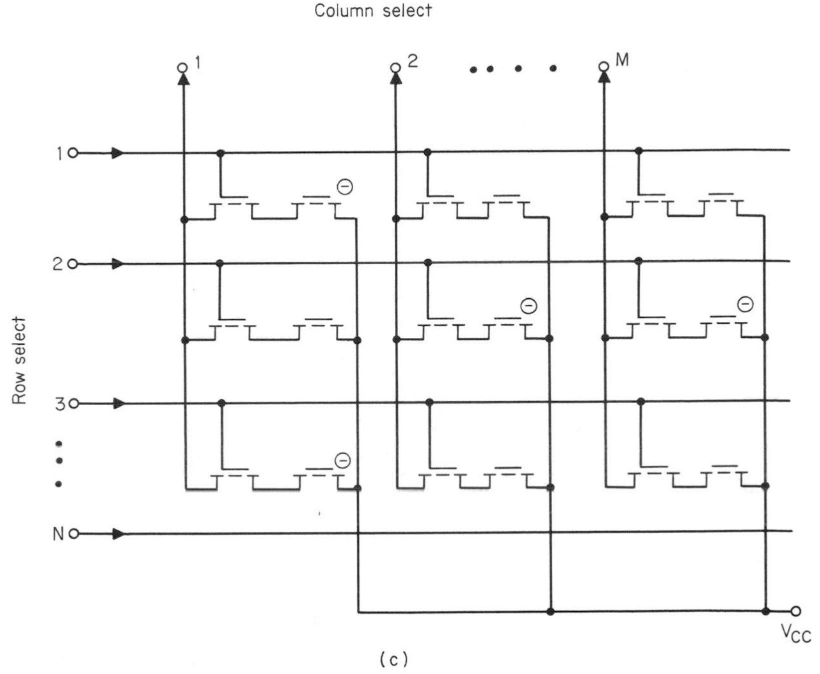

Fig. 7.3. Charge-storage MOS for electrically programing the ROM: (*a*) cell device equivalent; (*b*) geometry cross section; (*c*) matrix circuit schematic.

we find that column 1 is maintained at a V_{CC} voltage level corresponding to a positive-logic high. Thus, in the read mode, information in the selected memory cell is sampled by the output-sensing circuit.

Another type of ROM circuit includes the electrically fusible metalization pattern in which the metal interconnects can be either evaporated or migrated to obtain the desired open-and-closed circuit, respectively. The disadvantages of these fusible-link memories are that they cannot be reprogramed, do not have a high bit density, and may suffer problems of reliability. This technique, while effective for bipolar ROM, has not yet become useful for MOS ROM programing.

Yet another ROM programing technique is mechanical in nature. The mechanically programed ROM is basically a memory with 1s stored in each cell location. The 0s are written in by "scratching away" of the metalization stripe to electrically isolate the source or drain of the MOS device in any selected cell. If one mechanically isolates the gate for a standard MOS transistor ROM, the result is an erratic memory cell in which charge leaks unpredictably between the metalization lines and the electrically isolated metal gate. Therefore, the mechanically programable MOS ROM requires open-circuiting the source-drain and not the gate connection. The scratchable ROMs require larger cell area for two reasons: (1) the spacing must be large enough to permit a micromanipulator probe to position and mechanically remove metalization; and (2) the source or drain diffusion instead of a gate must be isolated for reliability. The mechanically programed ROM has not been used extensively. Existing circuits of this type generally use bipolar transistor or diode arrays instead of MOS devices.

These six techniques outlined have been used with varying degrees of success for MOS ROM programing and are listed in Table 7.1. Technique number 1 accounts for virtually all the present MOS ROM production. Numbers 2, 3, and 4 offer a reprograming feature and show promise for read mostly memory.

7.2.2 Code Conversion and Logic Applications

A rectangular ROM matrix can be used to program random, static logic. This point is most obvious when the circuit schematic for a single column of Fig. 7.1*b* is considered separately.

The column is seen to consist of nothing more than the circuit of Fig. 7.4*a* and is a negative-NOR or a positive-NAND gate with multiple inputs. Each separate input gate corresponds to a bit on separate row address lines. A ROM with N separate inverters is illustrated in Fig. 7.4*b*. Here the implementation, using p-channel enhancement-type devices, provides the Boolean functions

$$Q_1 = \overline{ABC} \tag{7-1}$$

$$Q_2 = \overline{A} \tag{7-2}$$

in positive logic. A simple weighted binary code is converted to excess-3 code through the ROM matrix of Fig. 7.5. When the weighted binary number is used for the row address, one obtains the corresponding excess-3 code converted with the corresponding column output B_1, B_2, etc. In this manner, a single ROM is used to implement two-level, random static logic. Code conversion is an important application for ROM.

The ROM matrix converting all possible combinations of an N-bit input code into an M-bit output code will require $M \times 2^N$ memory bits. The M output bit lines must permit all combinations of the input code, i.e., $N \geqslant M$. An example might be the conversion of the N-bit binary code to excess-3 binary. The output lines

Table 7.1. Techniques for Programing the ROM

1. Gate dielectric thickness control by photolithographic masking
2. Polarization of the MNOS and other dielectric structures using varying polarity gate potentials
3. Amorphous dielectric gate control
4. Avalanche injection of carriers to a floating gate
5. Electrically fusible pattern
6. Mechanical scratching of metalization

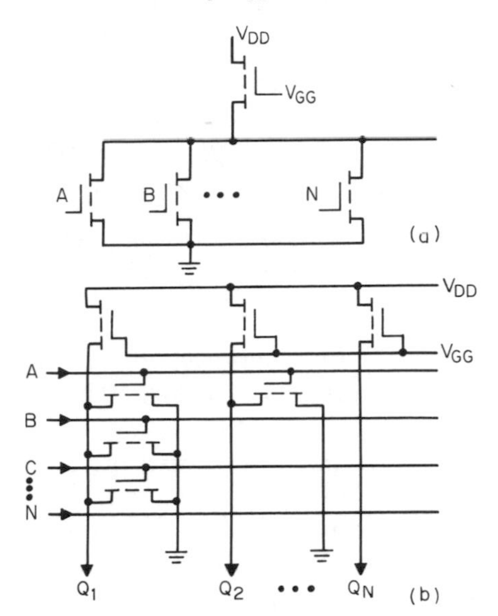

Fig. 7.4. ROM matrix provides single-level logic: (*a*) equivalent column circuit; (*b*) matrix for implementing negative logic. $Q_1 = A + B + C$.

must provide for the higher count and $M = N + 1$. Thus, this code conversion ROM implementation requires $N \times 2^{N+1}$ total bits. A direct layout would require a geometrical bit arrangement of N by $N + 1$ in size. Other ROM designs may require that the number of output lines differ greatly from the number of input lines. In that case, address decoding is used for the columns in addition to the rows to permit the actual geometry of the ROM to be relatively square. A square chip layout is advantageous for production and packaging economies. Figure 7.6 shows how an 8-bit address can be used to select any one of 256 different bits data output. The logical organization here is 256 × 1, and the geometrical organization within the ROM matrix itself is 16 × 16.

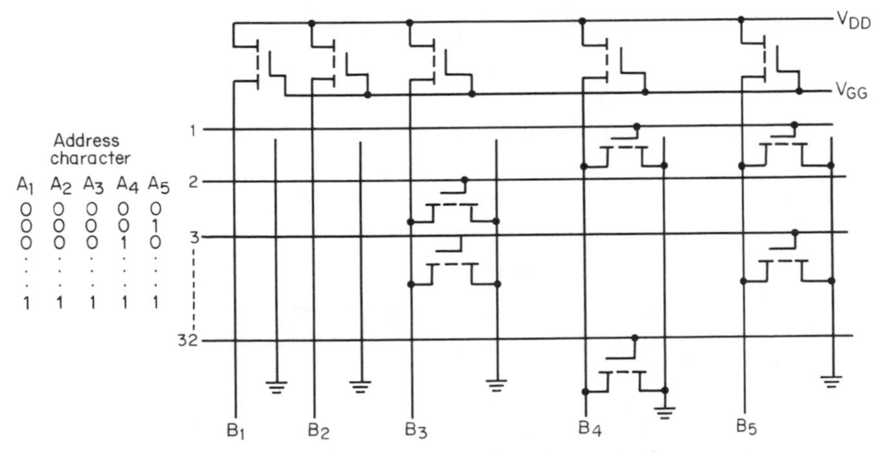

Fig. 7.5. Binary to excess-3 code conversion.

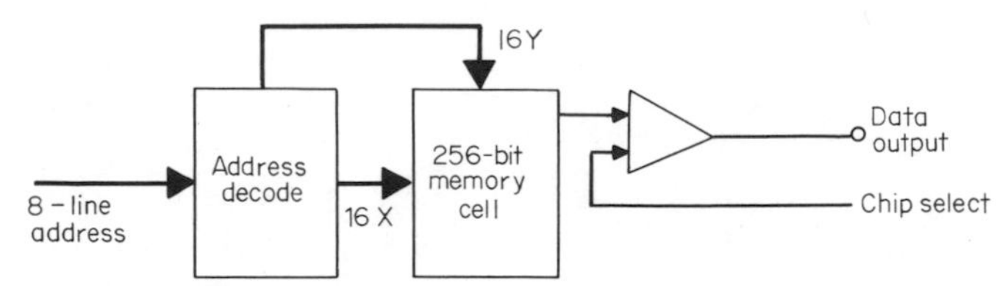

Fig. 7.6. Diagram indicating technique for typical code conversion with a 256-bit ROM.

A brute force approach to code conversion using a maximum number of ROM bits may often be avoided. There are algorithms for code conversions such as weighted binary-to-BCD decades that permit the use of ROMs and adders in a static circuit design. In this case, the ROMs are used to convert large binary numbers to a binary excess-M code which, when added from N-ROMs, provides an excess-MN code. A final level, subtraction, may be used to convert the excess-MN code into straight BCD. This technique permits a very fast code conversion using static logic and avoids the slower, clocked techniques that involve division and multiplication.

An example of efficient ROM usage is in converting Hollerith code to full 8-bit ASCII code. For Hollerith, no more than a single bit among the seven least significant bits is activated for any given character. Thus, these bits really form a one-of-seven code which can be compressed into a 3-bit binary code. With this compression, and the remaining 5 bits specified, the 256 Hollerith combinations here require an input total of 8 bits. Since the desired output format is 8-bit ASCII, the ROM should be arranged logically as 256×8 and contain a total of $2^8 \times 2^3 =$ 2,048 bits.

ROM may be used to provide more than two levels of random logic by stringing memories in cascade. This type of connection, illustrated in Fig. 7.7, permits the input data A_1, A_2, etc., to ripple through the cascade of three ROMs to provide the desired output function D_1, D_2, etc. Considerable delay time is inherent in the ripple-through approach, and one must be careful of the possibility of "glitches" appearing at the output during the transient which might provide undesired circuit operation. Any random, static logic circuit with N inputs and M outputs can be replaced by a ROM with N inputs and M outputs. The ROM contains a great amount of redundancy and is therefore not efficient for implementing relatively simple logic functions. The ROM comes into its own only when complex logic is desired.

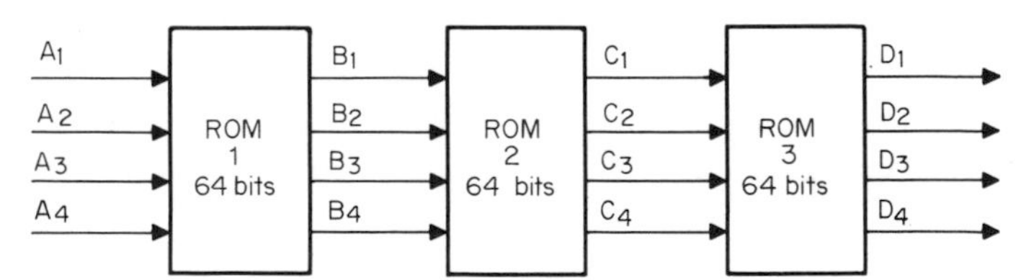

Fig. 7.7. Multiple levels of logic are obtained by stringing ROMs in sequence.

The ROM is also useful for providing the sequential instructions for digital control. For instance, the function of microprograming is an important application for the ROM. A simplified block diagram of a microprogramed digital processor is shown in Fig. 7.8. The processor contains two read only memories. One is the sequencing ROM, which is essentially a sophisticated counting circuit with flip-flops, and the other is the control ROM, which converts the sequence state into control commands to be used by the arithmetic unit. This circuit, providing a series of commands from its permanent store of information, guides the arithmetic unit through logical manipulations. A processor control of this type has many advantages. The arithmetic logic unit (ALU) is simplified since the number of circuit components is reduced. Sequence and control logic implemented with ROMs provides increased speed and reduced costs for processors that generally perform a certain repetitive function. The microprocessor is becoming an important feature within the central processing unit of various small- and large-scale computers. The MOS ROM is sometimes of marginal speed for microprocessor control in computers. In those applications bipolar ROM is used instead. There is an important class of ALU controllers that do use MOS ROM. These are desk calculators and slower control functions.

7.2.3 Character Generation for Displays

A very significant, and perhaps the best known, application of MOS ROM is for character generation. This includes display control for moving billboards, light emitting diode arrays, Hollerith card punching, paper tape coding, digital communications control, and most commonly, CRT display drivers.

Since the storage of significant alphabets and number digits for display involves a considerable number of memory bits together with logic gating, the MOS ROM

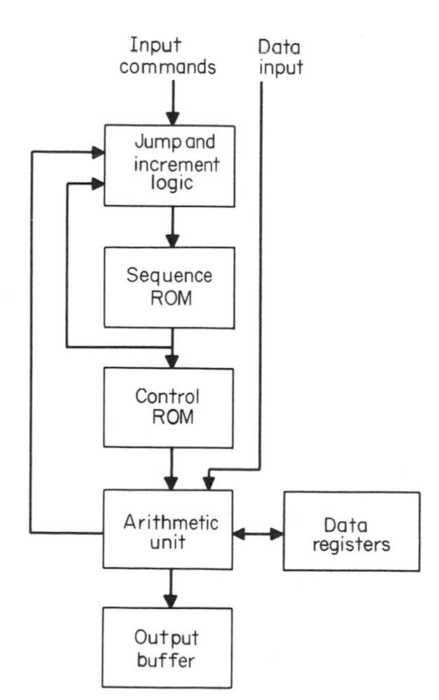

Fig. 7.8. Simplified processor using ROM control.

has become an important component. The digital character generators generally consist of the ROM in which a given input code produces a digital output coded for the character desired. This digital output is then gated for further processing into a CRT, moving billboard display, matrix printers, etc. There are many different output code schemes that might be designed into an ROM character generator. Naturally, the more elements in the display matrix, the more legible the characters can be to the human observer. The characters may be generated by using a dot-display matrix or an *xy*-stroke or vector generation scheme. A 5×7 dot matrix provides the minimum resolution for display of Western languages. One can anticipate, however, that larger matrix sizes, possibly up to a 12×16 matrix, will become practical with larger ROM character generators, permitting higher legibility and sharp definition, even with lower-case letters.

Let's consider a 5×7 dot matrix generator for displaying characters in raster-type display. In Fig. 7.9*a* the input and output functions for the ROM array are shown. The output from the ROM consists of 7 bits simultaneously. These 7 bits could be used to load control bits into a moving billboard display. A typical logical organization for an MOS ROM with a 5×7 dot matrix output and a 6-bit ASC II input code is $64 \times 5 \times 7$. If the character generator output is 7 bits, then the logical organization is 320×7. If the output is 5 bits, then the logical organization is 448×5. In both cases a total of 2,240 bits of ROM storage are required. Another common display is the 5×8 character dot format. This larger format requires 2,560 bits of memory. In general, a ROM with C characters, R rows, and N columns per character, will require CRN total storage bits. Since these logical organizations do not represent square geometrical layouts on the silicon chip, a column select circuit is used. An example of a column select circuit is shown in Fig. 7.9*b* and in this case is used to select one of the five column outputs in the circuit of Fig. 7.9*a*. By using the column select circuit, a 2,240-bit ROM is organized geometrically as 64×35. Similarly, the 2,560-bit ROM is organized geometrically in the matrix as 64×40.

The direct output from the character generator with column select in Fig. 7.9*a* might be used to directly control the video intensity for a horizontally scanning raster-type CRT display. With blanking on the retrace, the characters could be displayed as the beam moves from left to right. Without blanking on the retrace, we might display during both right-moving and left-moving beam scans. But, unfortunately, the present state of the art for MOS ROMs does not permit this direct control. At the moment such fast access times are not available in MOS and are the domain of bipolar ROMs only. Therefore, an economy design using MOS ROM uses a parallel-to-serial converter at the output of the character generator to obtain video modulation. If a parallel-to-serial converter is loaded with the 7 bits at the output in Fig. 7.9*a*, then a vertically scanned raster can be used to display the desired characters. The serial-to-parallel converter permits the character generator to operate with an access time at least seven times slower than that required for the direct video control case. For displaying the number 8, the MOS character generator supplies the selected column bits on five different occasions during the display frame. These bits are gated successively into the digital-to-analog converter, which in turn intensifies the beam to match the corresponding synchronized movement of the electron beam. In Fig. 7.9*c* a more complete character generation subsystem

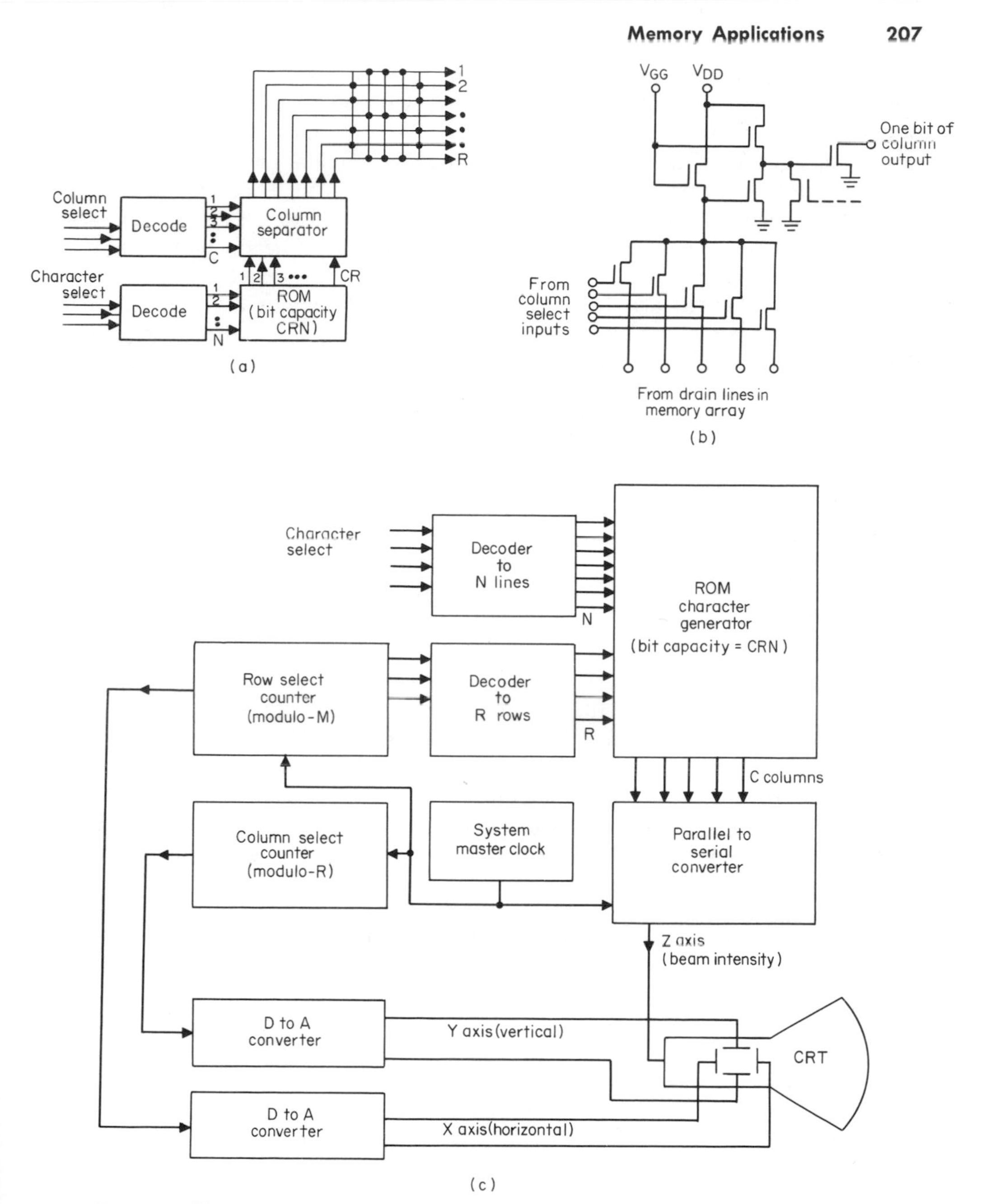

Fig. 7.9. Character generation: (*a*) ROM with output for a 5×7 line-raster scan; (*b*) one of the five column output circuits for (*a*); (*c*) system diagram.

is shown, including the ROM character generator with a bit capacity of $C \times R \times N$ bits. In this system a master system clock is used together with row and column select counters to provide a digital signal for the digital-to-analog converter. The converter in turn provides a saw-toothed voltage to the CRT for generation of the raster display matrix. The digital output from the ROM character generator is gated synchronously into a parallel-to-serial converter which in turn modulates the z axis of the beam intensity on the CRT. This parallel-to-serial conversion technique permits us to obtain what is essentially a multiplexing action, greatly increasing the total bit rate of data obtainable from the ROM generator. The flexibility of this particular display-refresh scheme is, of course, not limited to CRTs, but may also be used for driving the other types of displays such as print-heads, billboards, and even the microprograms and process control sequences mentioned earlier.

The MOS ROM may also be used to generate characters by a vector or stroke technique on an xy plotting display. Instead of using a dot matrix to represent the character, the image is formed from a series of vectors. Generally, the character generator will specify the beginning and end points of the vector. A display, such as the movement of a plotting pen, with proper raising and lowering for inking, results in the desired character. This approach can result in a high resolution for displayed characters, and indeed can permit random graphic generation. Control of the CRT beam by this technique results in vectors of varying intensity according to the vector length or time required for tracing. Additional circuitry may be used to sense the vector length and correct the beam intensity in an analog fashion to obtain a display with vectors of uniform brightness.

7.2.4 The Tree-type ROM

Another type of ROM using standard MOS technology can be obtained by arranging the devices in the form of a tree structure instead of the standard matrix. Figure 7.10 shows such a tree structure for decoding input data from a large number of addresses. For instance, the address $A_1B_1C_1$ with each corresponding line containing a negative-going voltage results in a zero voltage level at the output of this inverter combination. The load resistance R_L is a relatively small resistance on the order of 100 Ohms, providing a very small RC time constant for switching transients at the output node, and therefore producing a high-speed memory structure. The output voltages are relatively low—on the order of a few millivolts—and therefore require a bipolar sense amplifier for the data output.

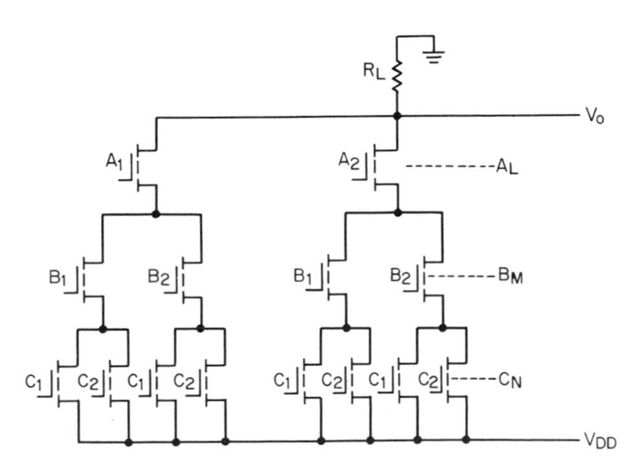

Fig. 7.10. Unconventional ROM circuit with a tree structure.

The tree-type ROM can provide data at the output node with access times on the order of a few tens of nanoseconds. The major problem associated with this type of ROM is that charges stored on internal nodes within the tree structure provide a voltage spiking at the output and a resulting high noise level. Since the voltage spiking can be larger than the actual signal voltage, it can prove to be very troublesome. The tree-type ROM is used only for those applications where a minimum access time is required.

7.3 RANDOM ACCESS MEMORY (RAM)

7.3.1 Static and Dynamic Configurations

The random access memory (RAM) represents a memory function that has acquired an unfortunate name. The RAM memory refers to a memory circuit in which one can both read and write the desired data with facility and speed. A better designator for the so-called RAM memory would be "read/write memory." Unfortunately, the designator "RAM" does not adequately differentiate the read/write memory, the subject of this section, from the previously discussed ROM. In any event, it is not the intent of this text to revise accepted nomenclature, and we will continue to refer to read/write memory and RAM memory as one and the same.

Random access memory devices made with MOS technology use two different techniques to store information. Depending upon the type of basic memory cell, MOS RAMs can be categorized as being either static or dynamic. Static MOS memories usually exhibit reduced speed and increased power dissipation compared with the dynamic circuits. Dynamic MOS RAMs make use of the very low leakage associated with the gate circuits and junctions of properly processed MOS devices. The leakage currents are small enough to permit the circuit's parasitic capacitances to exhibit time constants of several milliseconds. The long time constants may be used to provide temporary storage, which may be made permanent by appropriate cycling or refresh operations. Dynamic RAM circuits enjoy a wide application with reduced cost and high access speed.

MOS memory may be characterized along a spectrum of write-time. At one end of the spectrum are the standard RAMs, with very fast write times. At the other end of the spectrum are the gate-mask-programed, unalterable ROMs. At intermediate levels are the electrical and mechanically alterable ROM circuits. When considered along this spectrum, one can define the RAM as being "that type of random access memory that provides approximately equal read and write time." The read time for any of these memories is completely independent of write time, and in the case of the RAM, the read and write times will generally compare within a factor of two.

7.3.2 Static Circuitry

The static RAM memory cell consists essentially of the bistable storage element containing two cross-coupled inverters with appropriate signal-gating circuitry. The large static cell shown in Fig. 7.11 contains four signal lines in addition to the power-supply and ground lines. The static cell draws a relatively large current

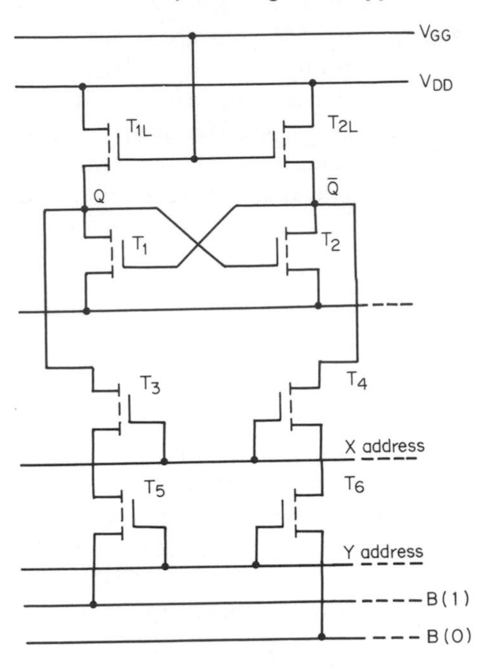

Fig. 7.11. Eight-transistor static RAM cell.

requiring that separate interconnects be used for the substrate V_{SS} lines. This cell consists of p-channel enhancement-type devices. Two of these devices, T_{1L} and T_{2L}, act as load resistors and are biased on by the V_{GG} supply. The two cross-coupled transistors T_1 and T_2 act as the bistable storage element and transistors T_3 through T_6 are gating devices permitting one to address properly each separate cell within a large matrix circuit using a random accessing scheme.

For instance, if we wish to read the contents of a cell at a particular *xy* address, we bring these two addressing lines into coincidence with a negative-going voltage. When the particular cell is interrogated, devices T_3 through T_6 are turned on, permitting us to sense the relative voltages on the two bit lines $B(0)$ and $B(1)$. These two bit lines are used to determine the logic state for each static cell. Each cell is randomly addressed through the *xy* address lines. Similarly, when we wish to write data into this cell, we address the cell in the same fashion and set or reset the desired voltages onto two bit lines $B(0)$ and $B(1)$, which in turn forces the bistable storage element into the desired state. The power dissipation for a typical cell of this type is approximately 1 mW per bit. This dissipation can be reduced to less than 0.5 mW per bit by pulsing the V_{GG} line with an appropriately small duty cycle. Pulsing is not, of course, required for normal operation, but offers a convenient method of conserving power when system dissipation must be kept to a minimum.

An alternate static storage cell containing only six devices is shown in Fig. 7.12*a*. The cell, similar to the previous case, consists of two cross-coupled transistors acting as a storage element with gating transistors T_3 and T_4 for read/write functions. In the storage mode, the word select line is in the zero voltage state, and transistors T_3 and T_4 are not conducting. As a result, the bistable element is isolated from

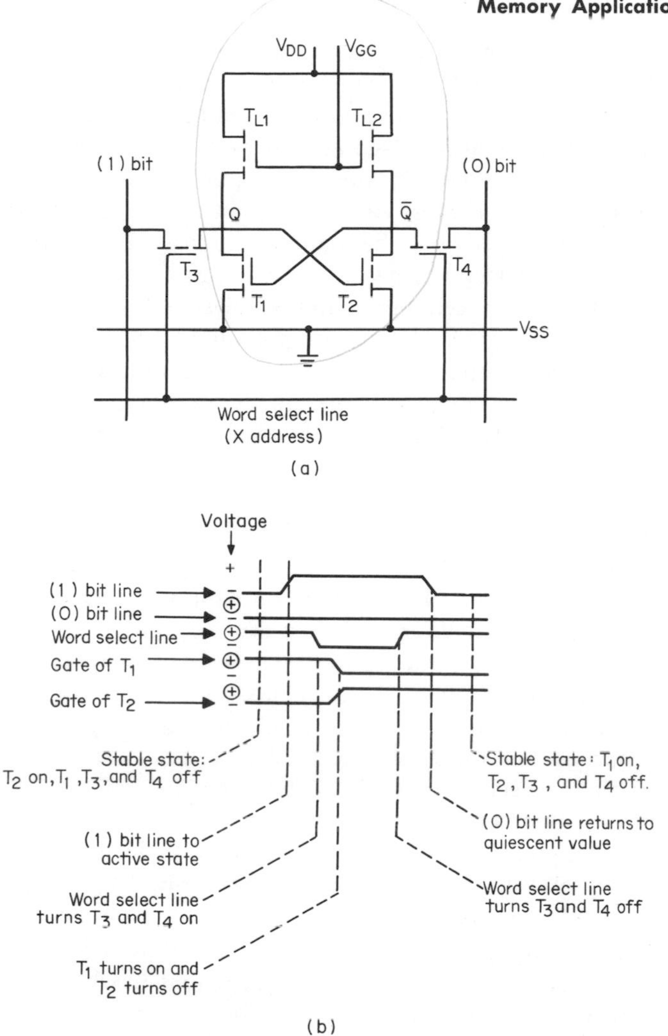

Fig. 7.12. Six-transistor static RAM cell: (*a*) circuit schematic; (*b*) timing for write operation.

the sense-digit lines. One of the possible stable states exists when the gate of T_2 is at a negative-voltage state. This means that T_2 is conducting so that its drain (Q) is at a zero-potential level. The difference in potential between the power supply (V_{DD}) and the drain of T_2 is dissipated across load device T_{L2}. The zero potential at the drain of the device T_4 is coupled to the gate of transistor T_1. This zero potential maintains T_1 in a nonconducting state and, as a result, the drain of this device is in a negative-voltage condition. There is essentially no current flow through load device T_{L1}.

To change the information stored in these basic static cells, the sense-digit lines are appropriately biased and the cell is addressed with the word selection line or with the *xy* address lines. A write operation for the circuit of Fig. 7.12*a* is illustrated in detail in Fig. 7.12*b*. The stable states existing between read/write operations find the bit lines in the negative-voltage state. These lines are normally held in a negative-voltage state since they can be switched faster for a positive-going transient as compared with a negative-going transient. To initiate the writing of a positive-logic 1 into this storage cell, consider that first the desired information must be placed on the bit line. Bit line $B(1)$ is brought into a positive-logic high state, connecting the bit lines directly into the drains of the cross-coupled transistors. Since this data input polarity places a zero voltage on the gate of transistor T_2, a switching action occurs and the storage cell toggles into the desired 1 storage condition. Next, the word select line is brought back to a zero-potential level, and bit line $B(1)$ returns to its normal negative-potential level. The write operation concludes with a return to stable and quiescent conditions. This state will be stored indefinitely in this cell as long as the power-supply voltages remain on, or the internal capacitances of the gate maintain their desired voltage levels.

To read the cell, the word line turns on again and the information contained within the flip-flop storage element is transferred directly to the bit lines. The bit lines, in turn, are connected to additional lines on the chip MOS decoding circuitry, or to external sense amplifiers. The two power-supply voltages can be identical, depending upon a particular design. When the V_{DD} and V_{GG} lines are the same, we can reduce the cell size. With separate power lines, however, the memory cell can be switched into a low-power standby mode during periods when the cell is not being accessed for data transfer. The substrate line (V_{SS}) is optional, and is added only to increase the switching speed for the cells. An alternate design would include an ohmic connection directly to the substrate within each cell, thereby eliminating the V_{SS} intercell bus.

An alternative design for static RAM cells uses CMOS devices. The CMOS RAM has the same very low dc power dissipation that CMOS logic gates provide. In Fig. 7.13*a* and *b* CMOS RAM cells with six and eight devices, respectively, are shown. Devices T_2 through T_5 in each circuit constitute a bistable latch or flip-flop. These four devices store the desired 1 or 0 that has been written from the bit lines (0) and (1). Figure 7.13*a* uses two logic steering devices T_1 and T_6 to permit access to an individual flip-flop from the bit lines (0) and (1). Figure 7.13*a* is a CMOS analog of the p-channel device circuit of Fig. 7.12*a*. Similarly, the CMOS cell of Fig. 7.13*b* is an analog of the p-channel device cell of Fig. 7.11.

In addition, it is possible to specify a class of CMOS dynamic memory cells which will provide both high bit density and greatly reduced power dissipation for clocking at rates below 100 KHz compared with other technologies. For our detailed discussions of dynamic RAM, we will, however, limit discussions to cells using p-channel enhancement devices.

7.3.3 Dynamic Storage Technique

The major disadvantages of the static flip-flop storage elements are that they dissipate higher power than other types, limiting the total number of bits that can be configured within a given package. Static storage occupies a large silicon area. Various techniques have evolved to combat these shortcomings for the double-inverter semiconductor memories. One method, mentioned previously, is to reduce

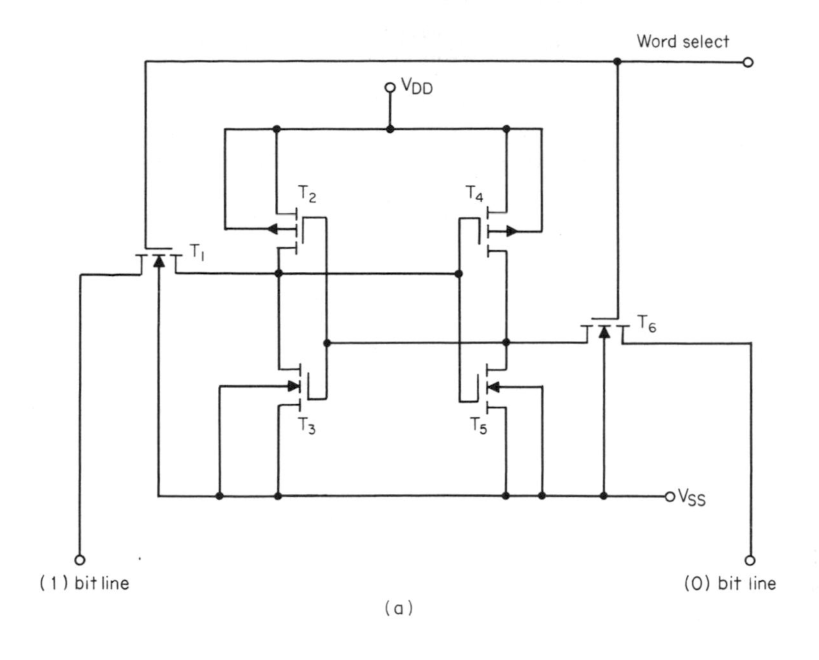

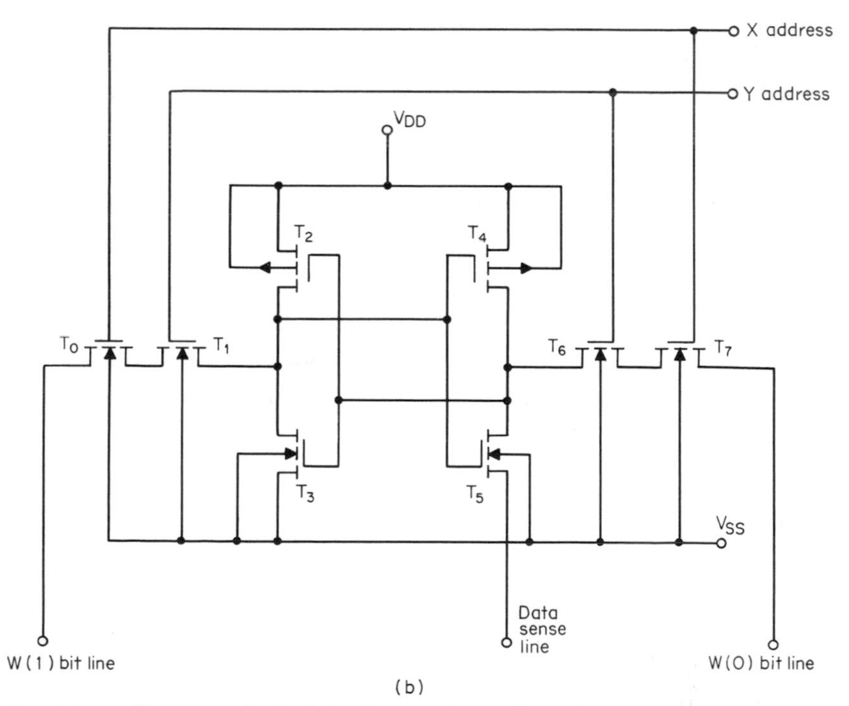

Fig. 7.13. CMOS static RAM cells: (*a*) six-device cell; (*b*) eight-device cell.

the power dissipation by clocking the V_{GG} supply line. During the period when the V_{GG} line is at a zero-voltage state, the charge stored in the gate capacitance of the inverters maintains the logic state for the double-inverter combination. Parallel with this gate capacitance, however, is the junction capacitance associated with the cross-coupled MOS devices. These junctions will provide a parasitic leakage path, through the leakage current of a reverse-biased junction, and therefore will limit the charge storage to a short interval. For instance, a 1-nanoamp current discharging a 1-pF capacitor results in a voltage change of 1 V per millisecond. One characteristic feature of all charge-storage memory cells is that periodic refreshing is required from control circuitry. In the case of the static element, this refresh is accomplished by periodically returning V_{GG} to its negative voltage, which activates the load elements and recharges the capacitors at the gate positions.

7.3.4 Four-transistor Cell

The device count for the memory cell containing the two cross-coupled inverters may be reduced to four devices with three connecting lines plus ground per cell, as shown in the circuit schematic of Fig. 7.14*a*. In contrast to the static cells, the word-enable transistors and the load devices T_3 and T_4 are one and the same here. To read data from the cell, the word line is brought to a negative-voltage condition, and the currents are sensed differentially between the two bit lines which are maintained in a negative-voltage state during the reading period. For instance, if T_2 is conducting, the bit line associated with device T_4 will carry current, and the other bit line will not. To write into this memory cell, we simply force the bit line into the proper states as for the static cell, and then transmit these data into the cell by enabling the word line. To refresh this cell, the bit lines and the word lines are all brought to a negative potential. A write function, followed by a read of

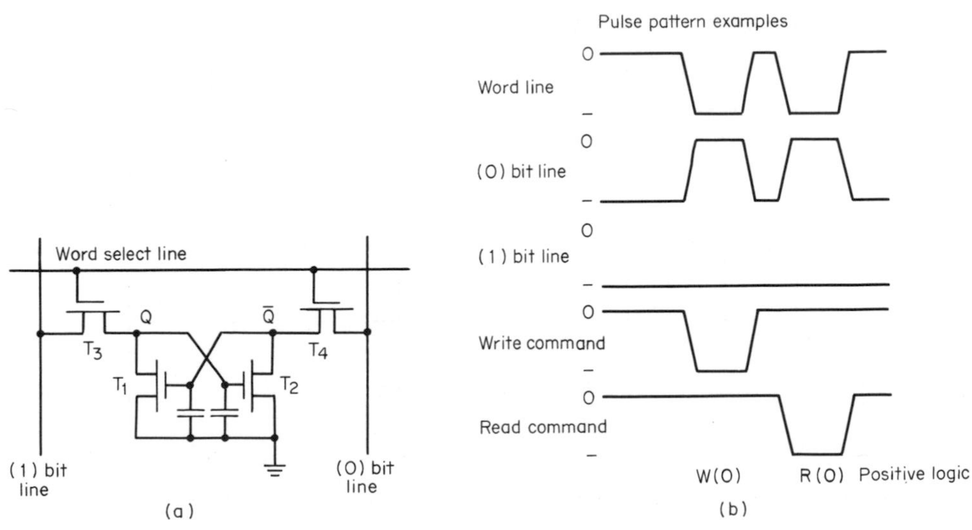

Fig. 7.14. Four-transistor dynamic RAM cell: (*a*) schematic; (*b*) pulse pattern example.

a positive-logic zero, is illustrated in Fig. 7.14*b*, which shows the pulse patterns. The read and write command pulses are shown, although they are processed externally to the storage cell in the control circuit. Note that the bit lines are normally maintained in a negative-voltage condition, and we observe the zero-voltage state only during those write and read timing intervals where the data requires.

7.3.5 RAM Cells with a Single Inverter (Half Flip-flop)

To store charge we actually do not need both coupled inverters. We can store charge with only a single inverter, or as we will see later, a single storage capacitor. In the case of a p-channel MOS transistor, the presence of a sufficient amount of negative charge on its gate will cause that device to conduct. An insufficient amount of voltage on the gate will turn the device off. There exists a series of three-transistor storage cells that contain only a single inverter together with a coupling transistor. Figure 7.15*a* contains such a cell, in this case, with four control lines. A pulse-pattern example for writing a positive-logic 0 followed by the reading of this same stored level is shown in Fig. 7.15*b*. The read and write data bit lines are normally maintained in a negative-voltage state. To write a logical 0 into this cell, we bring the write select line to a negative voltage level and maintain the read line at a zero level. The write data bit is maintained at the desired negative-voltage level and the result is a gating through transistor T_1 of the negative voltage onto node A and its corresponding storage capacitance. During the following read function, the read select line is brought to a negative potential and the write select line is maintained at zero level. The write data bit line is now isolated from node A. The read data bit line is discharged to a zero voltage state through the conducting path of T_2 and T_3. Thus, by interrogating the voltage on the read data bit line, one can determine the logic level stored in this cell. Note that data transfer through this three-transistor cell results in a logic inversion and therefore must be complemented externally.

A refresh action is required periodically for this cell. A circuit such as Fig. 7.15*c* will accomplish a refresh action. The nondestructive read time interval samples the charge on capacitor C_1 and discharges that capacitor if T_2 and T_3 are conducting. The read data bit line is inverted through devices T_5 and T_6 onto the write data bit line. During refresh cycling, the particular cell is addressed with the x and y appropriate coordinates and a read followed by a write command is given with the y-enable column control in a negative-voltage state. In this way, device T_6 provides the desired feedback function for effectively gating the read signal back into the write data bit line for the desired result. Here a read function is automatically followed by a refresh.

The refresh of a single cell is normally a periodic function controlled by a separate clocking, independent of normal read/write operation. For example, separate circuits may be provided to determine which cell addresses have not been refreshed by normal operation and therefore would not require a special refresh action. Such a determination results in complex control circuitry and is not economically effective. The memory refresh duty cycle is generally approximately 1 percent. Thus, even if all refresh cycles were to be saved by normal write/read cycling, the total represents only about 1 percent of the available memory time.

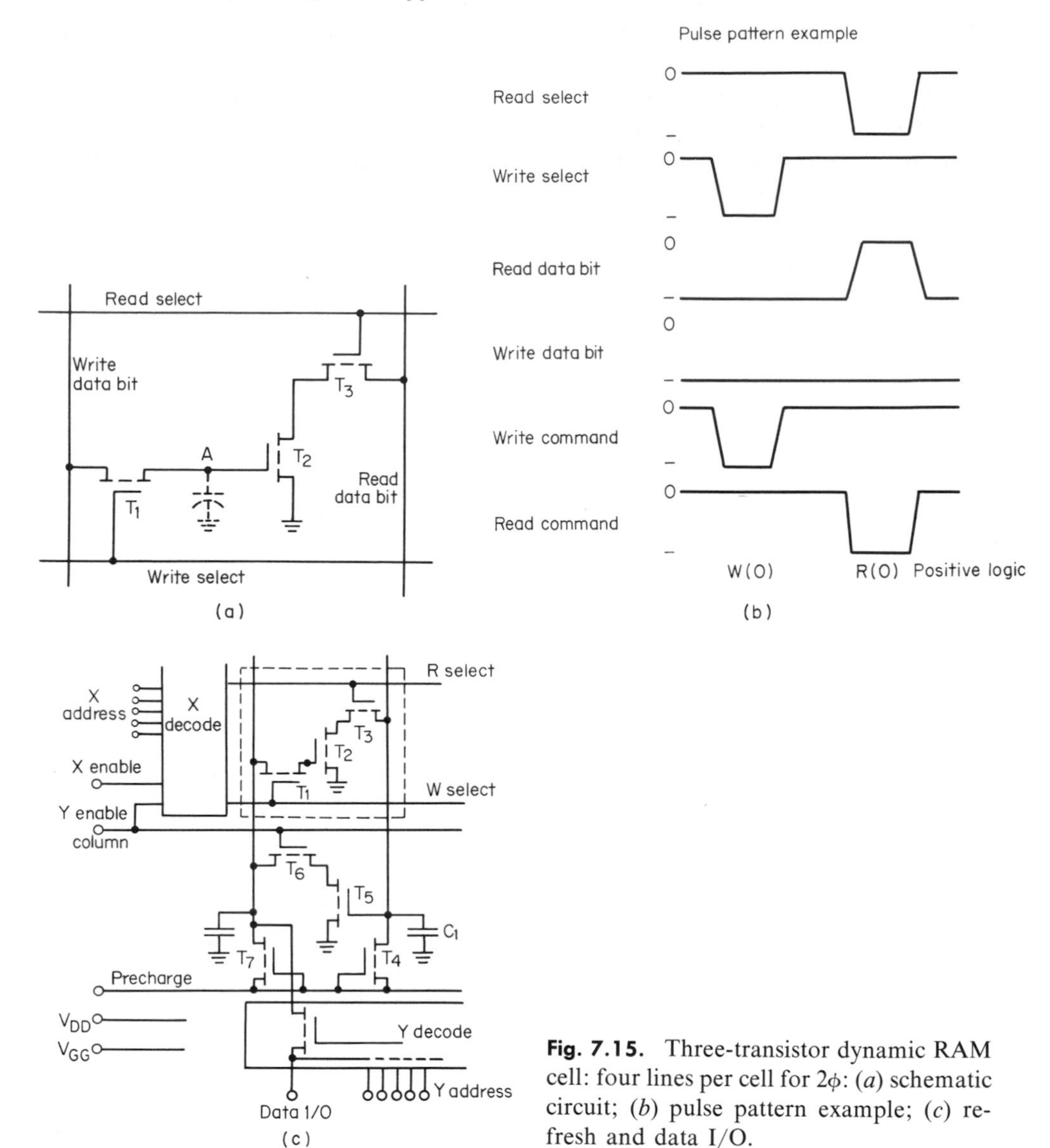

Fig. 7.15. Three-transistor dynamic RAM cell: four lines per cell for 2ϕ: (*a*) schematic circuit; (*b*) pulse pattern example; (*c*) refresh and data I/O.

7.4 THREE CONTROL LINES: SHARED DATA BIT LINES

Another three-transistor storage cell implementation shares the data bit lines and maintains separate read select and write select lines. This cell is shown in Fig. 7.16*a*, and a pulse pattern example is shown in Fig. 7.16*b*. To write a logical 0 into the cell, we maintain the data bit line at a zero level and bring the write select line to a negative voltage state. When reading from the cell, we bring the read select line to a negative voltage state and note that the data bit line is discharged to a zero potential value indicating a stored logic 0. For writing a logical 1 into the cell, we again bring the write select line to a negative voltage state while maintaining the read select line at zero potential. For writing a logical 1, we bring the data bit line to a negative voltage state, corresponding to a negative logic 1, and charge capacitive node A within the cell. This sets a logical 1 within the cell. For the read select function a moment later, the precharged data bit line is discharged through T_2 and T_3 to ground, indicating a stored logical 1.

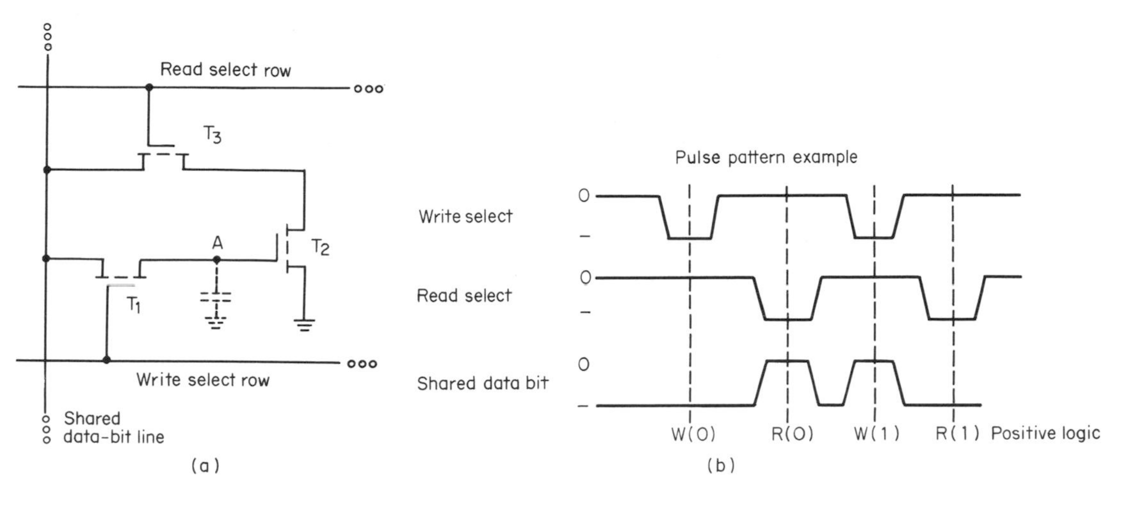

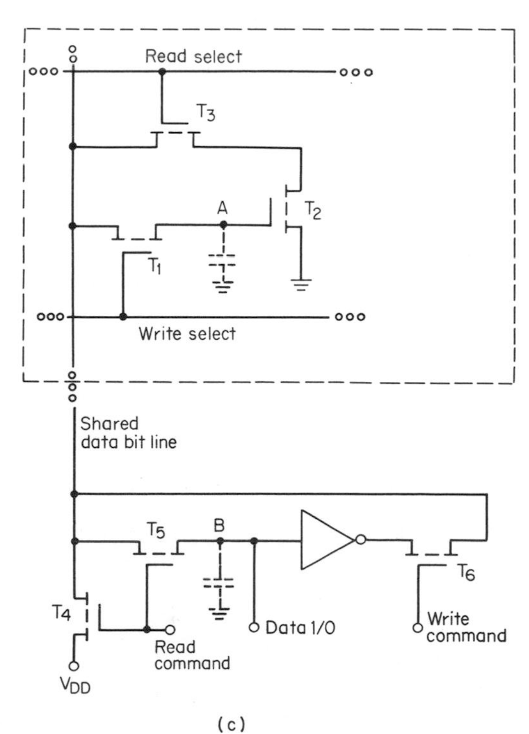

Fig. 7.16. Three-transistor dynamic RAM cell: shared data bit lines: (*a*) schematic circuit; (*b*) pulse pattern example; (*c*) refresh and data I/O.

This cell requires refreshing using data transferred in and out over the data shared bit line, as is shown in the circuit example of Fig. 7.16*c*. During the refresh cycle, the read command is always followed by a write command. The appropriate write select and read select coincident functions appear during the refresh cycle. During that brief time interval between read and write commands, the data for refreshing the cell is stored at capacitive node *B*.

7.5 THREE CONTROL LINES: SHARED R/W SELECT LINES WITH TRISTATE LEVELS

The read/write lines are combined into a single shared line in the cell memory shown in Fig. 7.17*a*. This cell requires a third voltage level in the read/write voltage levels for proper operation. When we wish to write data into node *A*, the R/W select line is brought to a full-negative-potential value. In this condition, both devices T_1 and T_3 are turned on. We neglect the operation of device T_3 during the write interval and depend only upon device T_1 to gate the desired data from the write data bit line onto node *A*, as shown in the pulse-pattern example of Fig. 7.17*b*. Here a precharged voltage is used to precharge these bit lines to a negative potential prior to the read or write function.

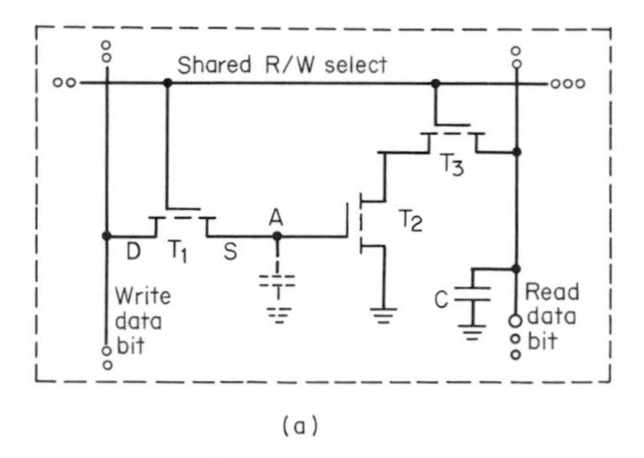

(a)

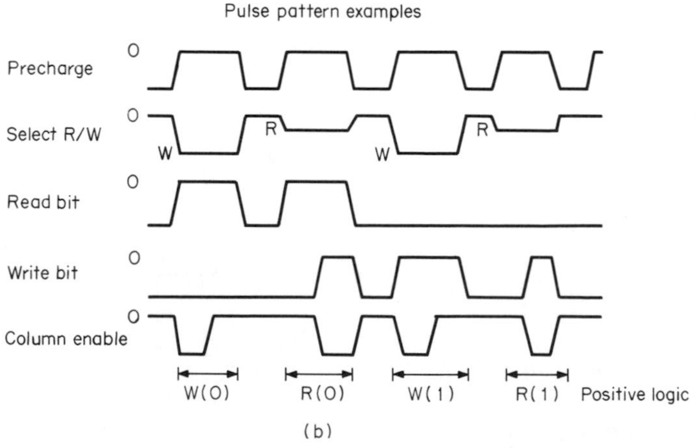

(b)

Fig. 7.17. Three-transistor dynamic RAM cell: shared R/W select lines: (*a*) circuit schematic; (*b*) pulse pattern example.

To read data from the new cell, the external precharge signal, which keeps both data lines at the precharge level, is removed from the read data line. The voltage level on the R/W line is then set at an intermediate level (a third voltage level), slightly above the threshold for device T_3. During this time period the read data bit line is discharged through the conducting path T_3-T_2 if device T_2 is conducting, corresponding to a negative logic 1 as shown in the pulse-pattern example. The important point to note here is that the R/W voltage at this intermediate level is not negative enough to turn device T_1 on. If node A carries a negative charge, it will be more negative than the R/W select voltage, and therefore device T_1 will never turn on. The effect at T_1 is the same as the substrate bias effect previously discussed. If, in the other instance, node A is at a zero potential level (not charged), then device T_1 may turn on only momentarily. In this case as soon as node A charges to within a threshold voltage of the R/W select line voltage, device T_1 will cut off. Thus, the voltage at node A does not permit device T_2 to change state during the read function.

The write function may be performed at maximum speed with the cell of Fig. 7.17. We can anticipate a slow read operation because device T_3 does not achieve full conduction when a limited voltage only is applied from the R/W line.

During a refresh operation, the circuitry will be similar to that of the dynamic cell with four lines. There is a refresh amplifier for each vertical column in this particular array, and data must be logically recomplemented during the refresh sequence. The series of amplifiers refreshes every cell in a selected row simultaneously, and successive rows in rotation. Since the refresh amplifiers also store data read from the cells for external use, they do the refreshing between external read and write cycles with a refresh interval that is fast enough to maintain data levels in the cells.

7.6 FOUR-PHASE MEMORY CELL

Another type of memory cell which uses a single inverter and a single charge storage capacitor is based upon a four-phase clocking scheme. A schematic for such a cell is shown in Fig. 7.18*a*. A four-phase clocking pulse is shown in Fig. 7.18*b*. Separate data bit lines for read and write, together with separate read and write select lines, are used for these cells. The basic cell consists of a half flip-flop for the four-phase dynamic shift register circuit. The charge storage node A is selectively charged or discharged during the write select time, through the write data bit line. Similarly, a read select operation is determined by selectively discharging the read data line and its corresponding capacitive node B through the series of combination T_2, T_3, and T_4 during a phase 4 time interval.

The precharging occurs during the first clock pulse, phase 3. Transistor T_1 is turned on and the data storage node A is charged to the input signal voltage level, which is either a logical 1 or 0. Simultaneously, T_2 is turned on and the node B is precharged to the supply voltage level V_{DD}.

During phase 4, a conditional discharging occurs. If the charge at node A is a negative voltage, then T_3 conducts and node B is selectively discharged. Similarly, if node A had been maintained at a zero potential, device T_3 would not conduct, and node B would retain its negative potential, indicating a positive logical 0. Figure

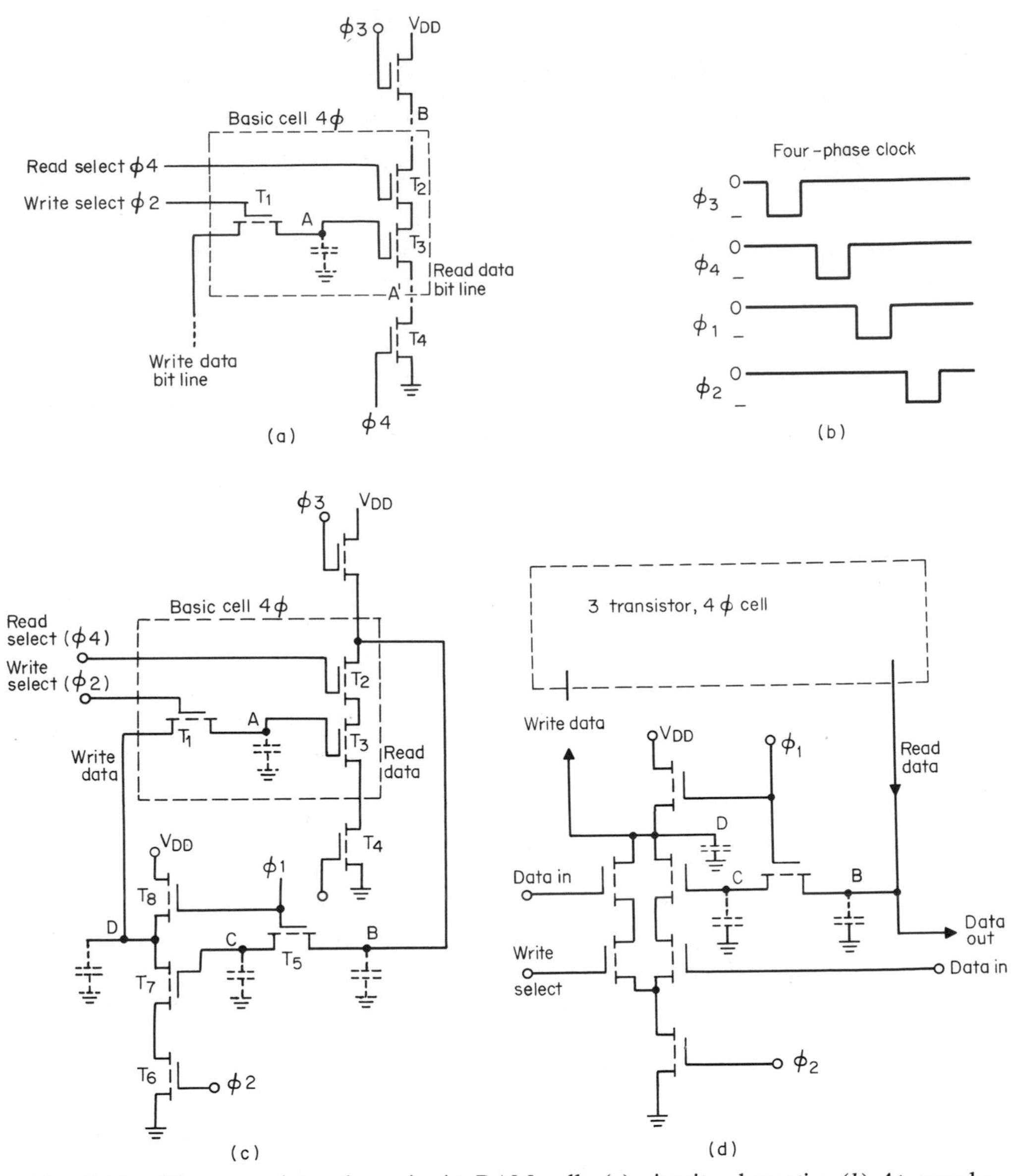

Fig. 7.18. Three-transistor dynamic 4ϕ RAM cell: (*a*) circuit schematic; (*b*) 4ϕ supply pulses; (*c*) refresh circuit; (*d*) data I/O added to (*c*).

7.18*c* indicates the next half-bit of four-phase shift register delay contained within the refresh portion of the circuit. Passing the signal from node *B* through the other half of the cell reinverts it and restores the signal level to the full voltage. During phase 1, node *C* is charged as a result of sharing with node *B*. During phase 2, the inverted signal at node *D* is written through the write data line onto node *A* to complete the refresh cycle.

The refresh circuit shown in Fig. 7.18*c* is used to refresh an entire column of memory cells. During successive clocking periods, the *xy* address controlling the read and write select lines transfers to different rows within a column and provides the desired refresh function. Note that node *A* is completely redundant with the phase 4 pulse and may be shared between adjacent memory cells. Therefore, the four-phase memory cell requires an average of 4½ connecting lines and three devices. The memory cell may be arranged with diffused data bit lines as rows. The power-supply lines are metalized and arranged in columns. A particular implementation of this cell required 9 square mils per memory cell, using standard p-channel MOS processing technology.

The data can be entered and taken out of the memory through the circuit in Fig. 7.18*d*, which also includes the refresh control. The circuit of Fig. 7.18*d* is derived directly from that of Fig. 7.18*c*, in which additional devices have been provided for the data in and data out functions. The actual circuit will correspond more precisely to Fig. 7.18*d*. The circuitry of Fig. 7.18*c* is shown for academic purposes to isolate the refresh function.

7.6.1 One-transistor Dynamic Cell

As we consider various storage cells with our goal of reducing the number of devices, the question arises, "Is it possible to fabricate a useful memory cell containing only a single device?" In Fig. 7.19*a* a single storage capacitance at node *A* may be used effectively as a dynamic storage element where charge is gated to and from this capacitance through the row address device T_1. A critical sharing of charge between nodes *A* and *B* during the read function determines the limiting speed and noise margin of this cell. Proper design of this storage cell requires the maximum capacitance at node *A* per unit area that is feasible for thin-oxide dielectrics at this writing. The total area for shunting diffused regions on device T_1 must be kept to a minimum to reduce the charge leakage from node *A*. With low leakage thin-oxide capacitance at node *A* coupled to a sensitive refresh amplifier, this storage cell can decay beyond normal digital logic levels and still permit logic-level discrimination. The read function occurs when the row and column address are in coincidence for the particular cell. This cell will normally require external sense amplifiers and cannot be used with all on-the-chip decoding. Thus, the read function may be slow. The write function for the one-transistor cell is fast. The parasitic capacitance C_2 at node *B* is generally somewhat greater in value than the capacitance C_1 at node *A*. A working design for this cell indicates that an area of 3.6 square mils is required per cell, using today's silicon-gate technology. The typical array layout containing the single device cell is shown in Fig. 7.19*b*. Here the data lines for read and write are shared, as are the read/write select lines. Each cell of the array shown in Fig. 7.19*b* contains two lines. A major problem in this design is in crosstalk between cells and processing nonuniformity.

7.6.2 RAM Summary

Many types of RAM cells have been described in this section. The number of devices per cell ranges from the maximum of eight for the CMOS cell, down to the single-transistor dynamic cell. A comparison of RAM cells is given in Table 7.2.

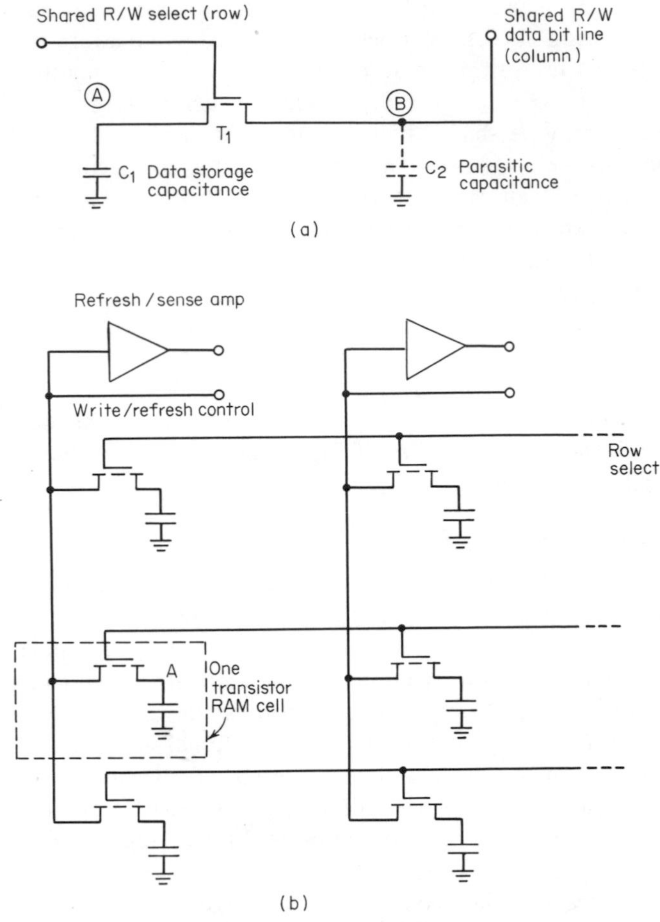

Fig. 7.19. One-transistor dynamic RAM cell: (*a*) circuit schematic; (*b*) matrix format.

Approximate areas for the various RAM cells based upon today's standard processing and the silicon-gate process are tabulated here with approximate values. It is interesting to note that the eight-transistor static cell requires approximately ten times as much area as the single-transistor dynamic cell. The noise margin of the larger static cell will, of course, be much improved over that of a one-transistor dynamic cell. Each of the dynamic cells will require a serial refresh circuit, with the exception of the four-transistor dynamic cell, which is self-refreshing through the word line. None of the static cells require refresh circuitry.

7.7 CONTENT ADDRESSABLE MEMORY (CAM)

The ROM and RAM are coordinate addressable memories, and data cannot be read or written until a particular, unique cell location is specified. In applications such as a search for a match to particular data, all cells of the memory must be

Table 7.2. Comparison of RAM Cells (Representative 1972 Values as Available)

RAM cells	Clocking	Connections per cell	Approx. area for standard process	Approx. area with silicon-gate process	Data refresh
6-transistor (CMOS)	1ϕ	5	.....	.....	No
8-transistor (CMOS)	1ϕ	7	.....	.....	No
8-transistor static	1ϕ	6	35 mil^2	mil^2	No
6-transistor static	1ϕ	4 to 6	.....	13	No
4-transistor dynamic	2ϕ	4	.....	.....	Parallel
3-transistor dynamic:					
4-lines	2ϕ, 3ϕ	4	11	9	Serial
Shared data bit lines	2ϕ, 3ϕ	3	8	6	Serial
Shared read select and write select lines	2ϕ, 3ϕ	3	8	6	Serial
$4\frac{1}{2}$ lines per cell	4ϕ	$4\frac{1}{2}$	9	.....	Serial
1-transistor dynamic	2ϕ, 3ϕ	2	.....	3.6	Serial

interrogated. There exists another class of memories which provides the necessary circuits for matching to a content specification instead of an address. This type of memory is termed, appropriately enough, content-addressable memory or CAM. These memories are also called associative memories. With CAM, the content desired for matching is address instead of location. We specify data desired by category, and not by specific coordinate address.

The CAM becomes effective when we must search a mass of data in a parallel fashion. This type of memory requires a certain amount of logic implemented within each CAM cell. In Fig. 7.20 a CAM cell with ten transistors is shown for illustration. This is a static cell and contains a double-rail bit line, a data summing line, and the one or two power-supply lines. Information can be written into this cell in the usual manner by bringing the word line to a negative potential and forcing the cross-coupled inverters into the desired logic state. Similarly, we can read information out of this cell in the same way as was done for the previously discussed static RAM cell by coordinate address.

Now let's consider CAM memory action in Fig. 7.20. Suppose we are looking for a negative logic 1, and that, instead, a 0 is stored by the flip-flop. In this case a zero voltage appears at node Q. To search for a stored 1, we bring bit line $B(1)$ high, the bit line $B(0)$ low, and the interrogate enable line (I) to a negative voltage state (high). The transistor quad T_5 through T_8 test for a matching function between the contents of the cell and the bit line voltages. In this case devices T_5 and T_7 are both conducting, and this provides a conduction path from the voltage line V_{11} (maintained at an intermediate voltage level) through device T_9 into the word line W. Thus, the current flow through the external resistor R is a maximum value for the case of a mismatch. The transistor quad T_5-T_8 provides the exclusive-OR logic function which is necessary for the logical comparison. If one looks for a best match (pattern recognition) situation, one will choose the row or word being interrogated, for which the current flow I_M is the smallest. The location of desired data can be found much faster using CAM than RAM.

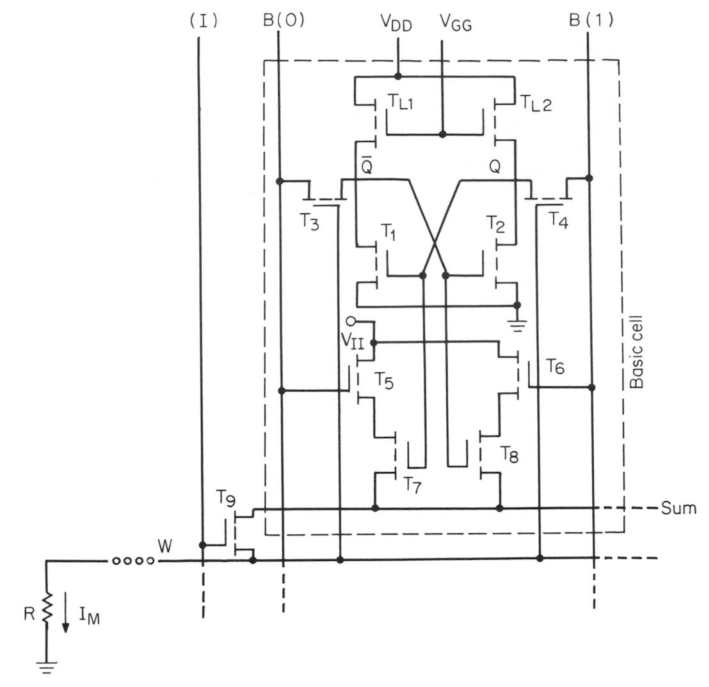

Fig. 7.20. Content-addressable memory (CAM) with ten transistors per cell.

In cases where there is a mismatch, the current I_M will be limited to the microamp range. In cases where the match is perfect for all the cells along a word line, the total current will reach many milliamps. The associative memory cell permits us to look for partial matching also. A partial match condition exists for intermediate current flow I_M and may be determined by measuring the magnitude of this current.

The operation of the associative memory system is shown in simplified form in Fig. 7.21. Here the desired word for which we are seeking a match is given as an input. A search register is masked in an appropriate manner to permit only those bits (for which we are interested in a match) to be transferred into the CAM master matrix. As the control circuit cycles through to determine which matches are best, the address decoding provides the address of these best match words. The output from the system comes through the address encoder, which provides the actual address for those words containing the best match, together with any additional matching signals desired.

The ability to search out or interrogate stored data on the basis of content can be a powerful asset in many applications. Another facet sometimes overlooked is that the CAM may form the basis for a truly general associative data processor. The simple matching process can be iterated to perform complex operations of obtaining best match, next best match, less than, greater than, etc., following a single memory interrogation. The computer does not need an address code to keep track of words stored in the associative memory. Instead, the computer specifies the *content* of the words it needs, and out come the words, wherever they are stored. The CAM provides the feature of processing several operations simultaneously, thus further alleviating the strain of designing higher-speed systems for processing more data in a serial fashion.

The cost-per-bit of associative memories is, of course, much greater than that of conventional memories. The associated address and coding circuitry, together with the considerably larger CAM cell, contribute this increased cost. In the past, this increased cost has made it prohibitively expensive to utilize core memory in associative designs to any extent. However, the advent of MOS now provides an intriguing possibility of economical associative memory processing. At the present time, very little is known concerning the programing that is needed for this type of computer hardware. As simple systems are introduced into the market place and computer scientists learn more concerning the manipulation of these systems, we can expect more useful applications.

7.8 SUMMARY

In this chapter we have been careful in making comparisons between the various circuit techniques and technologies. The circuits are presented with p-channel enhancement-type devices. A CMOS static cell is also shown for comparison. Not shown are the n-channel enhancement cells which would be identical with those p-devices presented except that a considerable increase in speed appears. It is

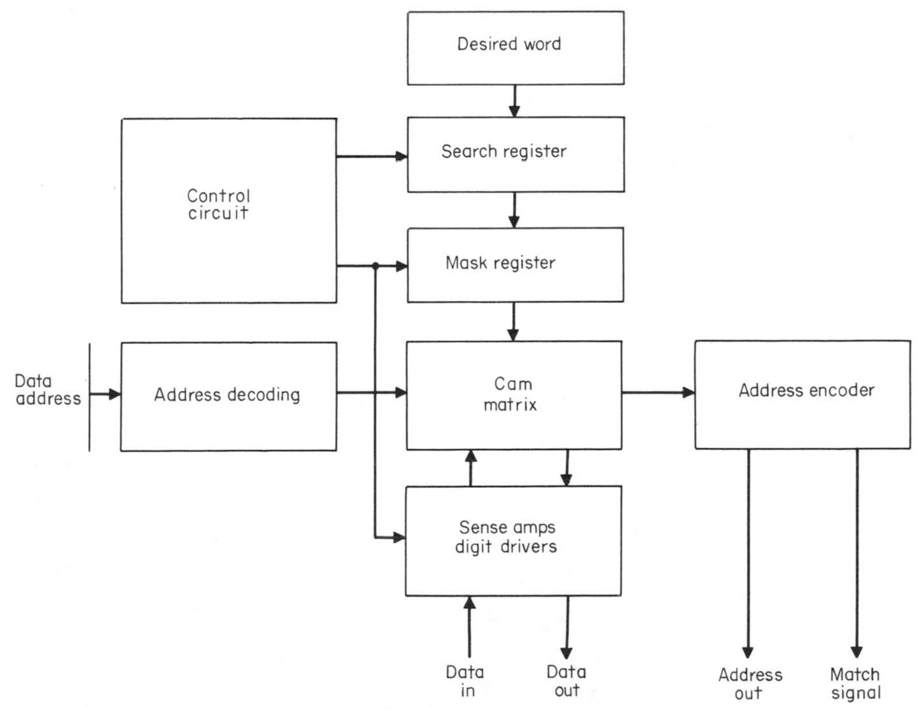

Fig. 7.21. CAM system configuration.

possible that the n-channel enhancement type memory will compete directly with the bipolar memory. For low power dissipation we have mentioned CMOS cells in particular. Also, the memory cells employing depletion-type load devices should be carefully considered for high-speed–low-power performance. There can be exceptions to almost every selection rule in the design of semiconductor memories. Therefore, one must be very careful to limit comparisons to specific memory cells and avoid jumping to broad conclusions based upon static versus dynamic, enhancement versus depletion, and other general categories.

MOS memory is being applied extensively to the design of MOS/LSI circuits in which control logic and memory are located on a single chip. Typical applications of this integrated logic/memory circuitry are found in character generators for CRT display terminals, and in the desk calculator. We can expect a wide variety of future applications of MOS/LSI designs following this same pattern of logic/memory integration.

Mass memory applications for MOS require that the semiconductor memory provide varying degrees of speed, cost, and reliability advantage over that of competing magnetic core, plated wire, and even the magnetic drum. MOS RAM memory has become a prime candidate for replacing magnetic core and certain high-speed drum memories. The MOS circulating memory with CCD or bucket-brigade circuits may prove to be strong competition for magnetic drums.

As we have observed with new semiconductor technologies in the past, the speed with which new technology replaces existing alternatives depends directly upon the ability of the process and design engineers to innovate and produce the cost-effective alternative. In this respect, the reader should recall that the track record for the semiconductor industry is impressive. In any case, it would be rash to be specific as to what future technologies will bring.

There are several interesting developments which may further expand the application horizons for MOS memories. One important example is that of ion implantation, which permits reduced capacitances throughout the circuit and also provides important possibilities of tailoring the threshold voltages for design optimization throughout the chip. Depletion-load inverting circuits utilizing an ion implantation technique appear to be one of many approaches we can anticipate providing a viable technology for the future.

In the area of specific circuit design innovation, many new developments are a certainty. We can add extra functions both within the control circuitry and within the memory cells. For instance, the latent image (the pattern assumed spontaneously by the static flip-flops each time the power is turned on) can be used to advantage. At the present time, this latent image is merely a nuisance and requires that one systematically initialize each memory cell prior to read/write cycling. But it is possible to program the mask, specifying this latent image, and provide bootstrap programs, microprograming instructions, and other desired listings for use in the central processor control in other functions. The latent image can be understood more precisely by referring to the memory cell of Fig. 7.22, which is a version of the six-device static cell. Here the time constant associated with capacitor C_1 and C_2 and load devices T_{L1} and T_{L2} determines the latent image. When the power is brought up and the word line is at a zero voltage potential, this cell will assume

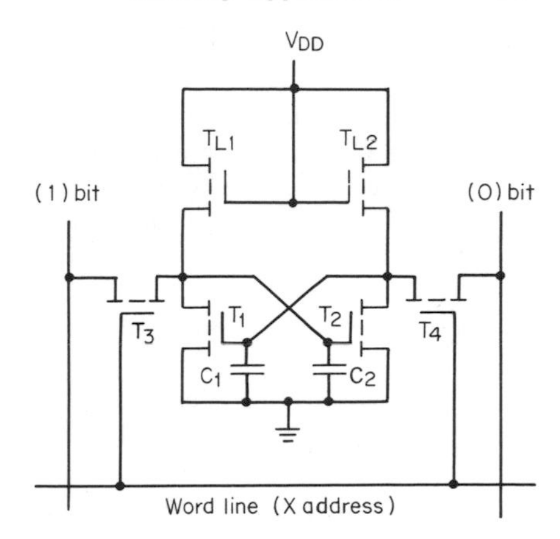

Fig. 7.22. Latent-image static cell contains asymmetry between L_1, C_1, and L_2, C_2.

a latent image. We could store an often-used program in the form of a latent image within a latent cell and interrogate this cell periodically following a special clocking cycle in which the V_{DD} supply level is reduced momentarily to substrate potential. The latent image permits some of the advantages of ROM in a RAM system.

Other applications, such as combining digital MOS memory with linear MOS circuitry on the same chip, have not been exploited at all at this writing, except for a few digital/analog conversion circuits. The content-addressable memory cell has not yet been developed in dynamic form. And still another evolving application based on MOS memory is that of the programable logic array, discussed in the next chapter.

REFERENCES

1. T. Muoio, Character/Symbol Generation by MOS ROM, *Proc. of the Soc. Info. Display,* **2** (1): 6–15, March 1970.
2. G. Carter and D. Mrazek, Better Way to Design a Character Generator, *Electronics,* **43:** 107–112, April 27, 1970.
3. G. B. Hoffman, "MOS Character Generators," Texas Instruments Application Report CA-145, January 1970.
4. C. C. Foster, "Computer Architecture," Van Nostrand Reinhold, New York, 1970.
5. J. S. Schmidt, Integrated MOS Random-access Memory, *Solid-state Design,* **6:** 21–25, January 1965.
6. P. Pleshko and L. M. Terman, An Investigation of the Potential of MOS Transistor Memories, *IEEE Trans. Elec. Computers,* **EC-15:** 423–427, August 1966.
7. H. G. Dill and T. N. Toombs, A New MNOS Charge-storage Effect, *Solid State Electronics,* **12:** 981–87 (1969).
8. S. Nakanuma, T. Tsujide, R. Igarachi, K. Onoda, T. Wada, and M. Makagiri, A Read-only Memory Using MAS Transistors, *ISSCC Digest of Tech. Papers,* **13:** 68–69, February 1970.
9. W. M. Regitz and J. Karp, A Three-transistor Cell, 1,024-bit, 500-ns MOS RAM, *1970 ISSCC Digest of Technical Papers,* **13:** 42–43, February 1970.
10. L. Cohen, R. Green, K. Smith, J. L. Seely, RAM Memory with Single-transistor Cell, *Electronics,* **44** (16): 69–76, August 2, 1971.
11. I. T. Ho and G. A. Maley, Latent Image Can Provide Chips with Built-in Control Memories (Bipolar), *Electronics,* **44** (16): 82–85, August 16, 1971.

12. F. Kvamme, Standard ROM for Complex Logic Design, *Electronics,* **43** (1): 88–95, January 5, 1970.
13. E. J. Boleky, J. R. Burns, J. E. Meyer, and J. H. Scott, MOS Memory Travels in Fast Bipolar Crowd, *Electronics,* **43:** 82–85, July 20, 1970.
14. W. A. Regitz and J. A. Karp, A Three Transistor Cell, 1024 bit, 500 ns MOS RAM, *IEEE, J. Solid-state Circuits,* **SC-5** (5) October 1970.
15. D. Frohman-Bentchkowcky, The MNOS Transistor—Characteristics and Applications, *Proc. of the IEEE,* **58:** 1207–1219, August 1970.
16. W. Frazee, MOS Storage Circuits Simplify Memory Design, *Computer Design,* 128–132, May 1971.
17. T. L. Chu, J. R. Szeden, and C. H. Lee, The Preparation and C-V Characteristics of Si-Si, N_4 and Si-SiO_2—Si_3N_4 Structures, *Solid-State Electronics,* **10:** 897–905 (1967).
18. F. W. Flad, C. J. Varker, and H. C. Lin, The Application of MNOS Transistors in a Preset Counter with Nonvolatile Memory, *1969 ISSCC Digest of Tech. Papers,* **12:** 46–47, February 1969.
19. J. R. Burns, J. J. Gibson, K. C. Ku, and R. A. Powlus, Int. Memory Using Complementary Field-effect Transistors, *1966 ISSCC Digest of Technical Papers,* **9:** 118–119, February 1966.
20. A. Apicella and J. Franks, A High-speed NDRO One-core-per-bit Associative Memory Element, *Proc. of the International Conference on Magnetics,* 14–15, April 1965.
21. K. E. Batcher, Sorting Networks and Their Applications, *AFIPS Conf. Proceedings, SJCC,* **32:** 307 (1968).
22. D. N. Leonard, MOS Content-addressable Memories, J. Eimbinder, ed., *Semiconductor Memories,* 69–74, Wiley-Interscience, 1971.
23. M. V. Hoover, Characteristics of Complementary MOS ICs, J. Eimbinder, ed., *Semiconductor Memories,* Wiley-Interscience, 1971.

8

Programable Logic Arrays

8.1 ROM LOGIC

In previous chapters we have implemented designs using the ROM as a source of systematically arranged data that can be accessed quickly. The applications discussed in Chap. 7 for the ROM each involve data retrieval. But another important feature of the ROM permits the implementation of random logic.[1]

An inspection of the devices connected to an output node in the ROM array reveals that these devices are actually part of a logic gate such as in Fig. 8.1*a*, which is rearranged into ROM format in Fig. 8.1*b*. Here, each gate provides the positive-NAND or negative-NOR function. Each of these separate gates constitutes a column within a complete ROM. One can easily locate the separate gates within the ROM array of Fig. 8.2*a*. Here, for instance, the Boolean functions P_1 and P_2 are as follows in positive logic:

$$P_1 = \overline{AB} \tag{8-1}$$

$$P_2 = \overline{\bar{A}\,\bar{B}} \tag{8-2}$$

The standard logic equivalent for Fig. 8.2*a* is shown in *b*. We will abbreviate our logic diagram further to the equivalent form termed *array logic* in Fig. 8.2*c*, for design work in this chapter. When the output of the ROM in Fig. 8.2 is used as the input for a second ROM, *B*, we immediately obtain a second level of functional logic. For instance, if the outputs of a second ROM, *B*, are specified as Q_1 and Q_2, and the P_1 and P_2 functions from ROM *A* are inputs to the second ROM, we have the circuit function and schematic of Fig. 8.3*a* and *b*, respectively. The P_1 and P_2 functions, from separate columns of ROM *A*, are used for the input to the second ROM. The resulting positive-logic Boolean functions are Q_1 and Q_2:

$$Q_1 = \overline{P_1 P_2} = AB + \bar{A}\bar{B} \tag{8-3}$$

$$Q_2 = P_1 = \overline{\bar{A}\bar{B}} \tag{8-4}$$

In this manner we can implement 1- and 2-level NAND logic. Note that the logic of Q_1 is equivalent to the AND-OR combination. The *A* and *B* input functions and their complements are ANDed in ROM *A*. ROM *A* is performing an ANDing

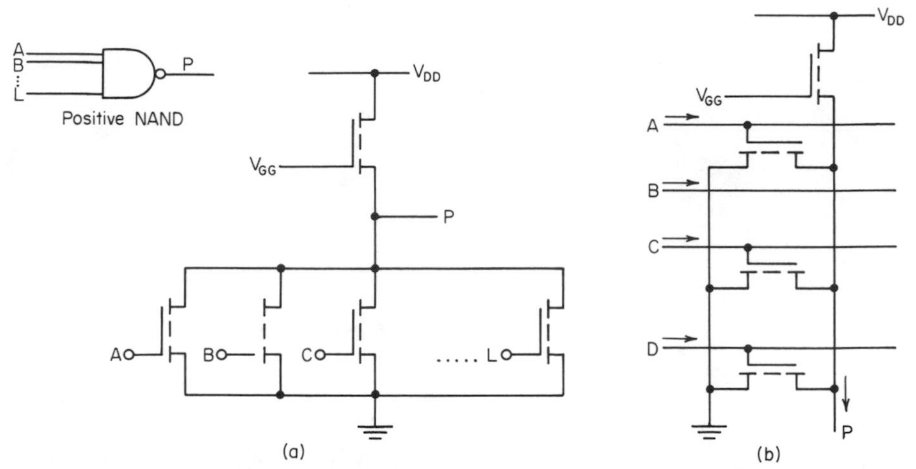

Fig. 8.1. Read-only-memory (ROM) logic: (*a*) basic gate element implements the function $P = \overline{ACD}$ with positive logic; (*b*) same basic gate for ROM matrix.

operation upon these two input functions. Similarly, the ROM *B* matrix is performing the OR function upon the two product terms P_1 and P_2 to obtain the output function Q_1. We can describe ROM *A* and *B* as the AND and OR matrix, respectively. Stated in another way, we can call ROM *A* a product generator and ROM *B* the summer. The ROM matrices coupled together as in Fig. 8.3*a* provide the possibility for implementing two-level, random and static logic. The coupled matrix array consisting of the two separate ROMs is presented differently in Fig. 8.3*c*, with the usual standard-logic schematic and matched to the array-logic schematic of Fig. 8.3*d*.

In Fig. 8.3 the product matrix contains eight product nodes and the summing matrix requires four nodes. The node structure for the coupled ROM arrays is shown clearly in the array-logic schematic of Fig. 8.3*d*. The dark circles within each ROM serve to indicate the presence of a gate for the transistor source-drain diffusions on the ROM chip. It is convenient to use an array-logic schematic similar to that of Fig. 8.3*d* for indicating random logic circuits, using the coupled ROM matrices.

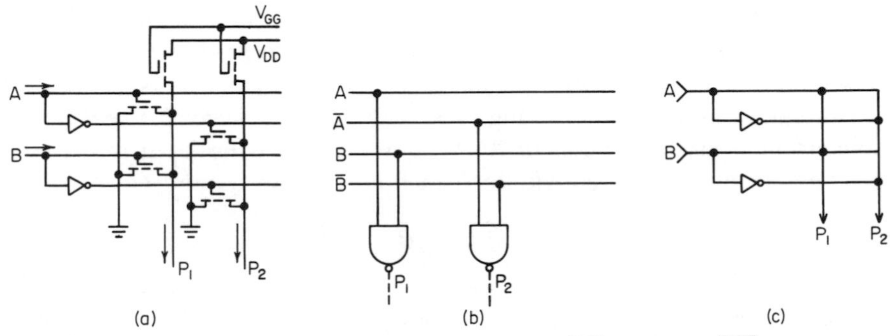

Fig. 8.2. ROM logic matrix for implementing $P_1 = \overline{AB}$ and $P_2 = \overline{\overline{A}\,\overline{B}}$. (*a*) Circuit schematic; (*b*) standard positive-logic equivalent; (*c*) array-logic equivalent.

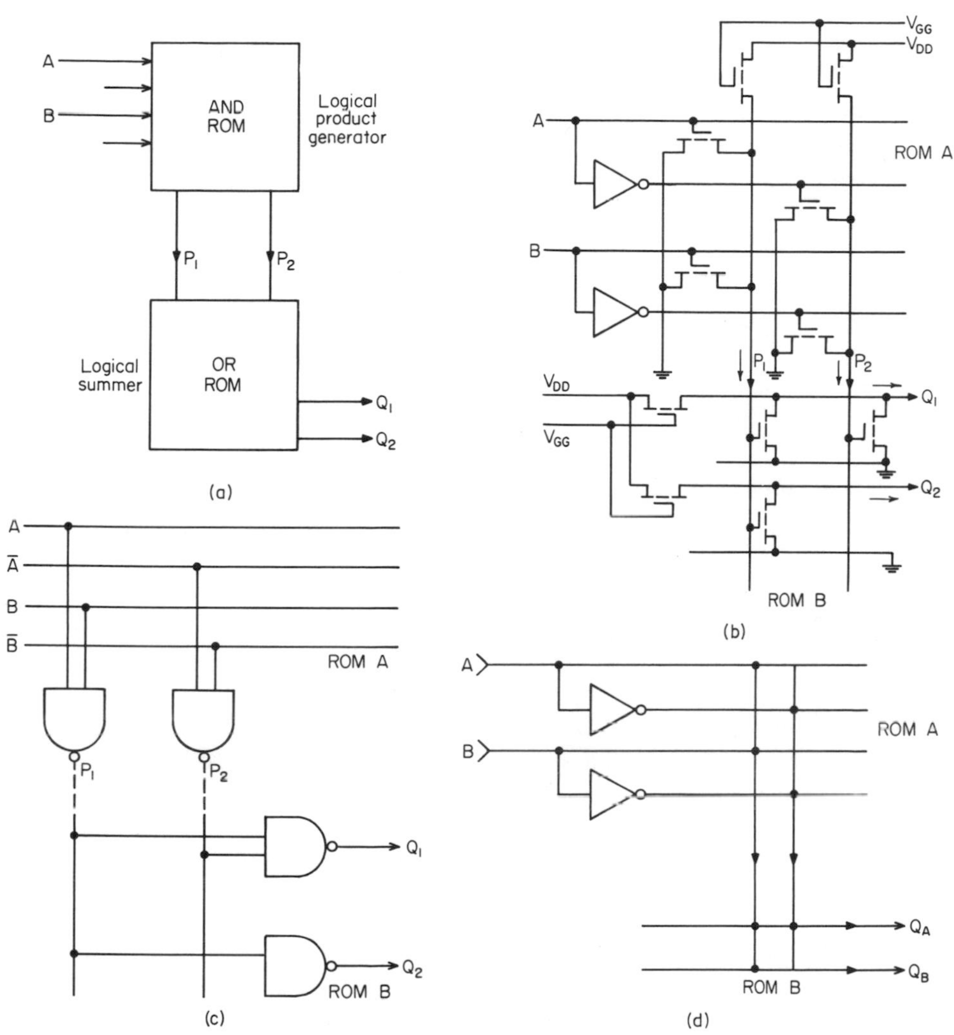

Fig. 8.3. Coupled arrays for implementing static logic: $Q_1 = AB + \bar{A}\bar{B}$ and $Q_2 = AB$. (*a*) Block diagram; (*b*) circuit schematic; (*c*) standard-logic schematic; (*d*) array-logic schematic.

A single simple logic function such as the exclusive-OR can normally be more conveniently implemented with random-type logic networks. The ordered arrangement of the ROM for implementing random logic becomes practical in more complex logic designs where automation can be used in design. Furthermore, when one desires to extend the application for these static gates into sequential designs, some form of temporary memory must be added. One approach is to add a flip-flop array into a feedback loop and use clocking to gate the data from ROM *B* back into ROM *A*. Figure 8.4 shows the flip-flop feedback added to static AND-OR ROM matrices. The resulting configuration of Fig. 8.4 we will now define to be the programable logic array (PLA).[2]

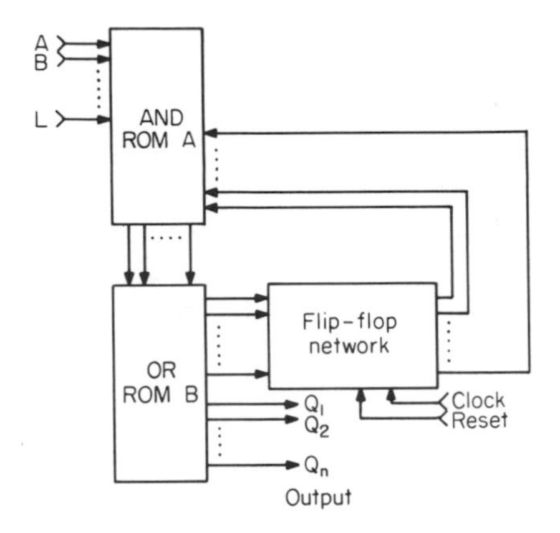

Fig. 8.4. Programable logic array (PLA) contains multiple ROMs and flip-flop feedback elements for sequential logic functions.

8.2 PROGRAMABLE LOGIC ARRAY

The simplest form of the PLA contains the dual ROM matrices with clocked flip-flops in the feedback path. The ROM array provides a matrix for design and production automation. A major advantage of the PLA results from the fact that a large variety of sequential combinational logic functions are economically obtained by designing for a single modification of the gate mask during circuit fabrication. Since ROM matrices may contain from 1,000 to 20,000 summing nodes, one can readily note that great logical complexity is possible using the PLA circuit system. In the PLA system the feedback loops must be clocked carefully. For instance, a race condition can develop if we feed the data back directly from the summing matrix into the product matrix. We therefore will generally require that clocked flip-flops be connected into all feedback paths from the summing matrix (ROM *B*) to the product matrix (ROM *A*) as in Fig. 8.4. We can control the reset of the flip-flops to initialize the logic. The designer is free to choose the number of matrix points and flip-flops in his particular design consistent with whatever circuit design automation techniques are used. To understand more fully the design procedures for PLA implementation of sequential logic, we will consider a series of design examples.

An example of two-level logic which includes an exclusive-OR data input is shown in Fig. 8.5*a*. Here the presence of true A or B levels as data inputs results in a true output Q^n level which is clamped into the logical high state while a clocking signal C remains high. The circuit function of Fig. 8.5*a* is one of pulse clamping with the exclusive-OR data trigger. The pulse length or duration of the high-level clamping function extends until the clock signal C returns to a logical 0. We can implement this same function with the PLA circuit system using the PLA-logic schematic of Fig. 8.5*b*. Here data inputs A and B with their complements are available together with the clock signal C in the product ROM. These signals are summed in the OR matrix and fed through a delay D-type flip-flop back into the product matrix to provide a logic match to Fig. 8.5*a*. Again, all product lines are summed to provide the desired output function Q^n. This same function is concisely shown in the array-logic diagram of Fig. 8.5*c*.

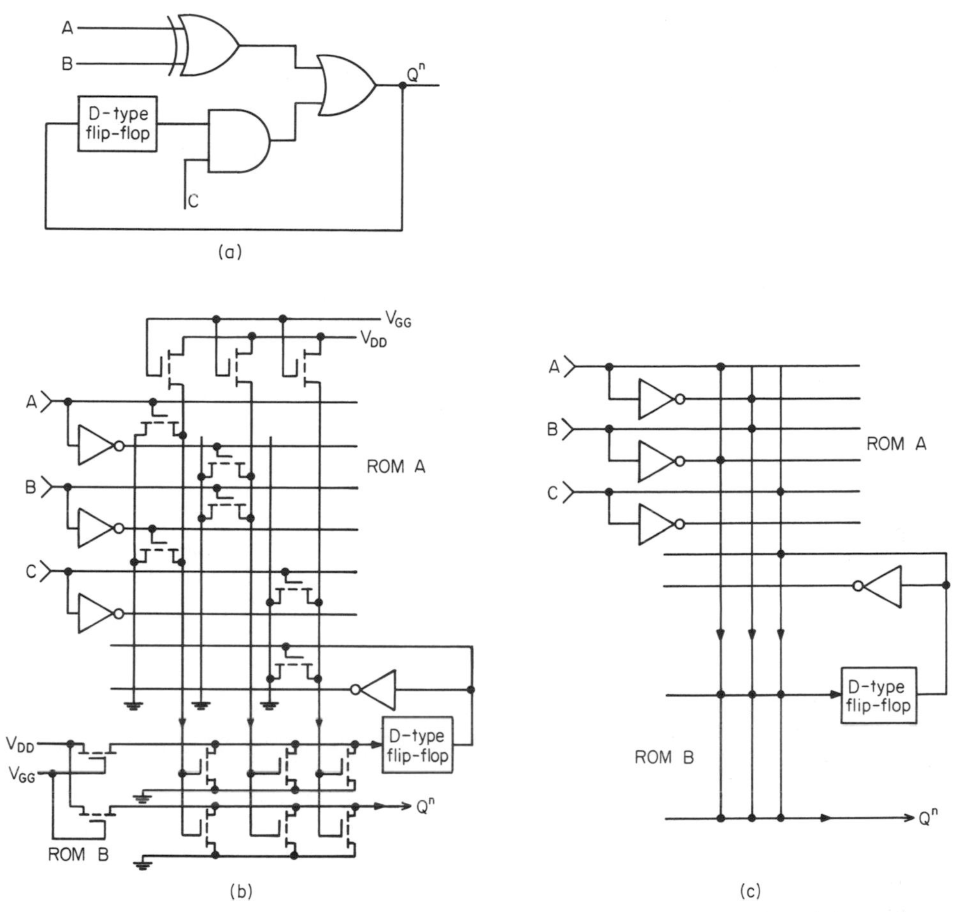

Fig. 8.5. PLA for implementing a pulse-clamping function with a clock and data trigger. An exclusive-OR ($A \oplus B$) data input is used as a trigger. (*a*) Standard-logic schematic; (*b*) circuit schematic; (*c*) PLA-logic schematic.

A second design example representing a further extension of Fig. 8.5 is shown in Fig. 8.6. Here the data input trigger is a more complex logic function $A \oplus B + D$. Again, the static logic portion of the circuit is included within the coupled ROM arrays. Specifically, the exclusive-OR function together with data input D and clock input C are shown as inputs to the product matrix. The lower summing matrix provides the desired output function and also the input to the feedback flip-flop. The D-type flip-flop in the feedback loop provides an additional data input into the product matrix.

8.3 PLA MASTER CHIPS

The uniqueness desired within the master PLA chip is often programed into the master chip by changing only the gate mask. The PLA used in this way often makes

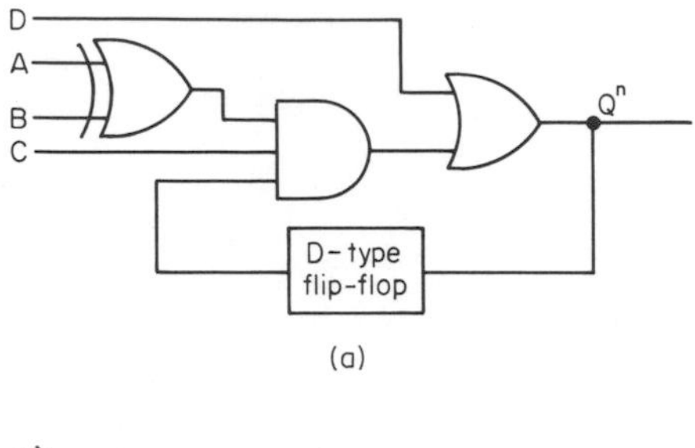

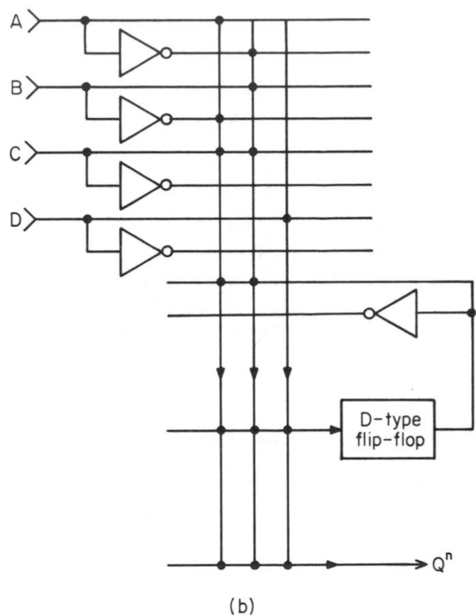

Fig. 8.6. PLA for implementing a clocked, pulse-clamping function with more complex data input trigger. The data trigger is $A \oplus B + D$. (*a*) Standard-logic schematic; (*b*) PLA-logic schematic.

possible the implementation of complex circuit functions at a reduced cost compared to designs where all of the masking steps are unique to the particular circuit function. The PLA design approach makes use of a generalized chip design in which only the gate mask is changed (programed) to obtain a unique and specific logic implementation. The gate mask is used to specify uniquely the product and the summing matrix only. Interconnections to and from the flip-flop array can also be fixed to match a large number of different circuitry requirements.

The complements of all data input signals feeding the product matrix are shown since, typically, the designer will be working with a standard ROM geometry which includes the complements. In Figs. 8.5 and 8.6 the data inputs to the product ROM matrix (including the outputs from the flip-flops) are all double rail. This generalized approach will be used throughout this chapter to indicate that, in a typical design using the PLA, one is designing with a master MOS/LSI chip which contains only one size ROM and flip-flop matrix array but permits the implementation of a variety of different functions. In typical PLA designs, one would have available a series of master PLA chips. Each of these chips would have available a certain size product matrix, a certain size summing matrix, together with a certain number of feedback flip-flops, which might be *D*, *R-S*, *T*, or *J-K* type. One selects the particular size master chip as a basis for a PLA design. The outputs from the flip-flops each are double-rail to accommodate a range of circuit design functions.

Next, let's consider a series of examples which specify the unique gate-masking arrangement required to implement various desired circuit functions.

8.4 PRINCIPLES OF PLA COUNTER DESIGN

In conventional counters a jump gate[3] is required where the modulo is not an integral power of 2. The jump gate detects the state corresponding to the counter modulo minus one and resets all the flip-flops for the following count. Conventional counters reset to the all-zero state and generally use random interconnect schemes. The jump-gate technique with arbitrary interconnection routing requires extensive redesign for extending the system function for different modulos and input control functions.

The counting function implemented using the PLA can be obtained for a variety of fixed or variable modulos by using the two-level combinational logic portion of the PLA to implement the jump gates. The PLA counter uses the product generator ROM section to sense the present state. The product ROM serves as a coincidence detector for set (or reset) levels into the PLA flip-flop array to determine the next system state. Inputs to the ROM section of the PLA may include external control lines in addition to those outputs provided by the flip-flops. One may use the external control lines to provide any desired counting sequence. In addition, the external control lines may be used to interrupt the normal counting sequence at any time and set to a predetermined state.

When the flip-flops are cascaded to implement a synchronous counter, suitable input functions for each flip-flop can be obtained by standard techniques to forward the counter in normal binary sequence. The counter may now be interrupted at a particular state by sensing with a jump gate. In the next state, the counter can be forced to any desired count by gating data to the flip-flops through external control inputs. From this desired count state, the counter can be allowed to continue incrementing in normal binary sequence.

The sequencing of the flip-flops requires two control functions: (1) the normal binary incrementing function and (2) the external control-data input function which forces the flip-flops to the desired state outside the normal binary sequence. The two control functions may be expressed as separate terms (TF and $\overline{T}f$) whose sum I is the required input function to each flip-flop:

$$I = TF + \overline{T}f \tag{8-5}$$

where T = minterm of the final state contained within the transition diagram

F = function which will force the counter to the desired state immediately following the final state

f = that function which moves the counter forward in a normal binary sequence

The input to each flip-flop is obtained from the summing OR ROM section in the PLA counter. The inputs to the AND ROM are (1) the output functions from the flip-flops and (2) the external control function. Standard techniques such as Karnaugh maps or Quine-McCluskey can be used to determine the minimum Boolean function for the input $\overline{Tf}$ in Eq. (9-5). The function F forcing the counter to the desired state at count-out is obtained in coincidence with control bits in the product-generator ROM.

For our design examples in this chapter, a transition map followed by multiple Karnaugh maps provides the desired reduced logic expressions. These Boolean expressions are used directly to determine $\overline{Tf}$ function for gate masking and flip-flop feedback requirements in the PLA counter. A modulo-10 counter has been selected, and design examples are given for circuits with feedback flip-flops of the *D*, *J-K*, *T*, and *R-S*, types. Finally, an arbitrary modulo counter is developed. By comparing the different approaches to PLA counter design, we can gain insight into the various trade-offs and details that are involved.

8.5 COUNTER DESIGN WITH *D*-TYPE FLIP-FLOPS

The transition sequence selected for a modulo-10 counter is shown in Fig. 8.7*a*. The counter counts in a simple weighted binary code, although the specific code selected for the PLA may, in general, be arbitrary. The counting code may be arbitrary because additional decoding is required where multiple output codes are obtained directly from the summing ROM. The transition diagram corresponding to the specified counting sequence is shown in Fig. 8.7*b*; it requires four binary digits. Six of the available states in the transition diagram are not used because the modulo is limited to 10. The arrows on the transition diagram between states indicate the desired sequence for counting. The counting loop is continuous since each output line is reset to zero following the ninth counting interval. The transition sequence and diagram are, of course, independent of the particular type flip-flop to be used in the PLA feedback circuit.

From the transition diagram and knowledge of the logic function of the *D*-type (delay) flip-flop, the Karnaugh map can be specified for each separate bit. Specifically, one Karnaugh map is used for each flip-flop in the feedback array. The entry in each map shows the input function required at each flip-flop to get the desired flip-flop output. These 4 bits are specified as *WXYZ* with the *Z*-bit being the least significant bit. These maps, shown in Fig. 8.7*c*, yield the desired minterm expressions that we will need for coding the PLA matrix.

Coding of the Karnaugh maps follows a standard approach. For instance, in the most significant bit, *W*, the time intervals 0 through 6 represent the case where the logic input to the *D*-type flip-flop occurs. During the seventh time period, the input to the *D*-type flip-flop according to the transition map must be a logical 1 to prepare for the delayed output of a logic true from the *W* flip-flop during interval 8. Similarly, during time period 8, the input to the flip-flop must remain in the logical 1 state since the delayed output during interval 9 must remain in a true or high level. During the time interval 9, the logic signal supplied to the *D*-type flip-flop returns to a logic 0 anticipating that a logic 0 will be required at the output of this flip-flop during interval 0. The Karnaugh map entries specify the inputs required for the particular flip-flop represented by that map.

The Karnaugh maps completed for the remaining three flip-flops complete our

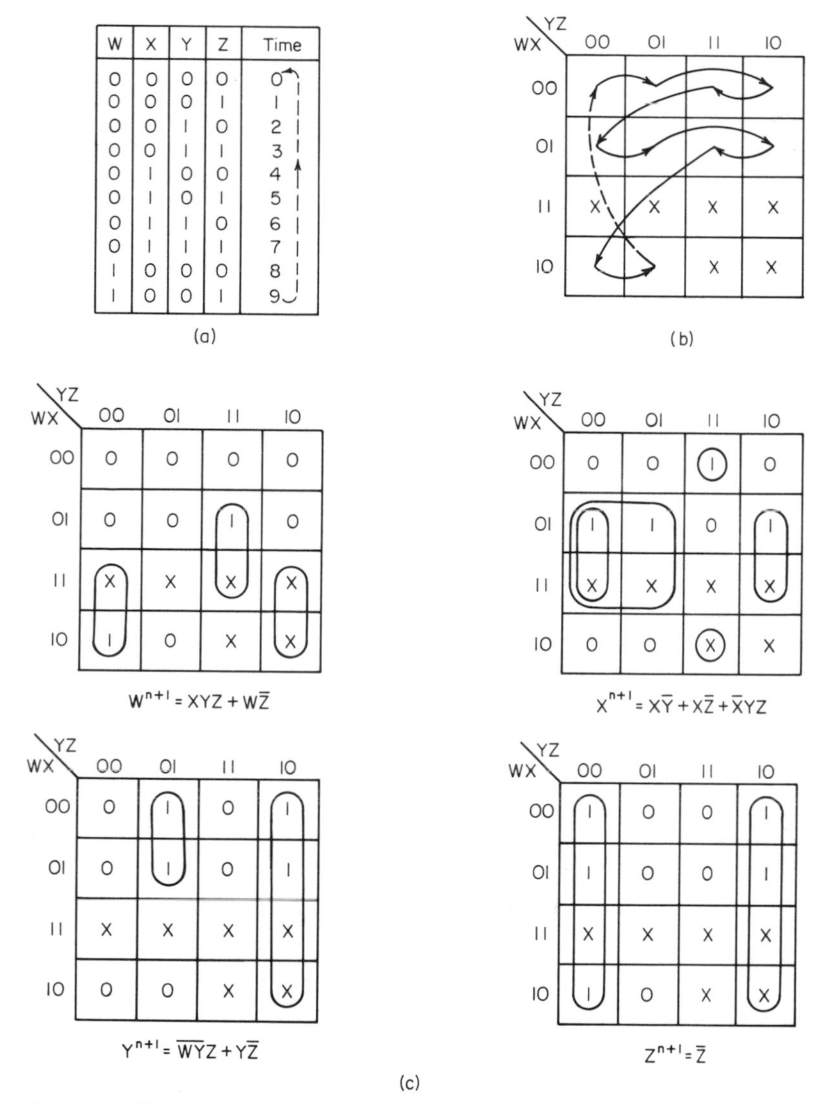

Fig. 8.7. Design sequence for a PLA counter of modulo-10 and using *D*-type flip-flops. (*a*) Transition sequence; (*b*) transition diagram; (*c*) multiple, 4-bit Karnaugh maps with logic functions for *D*-type flip-flops.

logic design necessary to specify the PLA matrix. For discussion purposes, the minterms representing next-state functions are as follows for counting with *D*-type flip-flops:

$$W^{n+1} = WXYZ + WXYZ \tag{8-6}$$

$$X^{n+1} = \bar{W} X \bar{Y} \bar{Z} + \bar{W} X \bar{Y} Z + \bar{W} \bar{X} Y Z + \bar{W} X Y \bar{Z} \tag{8-7}$$

$$Y^{n+1} = \bar{W} \bar{X} \bar{Y} Z + \bar{W} \bar{X} Y \bar{Z} + \bar{W} X \bar{Y} Z + \bar{W} X Y \bar{Z} \tag{8-8}$$

$$Z^{n+1} = \bar{W} \bar{X} \bar{Y} \bar{Z} + \bar{W} \bar{X} Y \bar{Z} + \bar{W} X \bar{Y} \bar{Z} + \bar{W} X Y \bar{Z} + W \bar{X} \bar{Y} \bar{Z} \tag{8-9}$$

The highest order of simplification is obtained when the logic 1 levels in the Karnaugh map are combined in as few groups as possible, each containing as many adjacent cells as possible. The equation above, representing the sum of the minterms from the Karnaugh map, can also be reduced by techniques more amenable to computer automation such as the Quine-McCluskey method. The resulting expressions containing a minimum number of terms are the desired result for PLA implementation with *D*-type flip-flops:

$$W^{n+1} = XYZ + W\bar{Z} \tag{8-10}$$

$$X^{n+1} = X\bar{Y} + X\bar{Z} + \bar{X}YZ \tag{8-11}$$

$$Y^{n+1} = \bar{W}\,\bar{Y}Z + Y\bar{Z} \tag{8-12}$$

$$Z^{n+1} = \bar{Z} \tag{8-13}$$

Further reduction beyond the simple approaches seldom reduces the PLA matrix significantly.

The expressions above represent next-state functions and permit one to code the PLA matrix very quickly. For instance, the input to the W flip-flop consists of two separate sum terms, specifically, $XYZ + W\bar{Z}$. Similarly, the next state input function to the X flip-flop consists of three sum terms, specifically, $X\bar{Y} + X\bar{Z} + \bar{X}YZ$. As you recall, each of these sums is determined by the placement of a gate at the appropriate node within the OR matrix. The product coefficients contained within these sum terms are obtained by masking the gates appropriately within the product matrix. Therefore, the product matrix contains the XYZ product and the $W\bar{Z}$ product. Figure 8.8*a* shows a PLA matrix from Eqs. (8-10) through (8-13) which will count sequentially in weighted-binary code with a modulo-10_{10}. The initial state is determined by the $WXYZ$ flip-flop matrix. If all flip-flops are reset to logic 0 initially, then the counting sequence will begin with all states at logic 0.

Another alternative would be to use the complete set of minterm expressions [Eqs. (8-6) through (8-9)]. These expressions given above are used to specify the PLA matrix of Fig. 8.8*b*. Notice that the next state inputs to the W and X flip-flops contain two and four separate sum terms, respectively. Inspection of the next state input to the Z flip-flop reveals that five separate product lines are required to obtain the five separate terms. A comparison of Fig. 8.8*a* and Fig. 8.8*b* reveals a reduction in number of product lines in every case except for the input to flip-flop W. Notice that here the minimum expression obtained by reducing the Karnaugh map was of no interest; only the reduction in number of product lines was of interest. We can conclude from these examples that good PLA design normally requires reducing the number of product lines and therefore the number of terms in the reduced expression. Further reduction of the product terms is generally not useful since the $WXYZ$ lines separately would all be required in any event.

Now that we have specified the static and active elements for the specific counter design, how are we to bring this information into the outside world or into bus lines which will drive additional circuitry? If the output from the flip-flops can be brought directly to the outside world, we have no need for additional gating or decoding. In general, the direct outputs from the flip-flops are of sufficiently high impedance that it is not good design practice to drive bus lines directly from these elements.

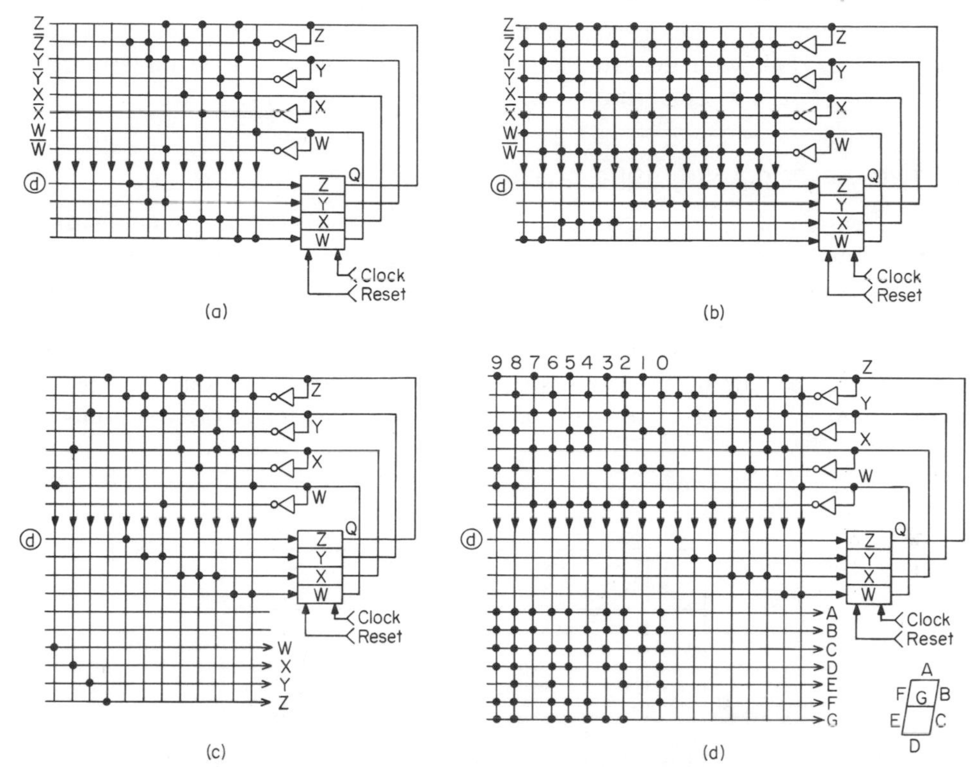

Fig. 8.8. PLA-logic schematic for modulo-10 counter using D-type flip-flops. (a) PLA design using the reduced logic from the Karnaugh maps of Fig. 8.7c; (b) PLA design without benefit of the reduced logic; (c) counter output decoded for binary code; (d) counter output decoded to drive a seven-segment display.

Furthermore, the particular code that the PLA is using for internal counting (in this case a weighted binary code) may not provide the total desired output. Therefore, one generally finds that the summing matrix should include additional elements to be used for decoding the product lines into the proper format for data output. The summing matrix can be used in this fashion to decode the product lines into the desired output function. In the simplest case where we desire the output function to remain a simple weighted binary code, the summing matrix should be coded to provide the output as shown in Fig. 8.8c.

If one further desires a PLA modulo-10_{10} counter with output in a form useful for driving a seven-segment decimal display, then the PLA coding of Fig. 8.8d is useful. The seven segments of the decimal display correspond to A through G as defined by the seven-segment numeral included in Fig. 8.8c. The proper decoding for a modulo-10_{10} requires 10 separate product lines, each of which includes the desired seven-segment combination. For instance, the count of 3 corresponding to the weighted binary code 0011 requires that segment outputs A, B, C, D, and G be selected and set to a logical 1. Similarly, we wish to display the decimal numeral 5 with sum lines A, C, D, F, and G each time the flip-flop output $\bar{W}X\bar{Y}Z$ (0101)

occurs. Thus the decoding format and specification of the PLA matrix is quite simple to implement. Following the same basic approach one can obtain, simultaneously, different codes such as Hollerith, Gray, etc., from a single PLA counter.

8.6 TWO-DECADE COUNTER USING *D*-TYPE FLIP-FLOPS

The extension of the design technique from a modulo-10_{10} to a modulo-100_{10} depends upon the counting code to be used. If we continue counting in weighted binary, the first decade of counting can be identical with the modulo-10_{10} counter already implemented in Sec. 8.5. The second decade of counting requires a coincidence (count-out) line C from the first decade. The transition diagram for the second decade containing the C-bit is shown in Fig. 8.9*a*. Note that the second-decade count advances by one step each clock pulse only when the count-out bit

WX \ CYZ	000	001	011	010	110	111	101	100
00								
01								
11	X	X	X	X	X	X	X	X
10			X	X	X	X		

(a)

WX \ CYZ	000	001	011	010	110	111	101	100
00	0	0	0	0	0	0	0	0
01	0	0	0	0	0	1	0	0
11	X	X	X	X	X	X	X	X
10	1	1	X	X	X	X	0	1

$W^{n+1} = W\bar{C} + W\bar{Z} + XYZC$

WX \ CYZ	000	001	011	010	110	111	101	100
00	0	0	0	0	0	1	0	0
01	1	1	1	1	1	0	1	1
11	X	X	X	X	X	X	X	X
10	0	0	X	X	X	X	0	0

$X^{n+1} = X\bar{C} + X\bar{Z} + X\bar{Y} + \bar{X}YZC$

WX \ CYZ	000	001	011	010	110	111	101	100
00	0	0	1	1	1	0	1	0
01	0	0	1	1	1	0	1	0
11	X	X	X	X	X	X	X	X
10	0	0	X	X	X	X	0	0

$Y^{n+1} = Y\bar{C} + Y\bar{Z} + \bar{W}\bar{Y}ZC$

WX \ CYZ	000	001	011	010	110	111	101	100
00	0	1	1	0	1	0	0	1
01	0	1	1	0	1	0	0	1
11	X	X	X	X	X	X	X	X
10	0	1	X	X	X	X	0	1

$Z^{n+1} = Z\bar{C} + \bar{Z}C$

(b)

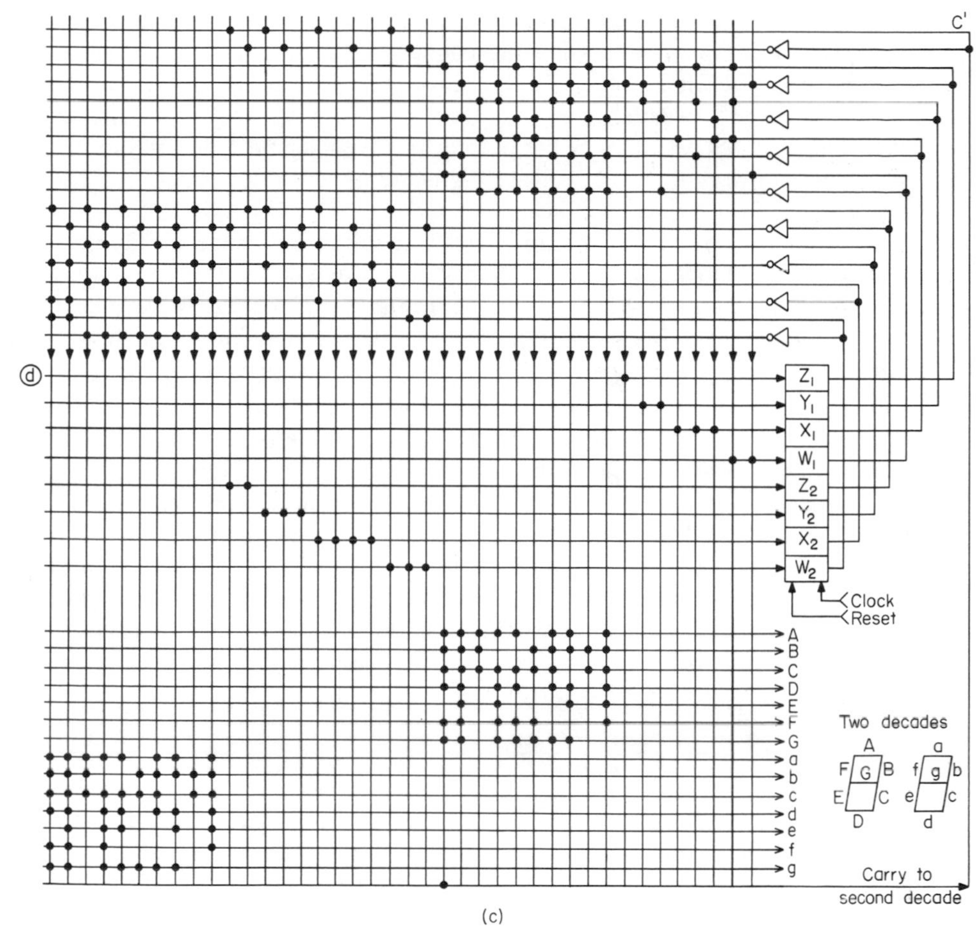

Fig. 8.9. PLA design sequence for the second digit of a modulo-100 counter in an application requiring a seven-segment decimal output. (*a*) Transition diagram for second decade; (*b*) multiple, 5-bit Karnaugh maps with logic functions for *D*-type flip-flops; second decade; (*c*) PLA-logic schematic with decoding for two (seven-segment) decimal digits.

C from the first decade is high. The Karnaugh maps for the 5 bits *WXYZ* and *C* in the second-decade have been developed in Fig. 8.9*b* for the *D*-type flip-flop function. Note that the minimum term expressions of Fig. 8.9*b* below

$$W^{n+1} = W\bar{C} + W\bar{Z} + XYZC \tag{8-14}$$

$$X^{n+1} = X\bar{C} + X\bar{Z} + X\bar{Y} + \bar{X}YZC \tag{8-15}$$

$$Y^{n+1} = Y\bar{C} + Y\bar{Z} + \overline{W}YZC \tag{8-16}$$

$$Z^{n+1} = Z\bar{C} + \bar{Z}C \tag{8-17}$$

reduce to the single (first) decade case of Eqs. (8-10) through (8-13) when $C = 1$.

The PLA array-logic of Fig. 8.9*c* now follows directly from Eqs. (8-14) through

(8-17) in the upper decade and Eqs. (8-10) through (8-13) in the lower decade. The carryout bit C is obtained when the count W, $\bar{X}$, $\bar{Y}$, Z (corresponding to the 9-count) occurs in the first decade. Each occurrence of the C-bit going high triggers a count in the second decade.

Figure 8.9*c* includes decoding at the output for seven-segment decimal numerals. The most significant and least significant digits are given in upper case (*ABC*, etc.) and lower case (*abc*, etc.), respectively.

Notice that two separate sets of sum lines are required to decode for the separate seven-segment decimal outputs. In Fig. 8.9*c* the Carry level (C) from the least significant decimal count is fed directly back into the product matrix without intermediate buffering. In this case it is fairly clear that a race condition should not develop under normal clocking situations. In more complex designs where the possibility of race conditions is now known, it is poor design practice to connect a node point in the summing matrix directly back into the product matrix. If, for instance, in the present situation there were a question of a race condition, the Carry-out bit C could have been derived from the product line corresponding to the eighth count of the least significant modulo-10_{10} counter, and buffered through a D-type flip-flop which would be in turn connected into the C' line of the product matrix. If one desires to implement PLA counters for more decades, then additional Carry lines between adjacent modulo-10_{10} counter sections should be accordingly added. Higher order counting decades would precisely match the PLA coding of the counting decade of Fig. 8.9*a* and *b*.

8.7 MODULO-10_{10} COUNTER DESIGN USING *J-K*-TYPE FLIP-FLOPS

Since the *J-K*-type flip-flop contains less logical redundancy than the *D*-type flip-flop, one might intuitively anticipate that PLA counters implemented with *J-K*-type flip-flops would provide more efficient utilization of chip area. It is instructive to implement the counter using various type flip-flops for comparison. In Fig. 8.10*a* the modulo-10 transition diagram is shown adjacent to the 4-bit Karnaugh maps (Fig. 8.10*b*) for ease in specifying the entries into the maps. In Fig. 8.10*b* the J and $\emptyset$ entries in the Karnaugh map correspond to the "don't care" states of the *J-K* flip-flop. The J entry indicates a "don't care" which is unique to the *J-K* function, and the $\emptyset$ entry indicates the "don't care" condition for the *R-S*-type flip-flop. This approach permits one readily to extend the design in Fig. 8.10*b* to the case for *R-S*-type flip-flops. The flip-flop designated X corresponding to the second most significant bit has all of its J inputs in the 0 state until time interval 3 appears, at which time the J input must go high in preparation for the output switching into the high state during the next time interval 4. Similarly, the K input to the X flip-flop during interval 3 receives a J (don't care) input, forcing the output of this flip-flop into the logic 1 state during the next time interval. The input bit K receives a "don't care" input during time intervals 0 through 3 and a logic input during intervals 4, 5, and 6. During time 7 the input bit K on flip-flop X must go to a logic high, forcing a transition to a logic 0 during the next time interval 8. This same flip-flop X during time intervals 8 and 9 remains in the "don't care" condition. The other Karnaugh maps are obtained with similar reasoning.

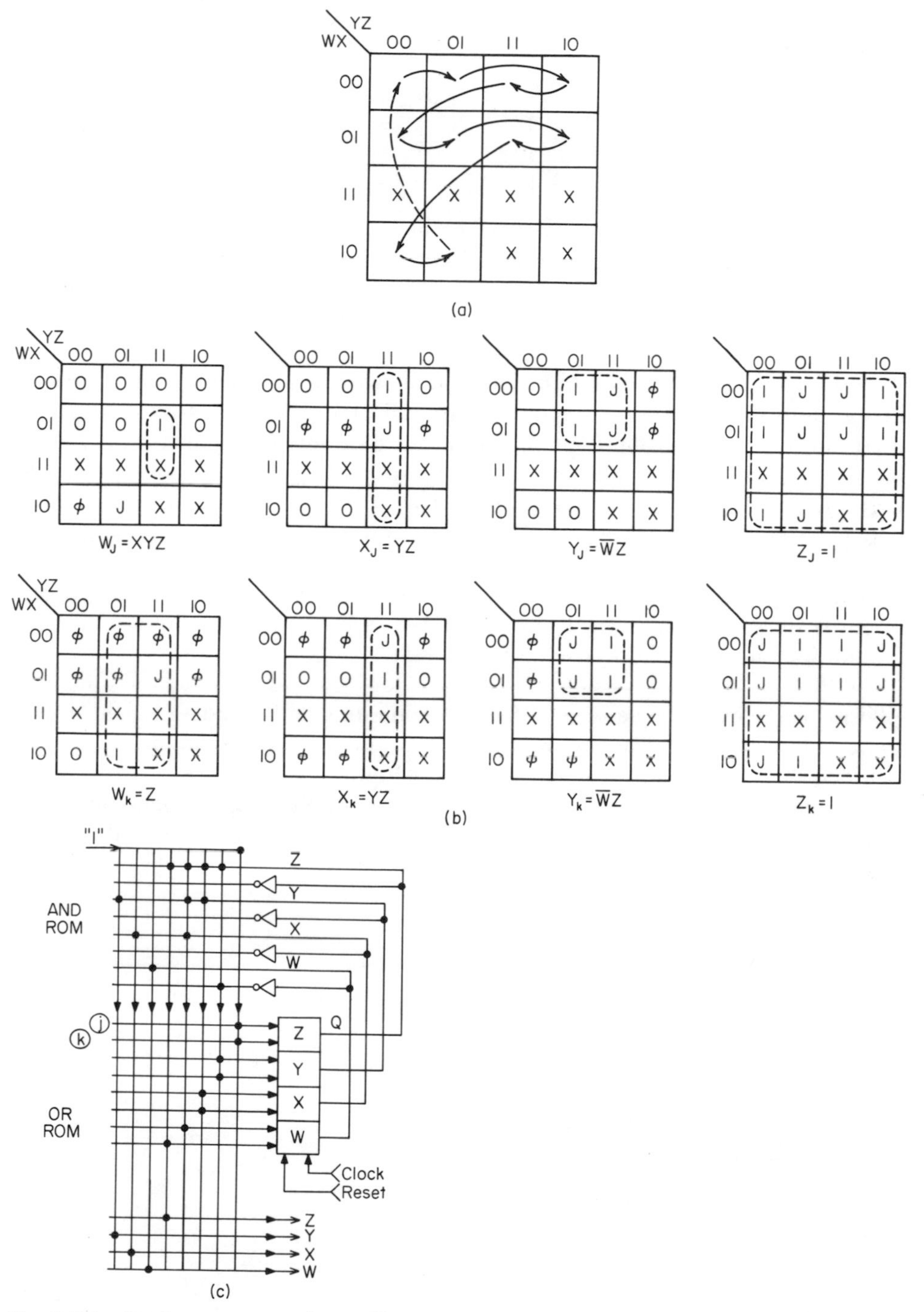

Fig. 8.10. Design sequence for a PLA counter of modulo-10 and using a *J-K*-type flip-flop. (*a*) Transition diagram; (*b*) 4-bit Karnaugh maps; (*c*) PLA-logic diagram with decoding for weighted binary output.

The reduced Boolean functions obtained from the Karnaugh maps of Fig. 8.10*b* are as follows:

$$\begin{aligned} W_J &= XYZ & W_K &= Z \\ X_J &= YZ & X_K &= YZ \\ Y_J &= \bar{W}Z & Y_K &= \bar{W}Z \\ Z_J &= 1 & Z_K &= 1 \end{aligned} \qquad (8\text{-}18)$$

The flip-flop Z requires that its input be maintained in a logic high state at all times. The straightforward implementation for this designed function is shown in the PLA-logic schematic of Fig. 8.10*c*, which includes the four separate *J-K*-type flip-flops. The total number of nodes or bits required in the dual ROM of Fig. 8.10*c* is 168.

Obviously the flip-flop Z could be eliminated, with the logic high input connected directly into the uppermost row of four gates in the product matrix. Also, it is interesting to note that flip-flops X and Y use data inputs for the J and K lines that are respectively identical. Therefore, flip-flops X and Y are each operated in the toggle mode and thus may be replaced by T-type flip-flops. The number of matrix nodes required for the modulo-10_{10} counter may be reduced significantly by eliminating the Z-type flip-flop and substituting a T-type flip-flop for bistable elements X and Y. The resulting PLA-logic schematic with decoding for a binary output is shown in Fig. 8.11. The product and sum matrices require a total of 64 and 56 discrete nodes, respectively, in Fig. 8.11.

The PLA counter design using *J-K*-type flip-flops may be extended for a modulo-100 total count by adding a second decade of counting and decoding for the appropriate output format. The two-decade PLA counter using *J-K* flip-flops is shown with decoding for a seven-segment display in Fig. 8.12.

The total number of matrix nodes required for the PLA-logic schematic of Fig. 8.10*c* requires 168 nodes for weighted binary output. If one desires the seven-

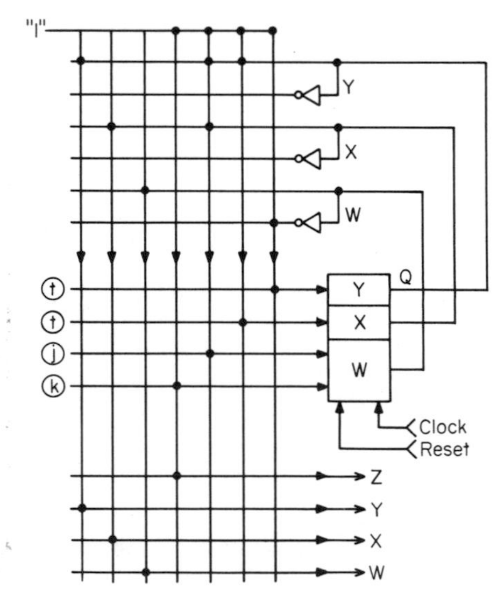

Fig. 8.11. Efficient modulo-10 counter using *J-K-* and *T*-type flip-flop combination in the feedback circuit.

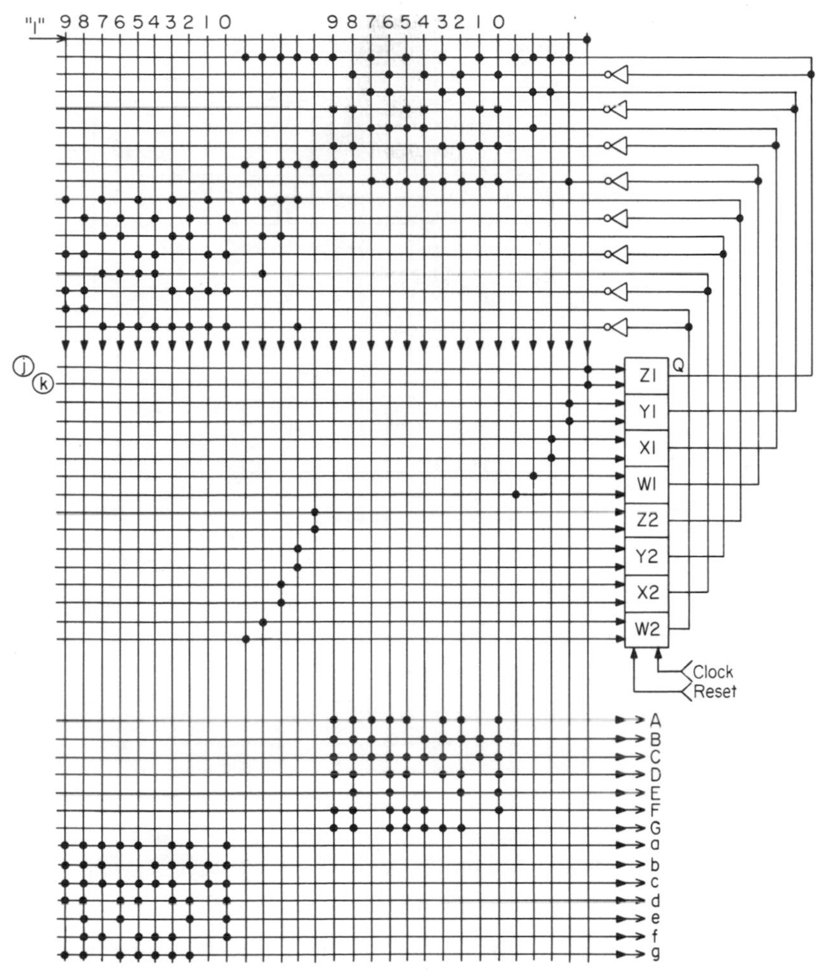

Fig. 8.12. Two-decade PLA-logic diagram with *J-K*-type flip-flops and decoded output for a seven-segment decimal display.

segment output for the modulo-10 counter with *J-K* flip-flops, a total of 360 matrix points are required. The 360-bit matrix for the PLA counter using *J-K*-type flip-flops and seven-segment output compares with 342 matrix points required for the same function using *D*-type flip-flops in Fig. 8.8*d*. It appears that the increase in logical efficiency obtained with the *J-K*-type flip-flops is offset by the increase in number of summing rows required to drive the *J-K* inputs as compared with the *D*-type implementation. When the number of circuit matrix node points is compared for the two-decade counter design in which seven-segment outputs are required, we again find that the *J-K*-type implementation is approximately equal to the *D*-type implementation. Specifically, the PLA-logic schematic requires ROM bit counts of 1,640 and 1,410 for the implementation with all *D*-type and all *J-K*-type flip-flops, respectively. Except for cases in which the *D*-type flip-flops require significantly

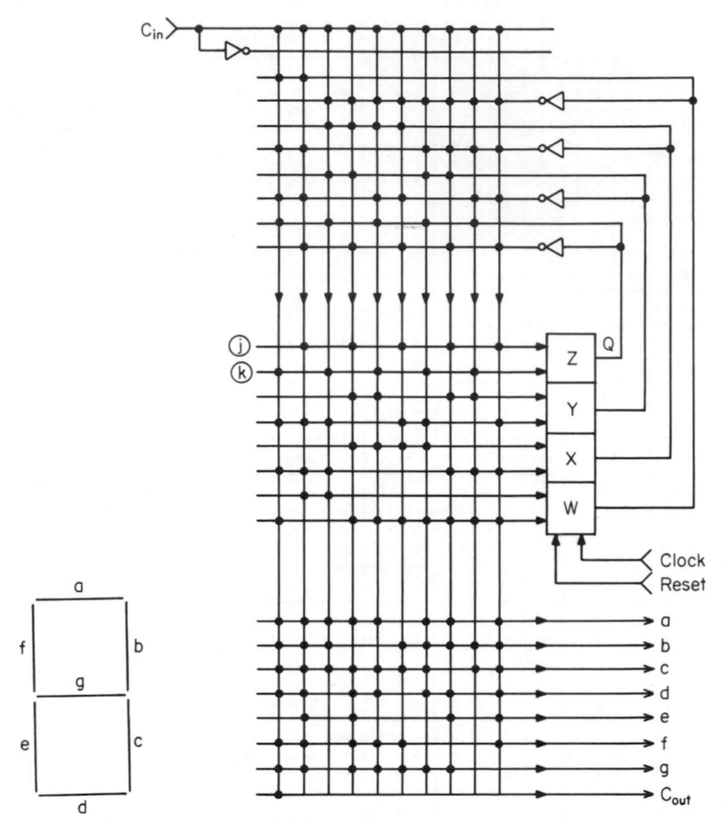

Fig. 8.13. Alternate design modulo-10 counter PLA array where the counting sequence is obtained directly in the seven-segment output code.

reduced area compared to the *J-K*-type flip-flop array, we can conclude that design efficiencies are approximately equal for implementation of the two-decade counter as specified in this case.

The counters above have all been sequenced through a simple weighted binary code. If only one code is desired at the output, it is sometimes advantageous to do the counting in that same code. If we desire only a seven-segment output, then we might wish to count directly in a decimal sequence instead of binary. Such a counter has been designed, with the result shown in Fig. 8.13. The decade counter sequencing through 10 separate states using *J-K*-type flip-flops requires 260 nodes in the ROM matrix. The single-decade counter which sequences in binary (Fig. 8.12) may be compared by adding the double-rail Carry-in C_{in} line to require 390 nodes. Thus, by counting directly with a decimal code, we reduced the number of ROM nodes by 30 percent.

8.8 DESIGN EXAMPLE: VARIABLE MODULO 4-BIT COUNTER

The complexity of the logic function performed using a basic 4-bit counting configuration can be greatly increased by the addition of a few control bits into the PLA matrix. For a design example, let's consider a controlled modulo counter. A controlled modulo counter can be implemented using the PLA scheme to result in an efficient design in which the total ROM matrix size is not increased much above that for the fixed modulo counter. Three specific counters will be designed using the jump-gate approach with (1) reset to the control input, and (2) reset to zero. In each case the modulo is determined by the 4 input bits $ABCD$ with A being the most significant bit. For the case (1) of reset to the control input $ABCD$, the modulo of the counter is the 2's complement of the variable input function $ABCD$.

8.9 VARIABLE-MODULO COUNTER: RESET TO THE CONTROL INPUT

The transition sequence for a variable-modulo counter in which the count advances between the input bit-lines and 1111 is shown schematically in Fig. 8.14*a*. For instance, with an input $A\,\bar{B}\,\bar{C}\,\bar{D}$ function, the counting modulo is 8_{10}. The input function $A\,\bar{B}\,C\,\bar{D}$ similarly results in a counting modulo of 6_{10}.

The variable-modulo design can be most easily understood by again referring to Eq. (8-5).

$$I = TF + \bar{T}f \tag{8-5}$$

The basic sequence (F) is determined by the modulo-16_{10} counter. The (F) functions using S-R-type flip-flops are obtained from the transition map of Fig. 8.14*b* and the Karnaugh maps of Fig. 8.14*c* in the usual manner.

$$W_S{}^{n+1} = \bar{W}XYZ \qquad W_R{}^{n+1} = WXYZ \tag{8-19}$$

$$X_S{}^{n+1} = \bar{X}YZ \qquad X_R{}^{n+1} = XYZ \tag{8-20}$$

$$Y_S{}^{n+1} = \bar{Y}Z + \qquad Y_R{}^{n+1} = YZ \tag{8-21}$$

$$Z_S{}^{n+1} = \bar{Z} \qquad Z_R{}^{n+1} = Z \tag{8-22}$$

$$T = WXYZ = 1\,1\,1\,1 \tag{8-23}$$

When $T = 0$, the counter sequences normally. When $T = 1$, the count-out has occurred ($WXYZ$ all high) and the next count to appear at the flip-flop output is the $ABCD$ input function. Therefore, the relationship of Eq. (8-5) becomes, for each flip-flop input,

$$W_S{}^{n+1} = \bar{T}\,\bar{W}\,X\,Y\,Z + TA \qquad W_R{}^{n+1} = T\,W\,X\,Y\,Z + T\bar{A} \tag{8-24}$$

$$X_S{}^{n+1} = \bar{T}\,\bar{X}\,Y\,Z + TB \qquad X_R{}^{n+1} = \bar{T}\,X\,Y\,Z + T\bar{B} \tag{8-25}$$

$$Y_S{}^{n+1} = \bar{T}\,\bar{Y}\,Z + TC \qquad Y_R{}^{n+1} = \bar{T}YZ + TC \tag{8-26}$$

$$Z_S{}^{n+1} = \bar{T}\,\bar{Z} + TD \qquad Z_R{}^{n+1} = \bar{T}Z + T\bar{D} \tag{8-27}$$

Equations (8-24) through (8-27) may now be used to specify directly the PLA variable-modulo counter array. This has been done in Fig. 8.15*a*. In addition, the $WXYZ$ bit lines are brought down to the sum matrix separately to provide the weighted binary output $WXYZ$. The count-out line T is brought from the sum to the product matrix directly. One could, if desired, change the coding slightly and bring the signal T through a fifth flip-flop instead. The circuit of Fig. 8.15*a* includes a total of 651 matrix nodes in the ROM section with the R-S-type flip-flops.

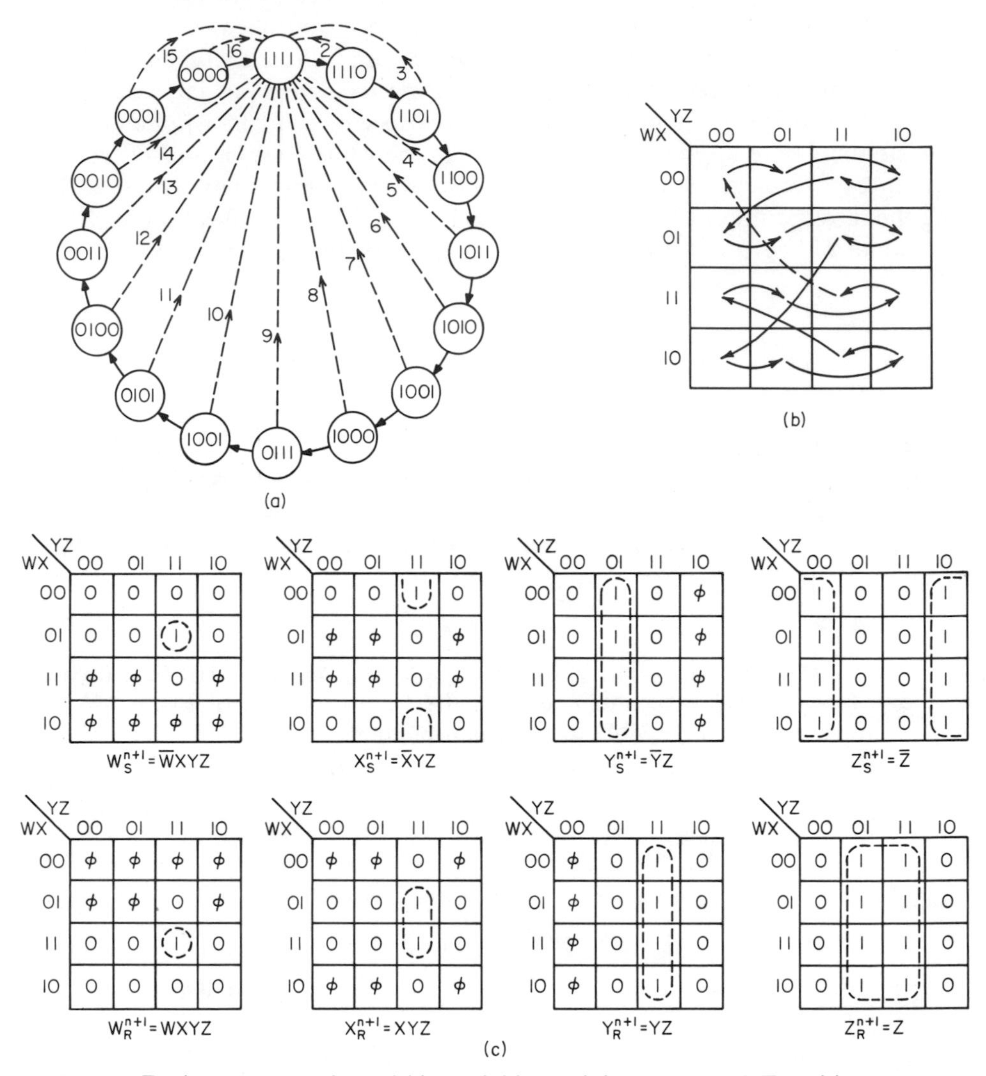

Fig. 8.14. Design sequence for a 4-bit, variable-modulo counter. (*a*) Transition sequence; (*b*) transition diagram for the modulo-16 counter; (*c*) multiple, 4-bit Karnaugh maps for modulo-16_{10} and *R-S*-type flip-flops.

The variable-modulo counter for *D*-type flip-flops requires control functions as follows:

$$W_d^{n+1} = \overline{T}W\overline{X} + \overline{T}W\overline{Y} + \overline{T}W\overline{Z} + \overline{T}\,\overline{W}XYZ + TA \tag{8-28}$$

$$X_d^{n+1} = \overline{T}X\overline{Y} + \overline{T}X\overline{Z} + \overline{T}\,\overline{X}\,YZ + TB \tag{8-29}$$

$$Y_d^{n+1} = \overline{T}\,\overline{Y}\,Z + \overline{T}Y\overline{Z} + TC \tag{8-30}$$

$$Z_d = \overline{T}\,\overline{Z} + TD \tag{8-31}$$

$$T = WXYZ = 1\,1\,1\,1 \tag{8-32}$$

The PLA array-logic for the D-type flip-flops is shown in Fig. 8.15b.

A simple derivation has been made for the same counting function using D-type flip-flops. The Boolean control is found to be

$$W_j^{n+1} = \bar{T}XYZ + TA \qquad W_k^{n+1} = \bar{T}XYZ + T\bar{A} \tag{8-33}$$

$$X_j^{n+1} = \bar{T}YZ + TB \qquad X_k^{n+1} = TYZ + T\bar{B} \tag{8-34}$$

$$Y_j^{n+1} = \bar{T}Z + TC \qquad Y_k^{n+1} = \bar{T}Z + T\bar{C} \tag{8-35}$$

$$Z_j^{n+1} = \bar{T} + TD \qquad Z_k^{n+1} = \bar{T} + T\bar{D} \tag{8-36}$$

$$T = WXYZ = 1\,1\,1\,1 \tag{8-37}$$

and the resulting array-logic schematic is shown in Fig. 8.15c.

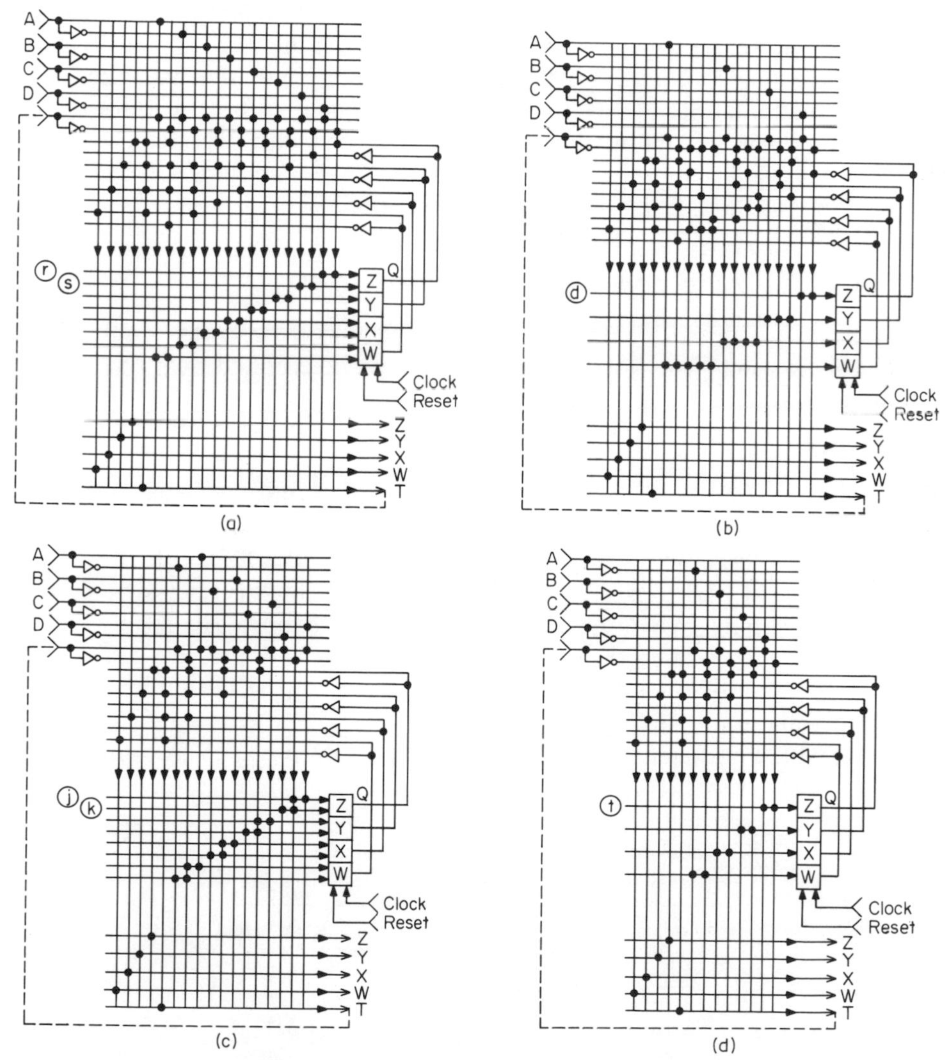

Fig. 8.15. Four-bit, variable-modulo counter with reset-to-input control bits. (a) S-R-type flip-flops; (b) D-type flip-flops; (c) J-K-type flip-flops; (d) T-type flip-flops.

The T-type flip-flop requires a simple Boolean control function

$$W_t^{n+1} = \bar{T}XYZ + T\bar{A} \quad (8\text{-}38)$$

$$X_t^{n+1} = \bar{T}YZ + T\bar{B} \quad (8\text{-}39)$$

$$Y_t^{n+1} = \bar{T}Z + T\bar{C} \quad (8\text{-}40)$$

$$Z_t^{n+1} = \bar{T} + T\bar{D} \quad (8\text{-}41)$$

$$T = WXTZ = 1\,1\,1\,1 \quad (8\text{-}42)$$

and thus may prove to be an efficient design. The PLA array-logic counter for the T-type flip-flops is shown in Fig. 8.15*d*.

Each implementation requires a different size ROM matrix. The differences are summarized in Table 8.1.

One might expect that the implementation using J-K-type flip-flops with no state ambiguity could be implemented with less complexity in the combinational circuitry and therefore would require a smaller ROM matrix compared with the D-type flip-flop implementation, but such is not the case. The T-type PLA ROM requires slightly more than half the total silicon chip area other flip-flops require; we can conclude that the design of this PLA counter with T-type flip-flops will be optimum for variable-modulo with reset-to-control input $ABCD$.

8.10 VARIABLE-MODULO COUNTERS: RESET-TO-ZERO

In the previous design examples the input function $ABCD$ was equated to the initial counter state. The final count-out state was determined by gate-mask programing and was fixed at $WXYZ = 1\,1\,1\,1$. The design with reset-to-control input would provide a binary output count which does not initialize to the all-zero state for all modulos.

It is often desired to initialize to an all-zero state and sequence forward by incrementing in a normal weighted binary count. For reset-to-zero, we desire each clocking period to compare the state of the counter ($WXYZ$) with the external data-control input ($ABCD$). The transition sequence for the reset-to-zero is presented in Fig. 8.16*a*. When the external input $ABCD$ coincides logically with the current state of the counter, the combinational logic network provides gating which resets all of the flip-flops. The modulo in this case will be the decimal value of the control-data input $ABCD$ plus one.

Since the PLA must sense for logical coincidence during each clocking interval,

Table 8.1. Variable-modulo, 4-bit Counters (Reset to Control Input)

Flip-flop	ROM size
S-R type	651
D type	513
J-K type	527
T type	351

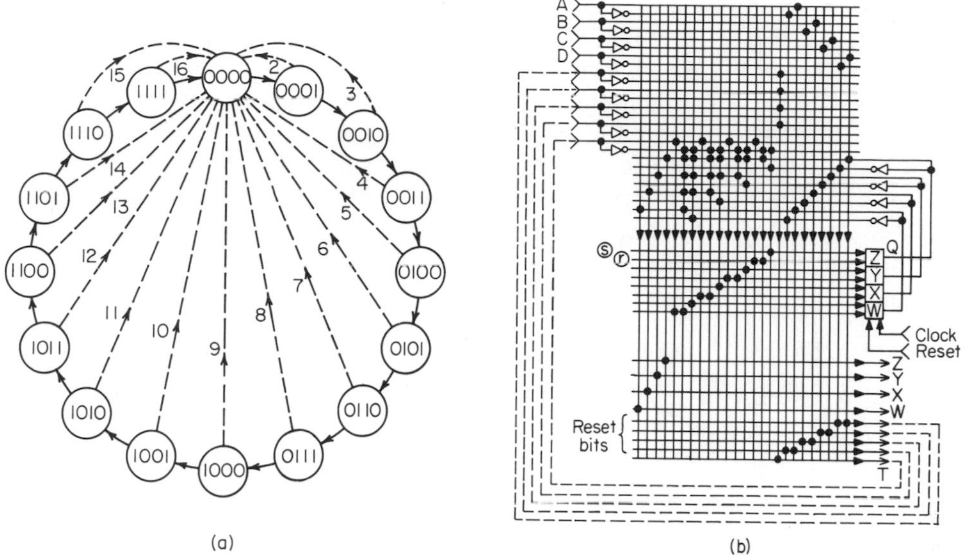

Fig. 8.16. Four-bit, variable-modulo counter with reset-to-zero and count-out determined by input control bits and feedback T. (*a*) Transition sequence; (*b*) S-R-type flip-flops.

we can anticipate an increased logic complexity for the ROM matrices as compared with the reset-to-control input case. The logic functions required for an implementation with S-R-type flip-flops in the desired counter are shown below:

$$W_s = \bar{T}\,\bar{W}\,X\,Y\,Z + T \tag{8-43}$$

$$W_r = \bar{T}\,W\,X\,Y\,Z + T \tag{8-44}$$

$$X_s = \bar{T}\,\bar{X}\,Y\,Z \tag{8-45}$$

$$X_r = \bar{T}\,X\,Y\,Z + T \tag{8-46}$$

$$Y_s = \bar{T}\,\bar{Y}\,Z \tag{8-47}$$

$$Y_r = \bar{T}\,Y\,Z + T \tag{8-48}$$

$$Z_s = \bar{T}\,\bar{Z} \tag{8-49}$$

$$Z_r = \bar{T}\,Z + T \tag{8-50}$$

$$T = (W\,A + \bar{W}\,\bar{A})(X\,B + \bar{X}\,\bar{B})(Y\,C + \bar{Y}\,\bar{C})(Z\,D + \bar{Z}\,\bar{D}) \tag{8-51}$$

The PLA implementation of the 4-bit counter is given in Fig. 8.16. The complexity results in the fact that the count-out T function becomes more complex. The four coincidence terms $W\,A + \bar{W}\,\bar{A}$, $X\,B + \bar{X}\,\bar{B}$, $Y\,C + \bar{Y}\,\bar{C}$, and $Z\,D + \bar{Z}\,\bar{D}$ are developed in one pass through the AND-OR ROM. These four terms are then fed back to the AND matrix with four separate lines, and the product desired is obtained on the second "ripple-through." The array-logic can be simplified considerably by adjusting the count-out T to its complement $\bar{T}$ to obtain:

$$\bar{T} = WA + WZ + XB + XB + YC + YC + ZD + ZD \tag{8-52}$$

which is equivalent to Eq. (8-51).

Using Eq. (8-56) for the count-out term $\bar{T}$ and complementing upon reentering the product generator to obtain T (true), we can simplify the PLA array-logic. The result is shown in Fig. 8.17*b*. Thus, for the variable-modulo design using *R-S*-type flip-flops, the change of count-out T to its complement reduces the ROM node count from 1,075 to 744. It should be noted that the count-out line T or $\bar{T}$ may also be used to reset the flip-flops to zero directly through the RESET line.

The design in Fig. 8.17*a* may be continued to obtain implementation with the *D*-, *J-K*-, and *T*-type flip-flops in Fig. 8.17*b*, *c*, and *d*, respectively. The Boolean control functions used are:

J-K-type

$$W_j = \bar{T} X Y Z \qquad W_k = \bar{T} X Y Z + T$$
$$X_j = \bar{T} Y Z \qquad X_k = \bar{T} Y Z + T$$
$$Y_j = \bar{T} Z \qquad Y_k = \bar{T} Z + T$$
$$Z_j = \bar{T} \qquad Z_k = \bar{T} + T$$

D-type

$$W_d = \bar{T} W \bar{X} + \bar{T} W \bar{Y} + \bar{T} W \bar{Z} + \bar{T} \bar{W} X Y Z$$
$$X_d = \bar{T} X \bar{Y} + \bar{T} X \bar{Z} + \bar{T} \bar{X} Y Z$$
$$Y_d = \bar{T} \bar{Y} Z + \bar{T} Y \bar{Z}$$
$$Z_d = \bar{T} \bar{Z}$$

T-type

$$W_t = \bar{T}XYZ + TW \tag{8-53}$$
$$X_t = \bar{T}YZ + TX \tag{8-54}$$
$$Y_t = \bar{T}Z + TY \tag{8-55}$$
$$Z_t = \bar{T} + TZ \tag{8-56}$$

The size of the various ROM matrices in Fig. 8.17 is given by Table 8.2.

The total ROM matrix size required is a minimum for the *T*-type flip-flop implementation. The *T*-type flip-flop implementation requires a minimum of 540 ROM matrix points. It is clear that the variable-modulo counter with reset-to-zero does require a more complex PLA implementation than the counter with reset-to-control input only. Also, the increased complexity of the counter with reset-to-zero will generally reduce the overall counting speed, because of capacitive loading within the ROM array.

8.11 ARBITRARY MODULO UP/DOWN COUNTER

Other counter applications sometimes require incrementing or decrementing according to a control-input function. A single control-input line designated as UP/DOWN line provides incrementing or decrementing control to the counter.

Table 8.2. Variable-Modulo, 4-bit Counters (Reset to Zero)

Flip-flop	ROM size
R-S type	744
D type	594
J-K type	620
T type	540

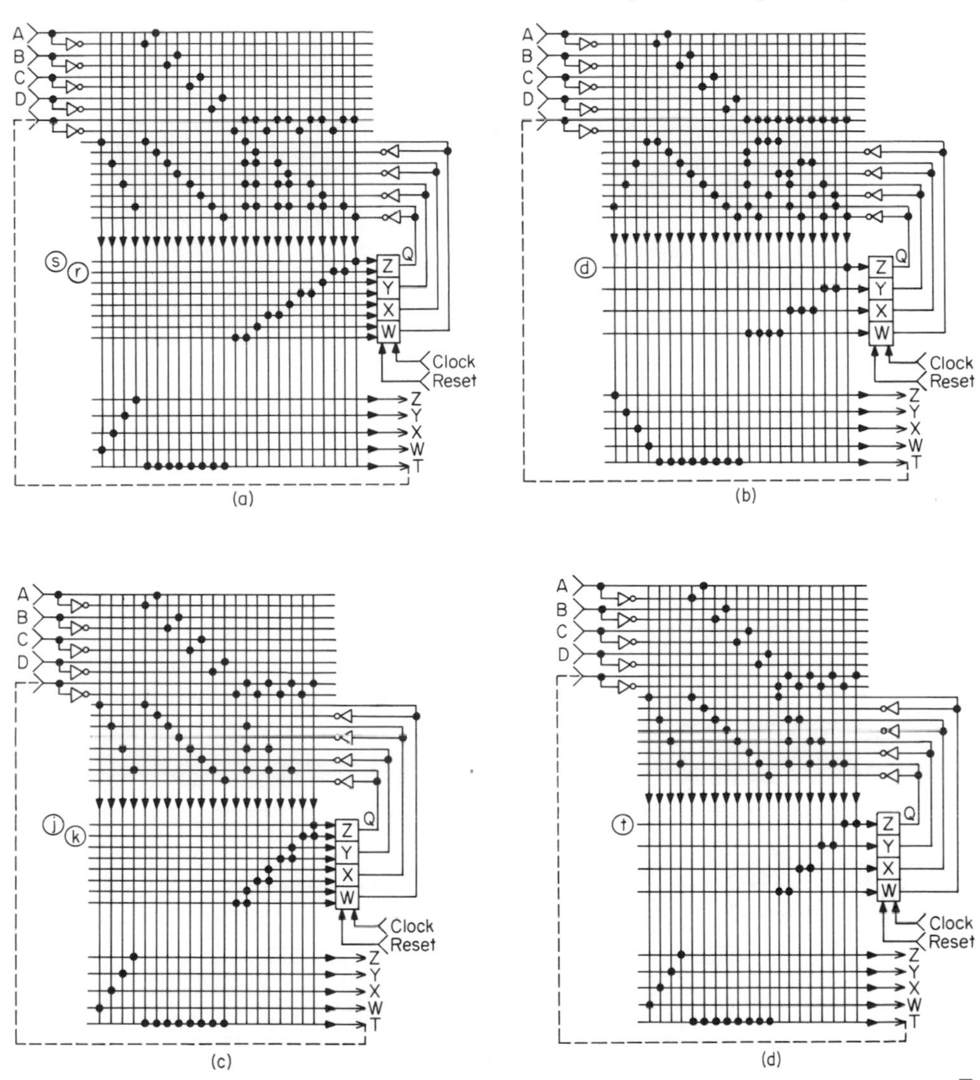

Fig. 8.17. Improved 4-bit, variable-modulo counter with reset-to-zero and feedback $\overline{T}$. (*a*) *S-R*-type flip-flops; (*b*) *D*-type flip-flops; (*c*) *J-K*-type flip-flops; (*d*) *T*-type flip-flops.

We will consider an UP/DOWN counter in which the transition count sequence is a simple weighted binary. When the UP/DOWN line is at logical 1, we wish the counter to increment in the forward direction. If the UP/DOWN line is at a logical 0, the counter should decrement from the initial state determined by *ABCD* down to the final state of 0000. Thus, with the UP/DOWN line in the logical 1 position, the modulo is the 2's complement of the control input *ABCD*. With the UP/DOWN line at a logical 0 state, the modulo becomes the decimal value of the control input *ABCD* plus one.

The logic functions required for implementation of the UP/DOWN counter using *J-K* flip-flops are shown below:

$$W_j = (\bar{T}_2 XTZ + T_2 A)\,U + \bar{U}\,(\bar{T}_1 \overline{XZ} + T_1 A)$$

$$W_k = (\bar{T}_2 XYZ + T_2 \bar{A})\,U + \bar{U}\,(\bar{T}_1 \overline{XYZ} + T_1 \bar{A}) \quad (8\text{-}57)$$

$$X_j = (\bar{T}_2 YZ + T_2 B)\,U + \bar{U}\,(\bar{T}_1 \overline{YZ} + T_1 B)$$

$$X_k = (\bar{T}_2 YZ + T_2 \bar{B})\,U + \bar{U}\,(\bar{T}_1 \overline{YZ} + T_1 \bar{B}) \quad (8\text{-}58)$$

$$Y_j = (\bar{T}_2 Z + T_2 C)\,U = \bar{U}\,(\bar{T}_1 \bar{Z} = T_1 C)$$

$$Y_k = (\bar{T}_2 Z + T_2 \bar{C})\,U + \bar{U}\,(\bar{T}_1 \bar{Z} + T_1 \bar{C}) \quad (8\text{-}59)$$

$$Z_j = (\bar{T}_2 + T_2 D)\,U + \bar{U}\,(\bar{T}_1 + T_1 D)$$

$$Z_k = (\bar{T}_2 + T_2 \bar{D})\,U + \bar{U}\,(\bar{T}_1 + T_1 \bar{D}) \quad (8\text{-}60)$$

$$T_1 = \bar{W}\,\bar{X}\,\bar{Y}\,\bar{Z} \qquad \text{count-down limit}$$

$$T_2 = W\,X\,Y\,Z \qquad \text{count-up limit}$$

The UP/DOWN line has been designated as U in these input functions.

The array implementation now includes the reset-to-control input $ABCD$. The UP/DOWN count is obtained by two separate counters, one for up-counting and the other for down-counting. The resulting PLA matrix for this function is shown in Fig. 8.18. A total of 1,080 ROM matrix points is obtained by using 30 product lines and 36 summing lines. The PLA output now includes the two T_1 and T_2 bit

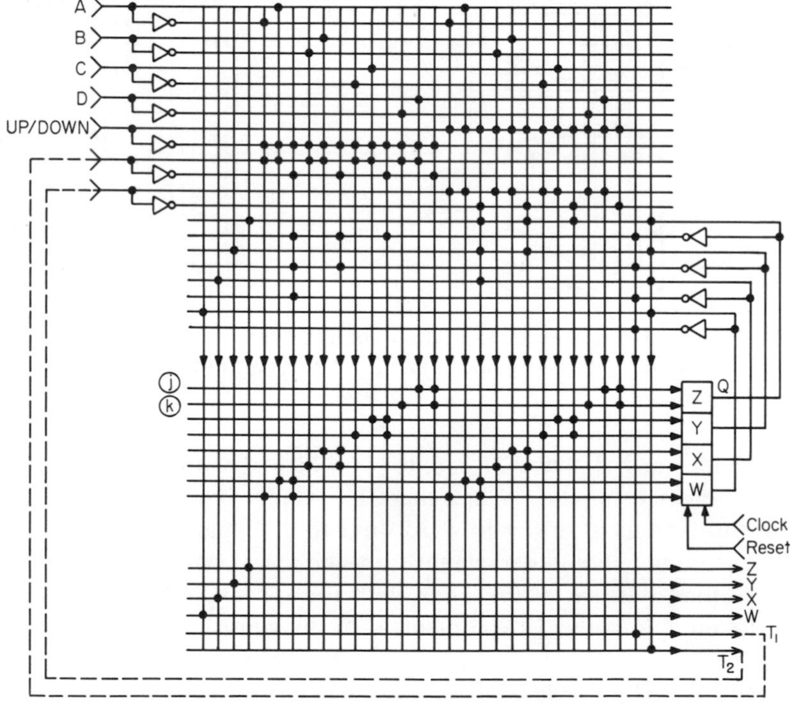

Fig. 8.18. Four-bit up/down, variable-modulo counter using PLA array.

lines which indicate (1) count-down with 0000 as the next state, and (2) Carry-out for all-ones, respectively. The T_1 and T_2 lines are connected back into the input of the AND matrix.

8.12 BIT-SERIAL ADDER/SUBTRACTER

The PLA really comes into its area of advantage when more complex functions are implemented. The adder/subtracter function is the most complex function that we can conveniently use for purposes of instruction. The bit-serial add/subtract function is shown in the standard-logic schematic of Fig. 8.19*a*. The PLA design is obtained by inspecting each logic functional component and adjusting the clocking for PLA implementation. The unit for binary-coded-decimal inputs is shown in Fig. 8.19*b*. Based on the original logic design derived by Irwin,[4] this requires two full-adders for serial implementation. The same function using a bit-parallel configuration requires seven full-adders. The circuitry for implementing the add/subtract function using BCD coding requires the addition of a correction flip-flop when a count greater than 9 is involved. The decimal coding requires two full-adders and a shift register to handle those additional counts associated with the hexadecimal code. The two data inputs are α and β ("sub" is the subtrahend when the unit operates as a subtracter). The ADD/SUBTRACT command is obtained with two input-control lines. The input data α and β are processed by the full-adder 1 and the result is temporarily stored in a 4-bit shift register.

The result of the add/subtract function in the shift register is examined during the fourth bit-time to determine if additional correction is necessary. The result during the following digit-time is modified by adding or subtracting 6, which is under the control of the correction flip-flop. Thus, the result of the serial addition will always be delayed by 4 bit-times. This principle has been implemented within the PLA format using *D*-type flip-flops.

The PLA design is shown in Fig. 8.19*b*; it requires a 32 by 51 ROM matrix with 1,632 matrix points. The circuit would normally require more than two-level logic. To circumvent this and use the standard two-level PLA format, three feedback direct connections are used from the OR matrix back into the AND matrix. These three lines include (1) the proper input of subtrahend to full-adder 1; (2) the carry signal generated by full-adder 1; and (3) correction input to full-adder 2 at proper time. This implementation uses a total of 1 1 *D*-type flip-flops.

The PLA approach is not limited to flip-flops within the feedback loop. The add/subtract unit with its 4-bit shift register provides a good example using other logic functions in the feedback loop. The design of Fig. 8.19*c* includes two separate 4-bit shift registers in the feedback loop. These registers are (1) *WXYZ*, which provides the temporary storage for the sum signal from the full-adder 1, and (2) *ABCD*, which produces the different bit-timing pulses T_1, T_2, T_3, and T_4. Both of these shift registers operate in serial fashion and require a single data input line. The use of the shift registers permits a reduction by 6 in the number of logic-summing rows for the implementation of Fig. 8.19*c* as compared with Fig. 8.19*b*. Other features in the two circuits are the same.

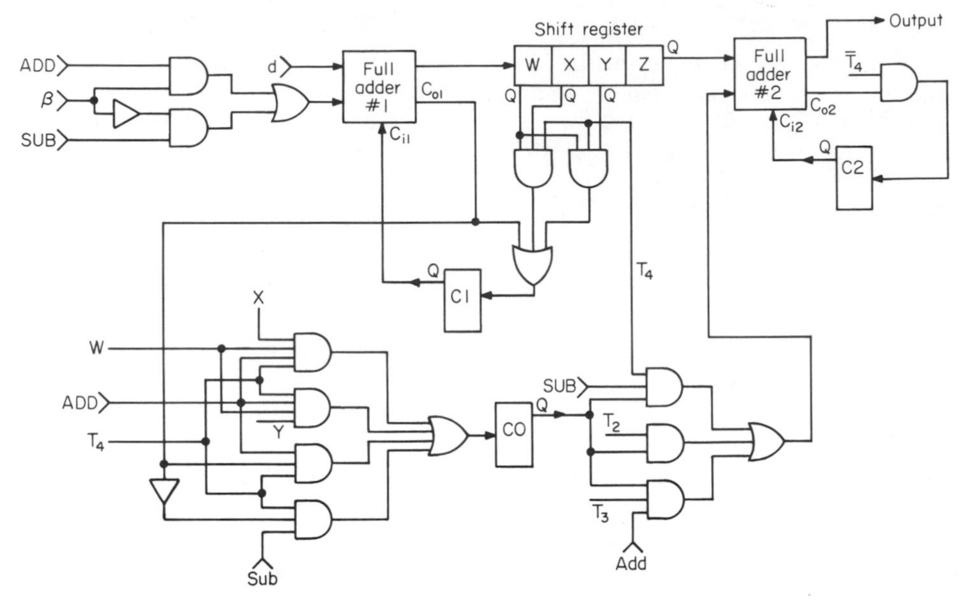

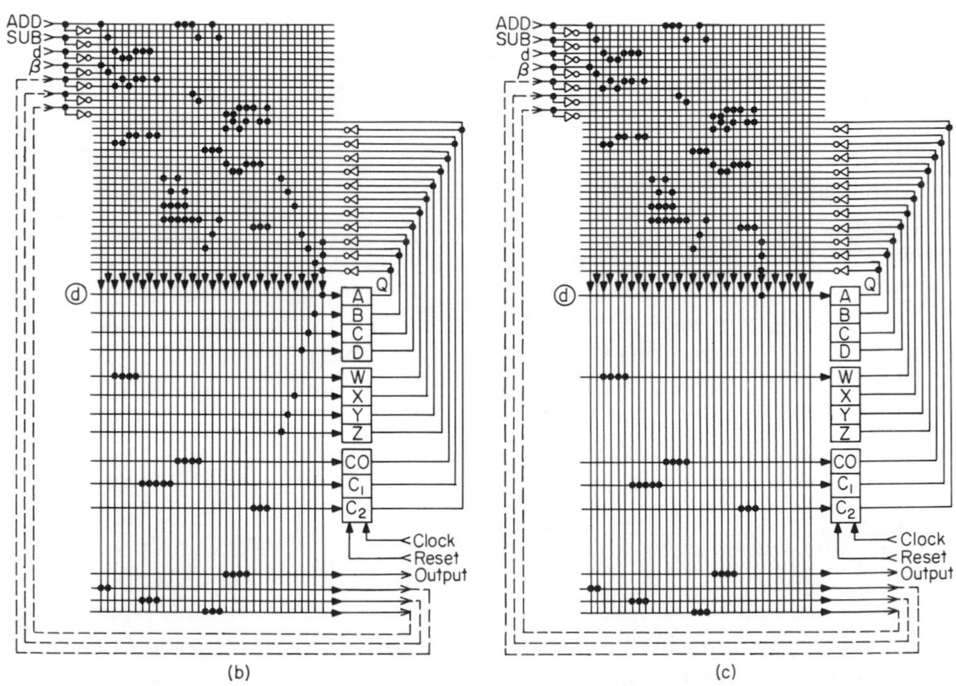

Fig. 8.19. Bit-serial adder/subtracter using PLA format (4-bits). (*a*) Functional logic net; (*b*) PLA implementation with *D*-type flip-flops only; (*c*) PLA implementation with shift registers substituted in the feedback circuit.

8.13 SUMMARY AND CONCLUSIONS

The PLA approach to sequential combinational logic design provides economic advantages over other design techniques such as geometrical layouts with random interconnects for production runs up to several tens of thousands. At some sufficiently large quantity of production, the "master slice" PLA concept may become more costly than designs in which each photomask of the process is uniquely specified. The advantages of PLA design increase with circuit function complexity. The design examples discussed in this chapter are indicative of design approaches useful for PLA implementation. The economic advantages of PLA increase further for circuit functions of higher complexity than those considered in the present chapter. The PLA approach is easy to automate with computer-aided design techniques. Logic designers can implement more easily with these arrays and check the design for errors. The ease of design may be compared with more typical random logic gate arrays in which the degree of geometrical ordering of components is considerably reduced and more difficult to check systematically for errors.

One can expect for the future that complex PLA arrays will evolve and find extensive application in MOS/LSI systems. The PLA approach provides complexities at the present time well beyond those conveniently handled through the existing logic design techniques.

The PLA configurations used in the design examples of this chapter do not show the full capabilities of the PLA approach. These examples are relatively simple and are chosen for the purpose of acquainting the reader with the design approach only. They are not illustrative of the great complexity that is possible. The maximum number of summing nodes contained in the examples of this chapter was 1,640. Since the ROM matrix section of the PLA even at this writing can contain well above 10,000 summing nodes, greatly increased logic complexity for specific designs is possible. Design techniques for the PLA system have not yet developed to maturity and one can anticipate very interesting developments as work progresses in this area.

When one requires more than two levels of random logic during a single clocking period, additional dual-level ROM matrices may be cascaded. An example of a four-level PLA system is shown in Fig. 8.20. Levels I and II of random logic are implemented in the initial PLA section with data feeding serially into a second PLA section which contains logic levels III and IV. The feedback clocking from the

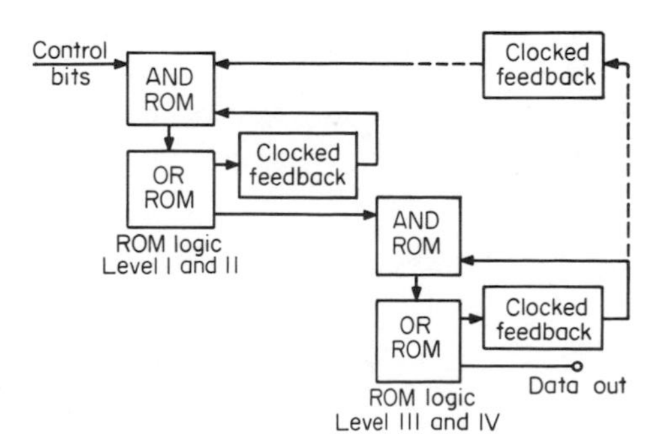

Fig. 8.20. PLA diagram for implementing four-level AND/OR logic in a complex sequential fashion.

flip-flop array can become quite complex, although a relatively simple feedback scheme is shown here. Logic designers have not developed techniques for implementing more complex PLA configurations of this type to date. The complexity of the logic function in which shift registers are introduced into the feedback loop can be increased further using the PLA basic format for the logic network.

The economy of PLA design is based on the facts that: (1) the master-chip approach permits inexpensive customizing by means of a single gate mask; (2) automated design techniques are possible, since the geometrical layout is an orderly array; and (3) the Boolean functions can be specified using automated techniques, in which a minimum number of logic terms instead of a minimum-term content is required. In applications involving a limited number of operational circuits, the PLA offers these above-mentioned economies. In large-scale production (at the present time, involving more than 50,000 units) one can anticipate that dedicated designs which are not based upon PLA layouts may provide the minimum-cost MOS/LSI design.

Logic designers can design easily into the PLA array, and further, the PLA array configuration is easier to check out for errors. Generally, the initial design in MOS/LSI suffers a high probability of prototype layout design errors. With the PLA approach it is easier to simulate the logic network using computer techniques and reduce the possibility of error during the initial mask design.

There are several techniques in which the gate mask may be effectively programed to implement the PLA-logic. The most common procedure is to design a custom gate mask for use in the photolithographic process during system fabrication. Other techniques include those of: (1) mechanically scratching the interconnects; (2) flashing the metalized interconnections electrically to the desired configuration; (3) avalanching and surface charge movement for positioning polarization charge semipermanently and thus fixing the gate threshold; and (4) ion implantation and high-energy excitation such as x-ray techniques for fixing the surface charge Q_{ss} or otherwise specifying the node points on the ROM matrices.

REFERENCES

1. F. Kvamme, Standard Read-only Memories Simplify Complex Logic Design, *Electronics,* **43:** 88–95, January 5, 1970.
2. K. Andres, MOS Programmable Logic Arrays, *Texas Instruments Application Report* CA-159, October 1970.
3. G. E. Goode, "Design of Synchronous Sequential Logic Circuits," to be published by McGraw-Hill Book Company, New York.
4. J. Irwin, MOS Shift Registers in Arithmetic Operations, *Electronic Engineer,* pp. 71–73, April 1970.

9

MOS Analog Circuitry

9.1 MOS-BIPOLAR COMPARISON FOR ANALOG APPLICATIONS

9.1.1 Comparison of g_m

A performance comparison of the MOSFET with the bipolar transistor in analog circuitry that yields hard-drawn conclusions is difficult to obtain. Problems arise in that the analysis often requires arbitrary choices of circuit configurations, voltage levels, power levels, and methods of circuit excitation from voltage or current sources. Probably the least ambiguous parameter for making the desired comparison is that of the transconductance g_m. The g_m concept has prevailed since the era of the vacuum tube and is well fixed in the system designer's repertoire, whether he is a newcomer to the field or a seasoned professional.

Let us begin then by examining the g_m of the bipolar transistor. By definition,

$$g_m \equiv \frac{\partial i_C}{\partial v_{BE}}\bigg|_Q = \frac{\partial(\alpha i_E)}{\partial v_{BE}}\bigg|_Q \approx \frac{\partial i_E}{\partial v_{BE}}\bigg|_Q = \frac{I_{EQ}}{kT/q} \tag{9-1}$$

where α is the forward current transfer ratio and is approximately equal to 1. The total instantaneous values of the collector current, emitter current and base-emitter voltage are i_C, i_E, and v_{BE}, respectively. I_{EQ} is the emitter quiescent current value. Thus for a small-signal bipolar transistor with $I_{EQ} = 10$ mA at $T = 300\,°\mathrm{K}$:

$$g_m = \frac{10 \cdot 10^{-3}}{0.026} = 0.38 \text{ mhos}$$

For a MOSFET operating in the saturation region with $I_D \approx 10$ mA, $W/L = 9$, $\mu = 150$ cm^2 per V-sec, $t_{\text{ox}} = 10^{-5}$ cm, and $|V_G - V_T| = 20$ V,

$$\begin{aligned} g_m &= \frac{\mu \epsilon_{\text{ox}} \epsilon_o W |V_G - V_T|}{t_{\text{ox}} L} \\ &= \frac{(150)(4)(8.85 \cdot 10^{-14})(9)(20)}{10^{-5}} \\ &= 0.00095 \text{ mhos} \end{aligned} \tag{1-117}$$

Thus on comparison of the two devices considered in this particular but general case, g_m for the bipolar transistor is a factor of $\approx$400 times larger than is g_m for the MOSFET.

If current is to be considered as the running variable on which a comparison of the g_m values of the two devices is to be made, then from Eqs. (9-1) and (1-117) and the expression for drain current for the MOSFET in saturation, i.e.,

$$I_D = \frac{\mu\epsilon_{\text{Ox}}\epsilon_o W}{t_{\text{Ox}}L}\frac{(V_G - V_T)^2}{2} \tag{1-108}$$

it follows that

$$\frac{g_m(\text{bipolar})}{g_m(\text{MOSFET})} \propto \frac{I_E}{\sqrt{I_D}} \tag{9-2}$$

Hence g_m values for the bipolar transistor become even more favorable in comparison to the MOSFET as conduction current increases. Furthermore it should be noted that although the g_m of the bipolar transistor is essentially fixed by nature, the g_m of the MOSFET is dependent on technology through W/L, μ, etc. It would appear, however, that no straightforward modification of MOSFET technology will serve to bring the g_m values of the MOSFET to a competitive position with those of the bipolar transistor—at least in the foreseeable future.

9.1.2 The *On* Voltage

The *on* voltage of an active device can be represented pictorially by the line *AB* of Fig. 9.1. Although *AB* exhibits a complex shape for both the MOSFET and bipolar transistor, we will for introductory purposes consider *AB* to be a simple straight-line function. Line *AB* is of importance in active-device operation since:

1. It is desirable to utilize as much of the I-V plane as possible and it is therefore

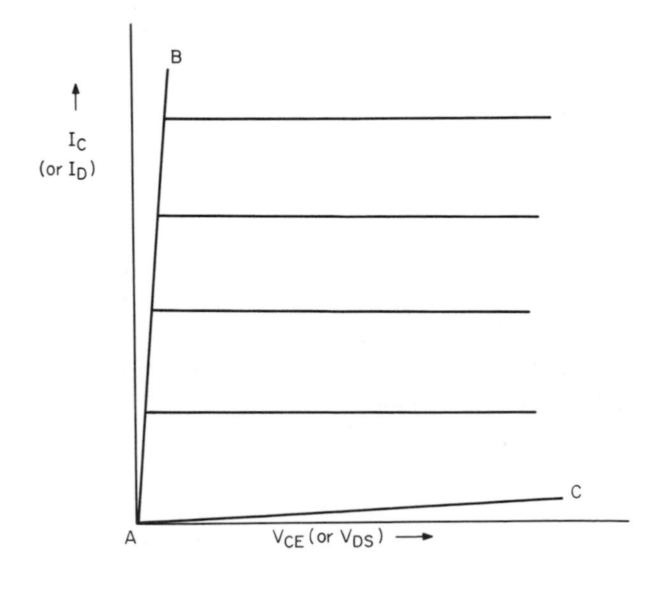

Fig. 9.1. Region of active-device operation in the I-V plane.

important that line AB be essentially vertical and of appreciable extent along the current axis.

2. A switching mode device should exhibit an *on* voltage ideally approaching zero volt in order that power dissipation in the *on* state will be low. Therefore it is again important that line AB be as close to the vertical as possible.

One of the most elegant features of the bipolar transistor is that in the presence of the proper external circuitry, the device will exhibit voltage saturation.[1] Voltage saturation is the condition in which both the emitter and collector junctions become forward biased and tend to cancel each other in voltage when the device is driven into the *on* condition (line AB). The voltage developed across the collector-emitter terminals of a medium-power transistor can typically be as low as 0.25 V at a collector-current value of 1 amp when the device is saturation.

The *on* voltage of the MOSFET as it enters its active region of saturation (note the essentially opposite meaning of the term "saturation" for the MOSFET and bipolar transistor!) is given by

$$|V_{D\,\mathrm{sat}}| = |V_G - V_T| \tag{1-107}$$

Thus for a p-channel enhancement-mode MOSFET with V_T value of -2.0 V and V_G value of -5.0 V, the *on* voltage corresponding to a point on representative line AB would be $|-3.0|$ V. The situation is not improved by utilizing a depletion-mode device. The *on* voltage could be lowered by reducing the V_G value in this example to correspond to nearly that of V_T if it were not for the fact that current flow through the device would then become negligible. It would thus appear that the *on* voltage of a MOSFET as it enters the saturation region will be at least a few volts. We have excluded from this comparison the *on* voltage of the MOSFET in the triode region. In the triode region the drain-to-source conductance appreciably shunts the output signal, thereby reducing the amplification. Therefore, only the saturation region will be treated analytically in this chapter on analog applications.

The above considerations point out that the *on* voltage of the bipolar transistor is an order of magnitude less than that of the MOSFET. The bipolar device consequently enjoys superiority with respect to this important parameter.

9.1.3 The Off Voltage

The *off* voltage of an active device can be represented pictorially by the line AC of Fig. 9.1. Line AC is of importance in active-device operation since:

1. It is desirable to utilize as much of the I-V plane as possible and therefore advantageous for AC to be essentially horizontal and of appreciable extent along the voltage axis.
2. Certain applications require a large *off*-state blocking voltage capability of the active device (e.g., sweep-drive circuitry for television receivers, 440-V motor control, ignition system control, etc.)

Modern construction of bipolar transistors employs the NPIN or PNIP configuration.[2] The I region (a high-resistivity extension of the collector body composing part of the collector-base junction) readily permits acceptance of a depletion-layer region. By this means, suitably designed bipolar transistors provide *off*-state blocking voltage capability of 1 to 2 kV with a corresponding region of active operation in the I-V plane. Little device design has been expended in extending the operation of the MOSFET to source-drain voltage levels greater than 100 V. It should be pointed out that the nature of field-plate overlap of the drain-substrate junction in the enhancement-mode device (Fig. 1.53) distorts the fringing field at the junction and thereby aggravates the problem of achieving high-voltage operation.

In summary of the *off*-state operation of the two devices, it is apparent that the bipolar transistor enjoys an order of magnitude operating voltage capability advantage over the MOSFET. The interaction of the field plate and drain-substrate junction in the present design of the enhancement-mode MOSFET does not provide inherent high-voltage capability for the device.

9.1.4 The Turn-on Voltage

By *turn-on* voltage we mean that voltage which must be applied to the emitter-base terminals of the bipolar transistor or the gate-source terminals of the MOSFET in order to place the devices at their desired quiescent points of active operation in the I-V plane. Reference to the literature[3] pertaining to bipolar-device operation reveals that the base-emitter turn-on voltage is readily controlled in device fabrication (similar to the turn-on voltage of the p-n junction diode). In integrated-circuit form the bipolar transistor turn-on voltage for adjacent devices matches to within a few millivolts. The MOSFET, however, represents an entirely different situation in that turn-on voltage (i.e., threshold voltage) is extremely process dependent. In practice it is difficult to match MOSFET threshold values of adjacent devices in integrated-circuit form to within ± 0.1 V.

The bipolar transistor thus enjoys almost a two-order-of-magnitude advantage with respect to turn-on voltage control in circuit fabrication over that of the MOSFET. This factor can be a significant detriment to the design and operation of conventional differential MOSFET amplifier configurations.

9.1.5 The Input Impedance

As a parameter, the input impedance points up the most striking difference between operation of the bipolar transistor and the MOSFET. The bipolar transistor is a current-controlled device; the MOSFET is a voltage-controlled device. In its most often used configuration (common emitter), the bipolar transistor exhibits an input impedance usually less than a few thousand Ohms. The MOSFET conversely has an inherent input impedance of $>10^{14}$ Ohms if biasing networks and high-frequency effects are neglected. At this point in the MOS-bipolar comparison we will merely note the unique input capabilities of both devices and use them accordingly in achieving useful circuit functions.

9.1.6 Summary

The above discussion brings to focus the striking advantages which the bipolar transistor enjoys in comparison to the MOSFET in conventional analog applications. The reader may well ask then, "Why continue the discussion of the MOSFET for analog circuitry?" We will see in the next section that certain unique features of the MOSFET and MOSFET circuit techniques offer promise for the device as a contender for applications in specific areas of linear and analog circuitry.

9.2 UNIQUE FEATURES OF THE MOSFET AS AN ANALOG DEVICE

9.2.1 The Input Impedance

(a) Introduction. One of the required properties of an "ideal" operational amplifier is that its input impedance be infinite. This is indeed a desirable feature to achieve in that the driving source will then not be affected by power drain at the input terminals of the amplifier. Fortunately the resistivity of the MOSFET gate insulating material SiO_2 is $>10^{16}$ Ohm-cm at $T = 300°C$, and it is not uncommon for the device to provide a dc input impedance $>10^{14}$ Ohms. Realization of the inherently high input impedance values of the MOSFET in practical circuitry does, however, pose major difficulties. The first problem encountered results from the presence of protective devices (Chap. 3) at the input gate of the MOSFET. These protective devices all employ p-n junctions in some form or other which shunt the gate-source input to the extent that their leakage currents exceed by a considerable margin that of the gate leakage current of the MOSFET. The protective devices must therefore be removed if the high input impedance of the MOSFET is to be enjoyed. This of course subjects the MOSFET to the hazards of gate-oxide rupture when the gate is subjected to excessive static charge. The protective devices can, however, be eliminated if proper attention is given to the design of the input driving circuitry and suitable precautions are taken in handling the unprotected MOSFETs (Chap. 3).

The remaining problem of achieving the high input impedance of the MOSFET in practical circuitry involves the required input biasing network for the device. The problem is best discussed by separately considering the situation as it pertains to the depletion- and enhancement-mode devices.

(b) Biasing the Enhancement-mode MOSFET to Achieve High Input Impedance. Since the enhancement-mode device is normally off, some form of biasing network must be utilized to place the device in saturation at the desired quiescent current value. Perhaps the most useful scheme for achieving these conditions is to employ the method of *diode biasing*.[4]

Diode biasing (Fig. 9.2) is easy to implement with existing MOS/LSI technology. Referring to Fig. 9.2, the parallel diodes D_1 and D_2 act as high-impedance elements in the *off* or non-forward-biased condition. The effective impedance of the diodes in the latter condition is thousands of megohms. Since, quiescently, zero current flows in the diodes, they establish the condition of $V_{GS} = V_{DS}$. The resulting quiescent point in the I-V plane is determined by the intersection of the load line with the I-V characteristic locus generated by $V_{GS} = V_{DS}$ as shown in Fig. 9.3. Since diode biasing causes the drain voltage to appear at the gate, the device is held in its saturation region of operation. The quiescent point in the I-V plane can be changed by changing the value of R_1 in Fig. 9.2.

For a single stage gain of ≈ 10 and with the input signal equal to a few millivolts or less, the diodes remain in their high-impedance state and negative feedback is

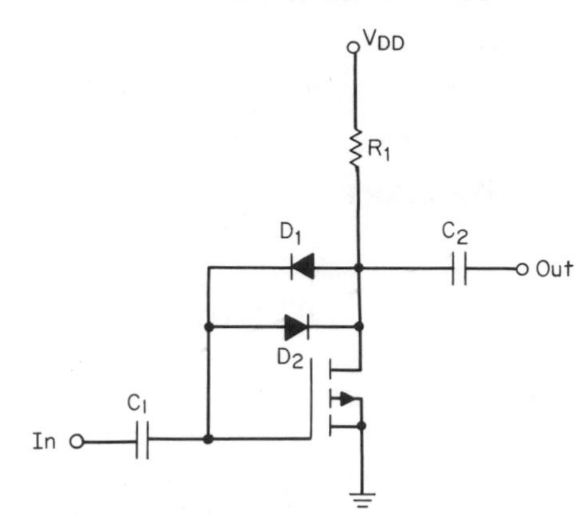

Fig. 9.2. Diode biasing the enhancement-mode MOSFET.

negligible. If however, power-supply transients occur which might tend to upset the quiescent bias point appreciably, the diodes can become active and will return the bias point to within a few hundred millivolts of the Q point.

It should be noted (Fig. 9.2) that diode biasing is applicable only to ac input coupling, thus precluding operation down to the zero-frequency level. The low-frequency cutoff point for amplifiers with diode-biased input stages has, however, been observed in the 1-Hz region for coupling capacitors of ≈ 20 pF. (A 20-pF MOS capacitor will occupy an area of $\approx 10 \times 10$ mils in integrated-circuit form.)

(c) Biasing the Depletion-mode MOSFET to Achieve High Input Impedance. Since the depletion-mode device is normally on, self-biasing schemes can be employed. In the self-biasing method, desired gate-to-source bias voltage for the device is developed by source current flow through a resistor placed in the source leg of the circuit. The scheme provides bias stability and also eliminates the power dissipation required to establish a fixed V_{GG} bias.

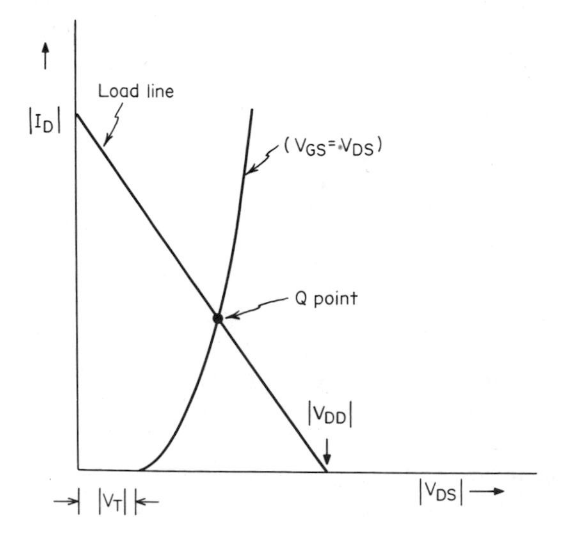

Fig. 9.3. Q point establishment with diode biasing.

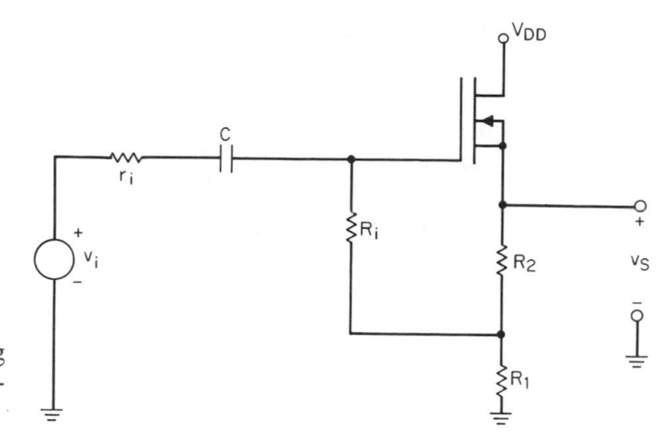

Fig. 9.4. The self-biasing method for the depletion-mode MOSFET.

One example of MOSFET self-biasing is shown in Fig. 9.4. Input impedance for the circuit of Fig. 9.4 is given by Eq. (9-3).

$$Z_i \approx \frac{R_i}{1 - \dfrac{u}{u+1} \cdot \dfrac{R_1}{1/g_m + R_1 + R_2}} \tag{9-3}$$

where $u = g_m r_{ds}$ and r_{ds} is the output impedance of the MOSFET. For operation of the circuit of Fig. 9.4

$$V_{DD} = V_{DS} + I_{DQ}(R_1 + R_2) \tag{9-4}$$

and the bias voltage is, assuming zero current flow in R_i,

$$V_{GSQ} = I_{DQ}R_2 \tag{9-5}$$

Typically V_{GSQ} will be only a few volts, whereas V_{SQ} will be roughly one-half of V_{DD}. This places the device in the desired saturation mode of operation and circuit parameters can be obtained by simultaneously employing Eqs. (9-4) and (9-5).

The self-biasing feature of the depletion-mode MOSFET thus affords a degree of freedom with respect to biasing that relieves some of the problems which conflict with achieving high input impedance. We shall find that the majority of present-day MOSFET linear and analog applications utilize depletion-mode devices.

9.2.2 The Driver-load Pair

(a) Introduction. The use of MOSFET load devices is not restricted to digital MOS circuitry. They are also employed in MOS analog circuitry. The MOSFET loads exhibit sheet resistance values at least 50 times greater than diffused resistors and can, in certain configurations, provide the ideal load characteristics of a constant-current source. These features ensure that high packing density as well as simplicity of fabrication with accompanying favorable economics will be enjoyed in MOS analog circuitry. The MOSFET driver-load pair also provides the features of gain linearity, inherent temperature stability, low-power dissipation and high gain in specific configurations. These advantages are best illustrated by separately considering the various modes of MOSFET driver-load circuitry.

(b) Enhancement-mode Driver—Enhancement-mode Load Pair. The enhancement-mode driver—enhancement-mode load configuration is shown in Fig. 9.5*a*. Typical drain characteristics of the driver device superimposed on the load line provided by the MOS load device are shown in Fig. 9.5*c*. The gain and linearity of the circuit can be obtained from the $V_{out} - V_{in}$ transfer characteristics derived in Chap. 4. It will be found that linearity over a large voltage range will be obtained when the load and driver devices are simultaneously operated in their saturation regions, because the nonlinearities of the devices will then cancel each other. This feature is established in the following derivation.

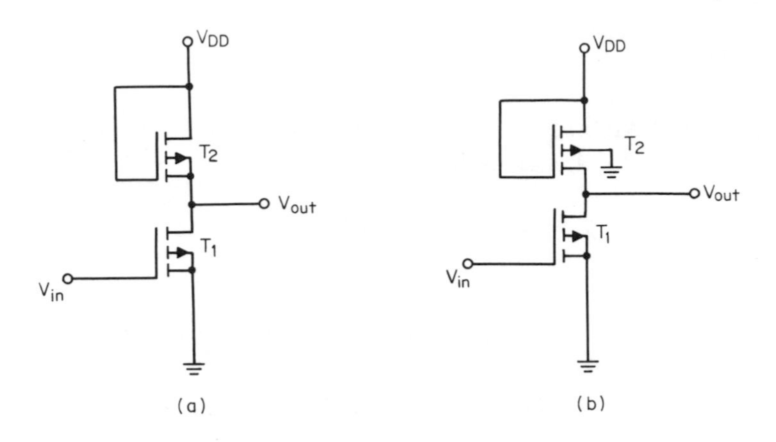

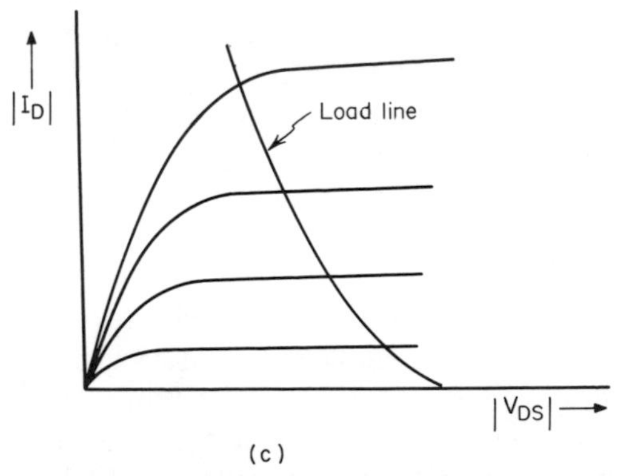

Fig. 9.5. (*a*) Discrete version of the enhancement-mode driver—enhancement-mode load configuration; (*b*) integrated circuit version of the enhancement-mode driver—enhancement-mode load configuration; (*c*) intersection of enhancement-mode MOSFET load line with driver device characteristics.

From reference to Fig. 9.5*a*, it is evident that load and driver device currents can be equated, i.e.,

$$I_{\text{driver}} = I_{\text{load}} \tag{9-6}$$

and since the devices are both postulated to be in saturation for this derivation, it follows that

$$\frac{\mu\epsilon_{\text{ox}}\epsilon_o W_{\text{driver}}}{t_{\text{ox}} L_{\text{driver}}} \cdot \frac{(V_{GS(\text{driver})} - V_T)^2}{2} = \frac{\mu\epsilon_{\text{ox}}\epsilon_o W_{\text{load}}}{t_{\text{ox}} L_{\text{load}}} \cdot \frac{(v_{GS\,(\text{load})} - V_T)^2}{2} \tag{9-7}$$

Solving Eq. (9-7) with $v_{GS\,(\text{load})} = V_{DD} - v_{\text{out}}$ as shown in Fig. 9.5*a*, we obtain

$$v_{GS\,(\text{driver})} - V_T = \sqrt{\frac{(W/L)_{\text{load}}}{(W/L)_{\text{driver}}}} \cdot (V_{DD} - v_{\text{out}} - V_T) \tag{9-8}$$

An expression for the small-signal voltage gain can be obtained by differentiating Eq. (9-8) and forming A_v, i.e.,

$$|A_v| = \left| \frac{dv_{\text{out}}}{dv_{GS(\text{driver})}} \right| = \sqrt{\frac{(W/L)_{\text{driver}}}{(W/L)_{\text{load}}}} \tag{9-9}$$

Eq. (9-9) has been derived under the assumption that the output impedances of the driver and load devices are infinite. The following important conclusions can be drawn from Eq. (9-9):

1. The temperature coefficients of both the load and driver device cancel each other. These coefficients have their origin in mobility and threshold voltage values. Gain variation is then less than 2 percent over a 105°C temperature swing.[5] This is a remarkable result when one considers that MOSFET device parameters are inherently temperature sensitive and also that no external compensating networks have been employed.
2. The nonlinearity of the load and driver devices operating in saturation has been effectively cancelled and linear amplification results.

It is to be further pointed out that practical considerations of *Q*-point establishment defined by Eq. (9-8) place constraints on the W/L values for the load-driver pair with the result that the maximum voltage gain per stage is only approximately 30.

Back-gate biasing effects (Chap. 1) must be taken into account for the load device of Fig. 9.5*b*. (Figure 9.5*a* shows the substrate returned to source as could be done when working with discrete devices.) Back-gate bias is a degenerative effect and results in gain reduction. The effect is not negligible and can reduce the inherent gain capability of the amplifier shown in Fig. 9.5*b* by a factor[6] of 2.

(c) Enhancement-mode Driver—Depletion-mode Load Pair. If the MOSFET load device is of the depletion-mode variety (Fig. 9.6*a*), it can offer essentially an infinite load impedance to the driver device without at the same time requiring an infinite supply voltage to establish a *Q* point! Another way of stating this condition is that the load device will function as a constant-current source if the *Q* point is properly established. An amplifier stage using a MOSFET depletion-mode load will in

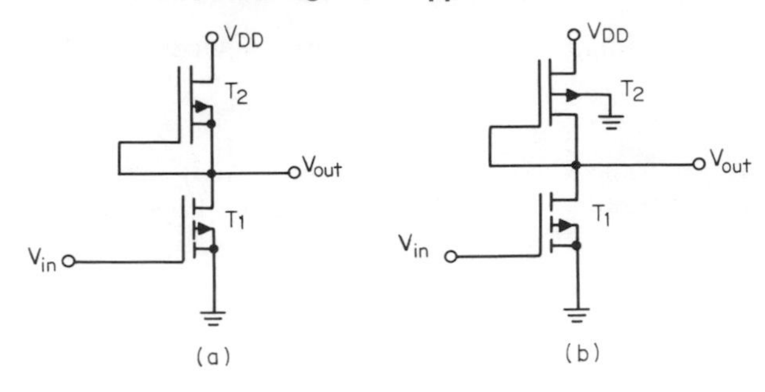

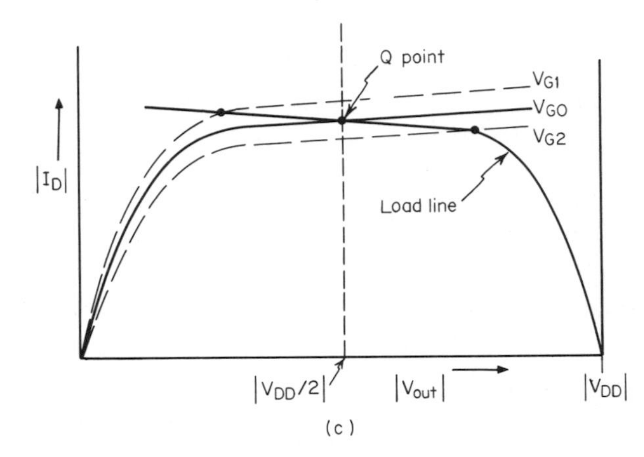

Fig. 9.6. (*a*) Discrete version of the enhancement-mode driver—depletion-mode load configuration; (*b*) integrated circuit version of the enhancement-mode driver—depletion-mode load configuration; (*c*) intersection of depletion-mode load line with driver device characteristics.

general exhibit a higher gain than that employing an enhancement-mode load. This results from the very high impedance presented by the load as shown in Fig. 9.6*c*. The load gate bias is kept constant and at zero volt by connecting the gate and source terminals together as shown in Fig. 9.6*a*.

The gain of the stage operating at point Q of Fig. 9.6*c* is given by:

$$A_v = g_{m1} \cdot \frac{r_{ds1} \cdot r_{ds2}}{r_{ds1} + r_{ds2}} \tag{9-10}$$

Since gain depends on the output impedance values of T_1 and T_2, we should therefore recall that the output impedance of the MOSFET is inversely proportional to the drain current. Low power dissipation and high gain are therefore compatible for this circuit. Typical voltage gain for the stage of Fig. 9.6*a* is, very approximately, 200. Here again the effect of back-gate bias on an integrated stage load device (Fig. 9.6*b*) can reduce the inherent gain of the configuration by as much as a factor[6] of 5.

(d) Depletion-mode Driver—Depletion-mode Load Pair. Consider for example the n-channel depletion-mode—driver-load MOSFET amplifier shown in Fig. 9.7. An

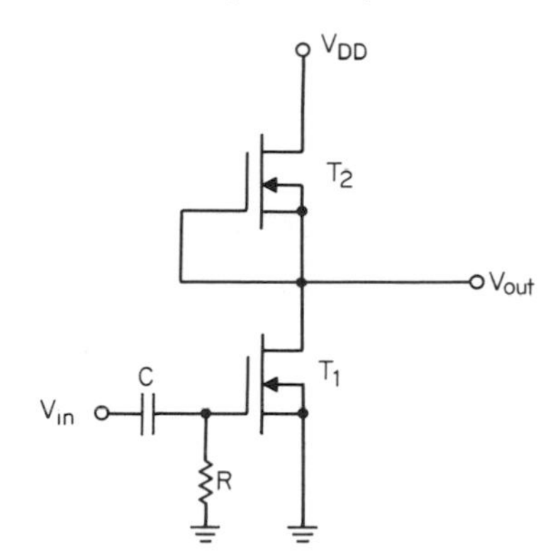

Fig. 9.7. The n-Channel depletion-mode driver—depletion-mode load configuration.

n-channel depletion-mode stage is chosen for discussion only because it is more prevalent than its p-channel analog. Assume then that T_1 and T_2 of Fig. 9.7 can be fabricated to have completely identical characteristics (g_m, r_{ds}, V_T, W/L, etc.). Under these conditions, the supply voltage will divide equally across the load and driver devices, and the voltage Q point will be located propitiously at $V_{DD}/2$. Quiescent current flow through load and driver devices will be given by:

$$I_D = \frac{\mu\epsilon_{\text{Ox}}\epsilon_o W}{t_{\text{Ox}} L} \cdot \frac{(0 - V_T)^2}{2} \tag{9-11}$$

The quiescent gate-source voltage of T_1 and T_2 is zero volt. Both devices will be in the saturation region since it is postulated that

$$\left|\frac{V_{DD}}{2}\right| > |0 - V_T| \tag{9-12}$$

Since the gate-source voltage of the load device cannot vary from zero volt, essentially an infinite load impedance is offered to the driver device (i.e., the load device serves as a constant-current source).

With T_1 and T_2 identical, the voltage gain of the stage is given by

$$A_v = \frac{g_{m1} \cdot r_{ds}}{2} \tag{9-13}$$

where $r_{ds} = r_{ds1} = r_{ds2}$

The appreciable voltage gain achievable with this configuration ($\approx$200) is readily understood by referring to Fig. 9.6*c*. To maximize the gain, the r_{ds} values must be made as large as possible; this advantageously constrains the amplifier to low-current, low-power operation. It is of further interest to note that the electrical characteristics of this gain stage (Fig. 9.7) are subtly but markedly different from those of the enhancement-mode—driver-load counterpart of Sec. 9.2.2*b* where, if T_1 and T_2 were identical, the voltage gain would be unity!

The optimum gain of the depletion-mode—driver-load pair is achieved by employing identical devices to set the Q point at $V_{DD}/2$. Since identical devices do not exist in practice, negative feedback must be employed to ensure that the Q point

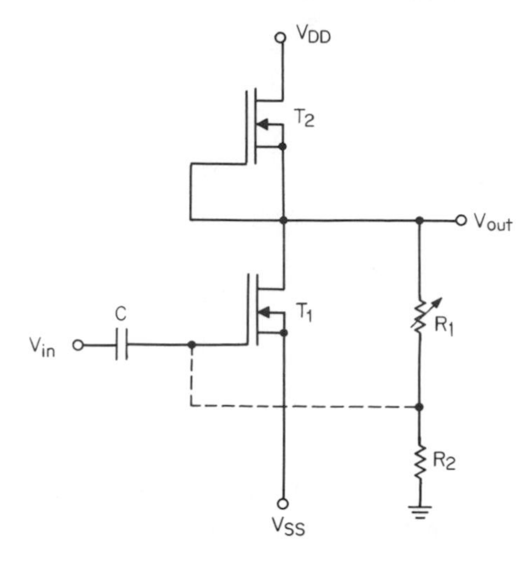

Fig. 9.8. Q point adjustment with negative feedback.

rests very nearly at $V_{DD}/2$. The configuration of Fig. 9.8 could be employed to set up the appropriate negative feedback bias voltage. Omitted for simplicity in Fig. 9.8 is the required source-follower to provide the proper impedance transfer between the feedback loop and amplifier.

Another condition which must be met in obtaining the optimum gain from the integrated-circuit version of the depletion-mode—driver-load stage is that back-gate bias effects in the load device must be eliminated. To accomplish this, the load device must be connected with gate, source, and substrate common, and of course above ground potential. This may be accomplished by diffusion isolation on epitaxial material, but this involves expensive material and process steps, tending to nullify the favorable economic aspects of MOS/LSI.

(e) The Complementary Pair. A complementary common-source MOSFET amplifier which uses enhancement-mode devices is shown in Fig. 9.9. The allowable operating limits for linear applications can be studied by superimposing the drain characteristics of the two matched complementary devices as shown in Fig. 9.10. The transfer characteristics of the complementary devices reinforce rather than cancel

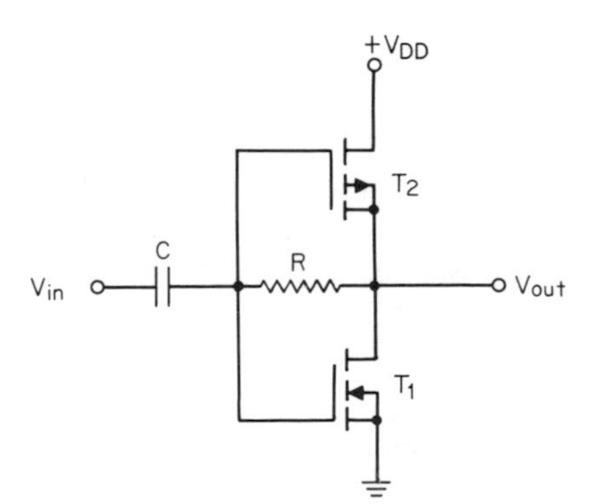

Fig. 9.9. The complementary common-source MOSFET amplifier.

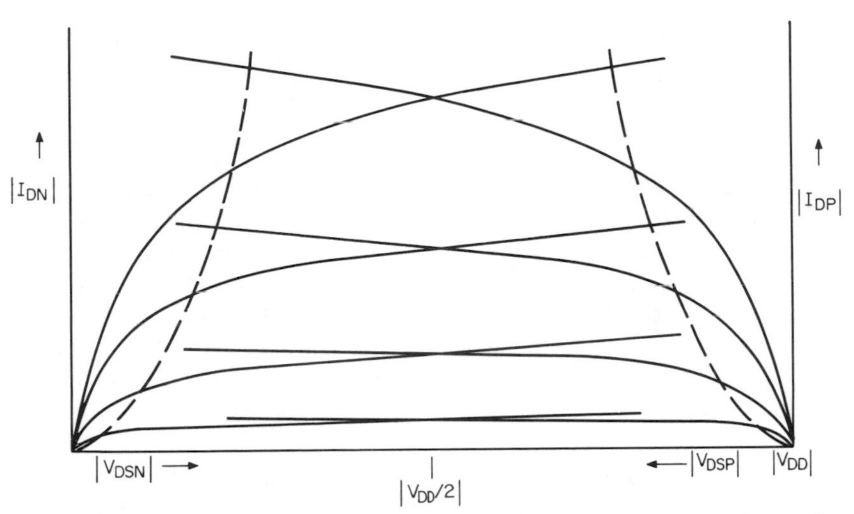

Fig. 9.10. Intersection of complementary MOSFET load lines with active device characteristics.

each other, thereby resulting in a high-voltage-gain configuration. The superposition graphically represents the relationship:

$$V_{DSP} = -(V_{DD} - V_{DSN}) \tag{9-14}$$

Note that the output voltage of a complementary-symmetry amplifier cannot swing through the full supply voltage scale without entering the triode region of one or both of the MOSFETs. For truly matched complementary devices, the drain characteristics and parameter values will be identical except for sign differences, and swing will be symmetrical about $V_{DD}/2$.

The transfer characteristic for the complementary configuration is shown in Fig. 9.11. The gain of the amplifier is equal to the slope of the transfer characteristic at the bias point. The bias point should be chosen such that $V_{out} = V_{DD}/2$ in order

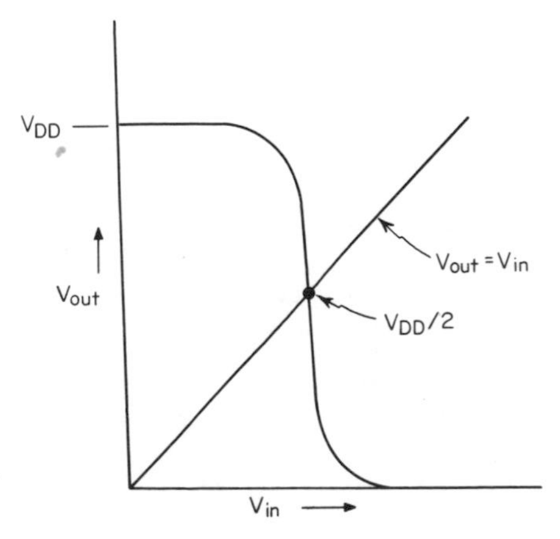

Fig. 9.11. Transfer characteristics of the complementary MOSFET configuration.

to obtain maximum gain and linearity. If the two devices are not exact complements of one another, then it is difficult to set the bias at $V_{out} = V_{DD}/2$ with the bias method of Fig. 9.9, and somewhat more complex circuitry must be employed for bias-point balance.[7]

The complementary pair has an extremely sharp transfer characteristic in its linear region. The steepest slope and therefore highest gain occurs when $V_{DD} = |V_{TP}| + V_{TN}$, where V_{TP} and V_{TN} are the p- and n-channel threshold voltages, respectively. The configuration offers low power, good linearity, and temperature stability plus an approximate voltage gain of >500 per stage. Finally, the fabrication process for complementary MOS (CMOS), with its associated isolation diffusions, permits operation without the deleterious effect of back-gate bias (Fig. 9.9). Thus CMOS analog integrated circuitry approaches the ideal, and its only serious disadvantage is the manufacturing costs that accompany the sophisticated technology required for circuit fabrication.

(f) Driver-load Pair for the Source-follower Configuration. To this point we have discussed MOSFET driver-load pairs that achieve voltage gain per stage greater than unity. It is often desirable, however, in circuit synthesis to utilize the classical source-follower configuration (the vacuum-tube version would be a cathode follower, and the bipolar version would be an emitter follower) for active impedance transformation. In the source-follower configuration the gain is less than unity.

The ideal source-follower configuration is best approached with an enhancement-mode driver T_2—depletion-mode load T_1, as shown in Fig. 9.12. An n-channel configuration is chosen for the discussion. The circuit can be physically designed (V_T, W/L, V_{DD}, etc.) so that the load device in the source leg of the driver very nearly presents the ideal "horizontal" load line (i.e., functions as a constant-current source of high impedance) for the driver. A stage gain approaching unity can thereby be achieved for the configuration.

Conversely, if the depletion-mode load device T_1 of Fig. 9.12 is replaced by an enhancement-mode load as shown in Fig. 9.13*a*, the gate-source voltage will vary in a degenerate fashion with input signal. This will result in a loss of voltage gain for the enhancement-load circuit. The degenerative effect in the enhancement-load configuration of Fig. 9.13*a* can be eliminated by employing a circuit of the type

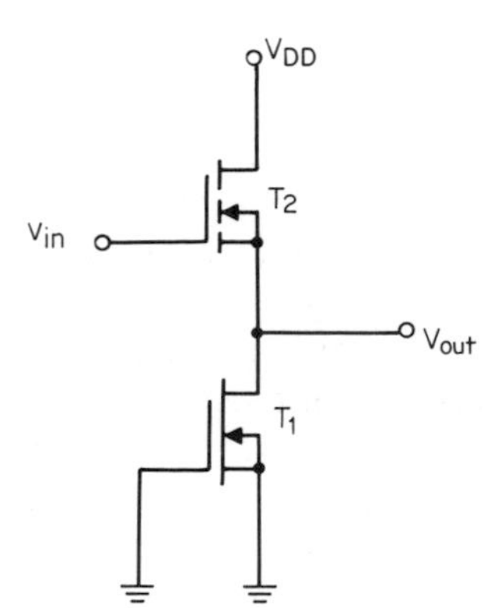

Fig. 9.12. MOSFET source follower with enhancement-mode driver and depletion-mode load.

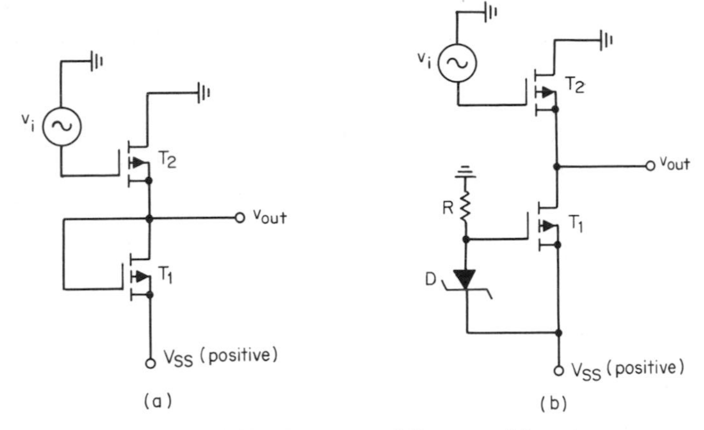

Fig. 9.13. (*a*) MOSFET source follower with enhancement-mode driver and depletion-mode load; (*b*) Zener diode biasing of enhancement-mode load for the MOSFET source follower.

shown in Fig. 9.13*b*. It will be noted in Fig. 9.13*b* that the Zener diode bias supply provides a constant gate-source voltage for the enhancement-mode load, thereby removing the bias degeneracy. The added complexity of the circuit of Fig. 9.13*b* places it at a disadvantage, however, with respect to the depletion-mode circuit of Fig. 9.12.

Back-gate bias effects in the driver device have a deleterious effect on the gain of all non-diffusion-isolated integrated-circuit versions of the source followers discussed in this section. In the integrated-circuit versions, a gain reduction as high as 50 percent can be experienced in comparison to the discrete MOS source-follower circuits of Figs. 9.12 and 9.13*a* and *b*, where the substrate of the driver device is returned directly to the source terminal of the driver device.

9.2.3 Square-law Device

(a) Introduction. The current-voltage relation for the MOSFET in saturation is ideally given by

$$I_D = \frac{\beta(V_G - V_T)^2}{2} \tag{1-116}$$

where all terms have been previously defined. In this mode of operation, the MOSFET is referred to as a *square-law* device since change in output voltage (generated by change in output current) is proportional to the square of the input voltage. An interesting but somewhat limited application for the MOSFET is thus found in circuitry which performs the *voltage-squared* function. Later in this chapter we will consider a MOSFET circuit which performs the voltage-squaring function. A more important area of application, however, which employs the square-law feature of the device is that of linear circuitry. This statement may be found somewhat surprising in view of the fact that a square-law device is basically nonlinear. It is through the square-law characteristic, however, that third and higher harmonics leading to cross-modulation and intermodulation distortion in amplifiers are eliminated, thereby resulting in utility of the MOSFET in various linear and analog circuitry. As an introduction to these concepts let us first consider the basic distortion found in the square-law device.

(b) Distortion. The total instantaneous current for the MOSFET in saturation is ideally given by

$$i_D = \frac{\beta(v_{GS} - V_T)^2}{2} \tag{9-15}$$

where V_T = threshold voltage
v_{GS} = total instantaneous value of the gate-to-source voltage

For this discussion v_{GS} is given by

$$v_{GS} = v_{GSQ} + V_{im} \cos \omega t \tag{9-16}$$

where V_{GSQ} = quiescent bias voltage
V_{im} = maximum value of time varying input voltage

Substituting Eq. (9-16) into Eq. (9-15) yields

$$i_D = \frac{\beta(V_{GSQ} + V_{im} \cos \omega t - V_T)^2}{2} \tag{9-17}$$

And then simplifying and using the identity

$$\cos^2 \omega t = \frac{1 + \cos 2\omega t}{2}$$

yields

$$i_D = \frac{\beta}{2}\left(V_{GSQ}^2 + V_T^2 - 2V_{GSQ}V_T + \frac{V_{im}^2}{2} + 2V_{im}V_{GSQ} \cos \omega t - 2V_{im}V_T \cos \omega t + \frac{V_{im}^2 \cos 2\omega t}{2}\right) \tag{9-18}$$

The last term of Eq. (9-18) indicates the presence of a second harmonic component in the output signal developed in a linear resistive load. The second-harmonic distortion D_2 can be expressed as the ratio in decibels of the amplitude of the second harmonic to the amplitude of the fundamental, and hence

$$D_2 = 20 \log \frac{V_{im}^2/2}{2V_{im}V_{GSQ} - 2V_{im}V_T} \tag{9-19}$$

Therefore

$$D_2 = 20 \log \frac{V_{im}}{4V_{GSQ} - 4V_T} \tag{9-20}$$

For an n-channel depletion-mode device with $V_T = -1$ V and $V_{GSQ} = 4$ V, D_2 of Eq. (9-20) becomes

$$D_2 = 20 \log \frac{V_{im}}{20} \tag{9-21}$$

and hence the input signal amplitude in this example must not exceed 0.2 V if the second-harmonic distortion is to be less than −40 dB.

(c) Cross-modulation and Intermodulation Distortion in the Narrow-band RF Amplifier. The square-law characteristics play a most important role in keeping the level of cross-modulation and intermodulation distortion in tuned RF amplifiers at a minimal level. In fact, these two types of distortion are essentially absent in narrow-band RF amplifiers where the ideal square-law device is employed with a linear resistive load, as will be seen from the following derivations.

Cross-modulation distortion in an amplifier is caused by the interaction of a large-amplitude interfering signal with the desired signal. Cross modulation is then, in essence, the transfer of the modulation of an undesired or interfering signal to the carrier of the desired signal. Let us examine the likelihood of this taking place in a narrow-band MOSFET amplifier stage tuned to frequency ω_1 where the desired carrier signal given by $V_1 \cos \omega_1 t$ and a large amplitude interfering signal given by $V_2(1 + m_2 \cos \omega_{m2} t) \cos \omega_2 t$ are simultaneously applied to the input of the amplifier, i.e.,

$$v_{GS} = V_{GSQ} + V_1 \cos \omega_1 t + V_2(1 + m_2 \cos \omega_{m2} t) \cos \omega_2 t \tag{9-22}$$

where V_1 = carrier amplitude of desired signal
V_2 = carrier amplitude of undesired signal
m_2 = modulation depth of interfering signal
ω_1 = carrier frequency of desired signal
ω_2 = carrier frequency of interfering signal

Substituting Eq. (9-22) into Eq. (9-15) yields

$$i_D = \frac{\beta}{2}[V_{GSQ} + V_1 \cos \omega_1 t + V_2(1 + m_2 \cos \omega_{m2} t) \cos \omega_2 t - V_T]^2 \tag{9-23}$$

Simplifying Eq. (9-23) by collecting terms in $\cos \omega_1 t$ and using the identity $2 \cos \omega_1 t \cos \omega_2 t = \cos (\omega_1 + \omega_2)t + \cos (\omega_1 - \omega_2)t$ yields

$$\begin{aligned} i_d = \beta(V_{GSQ} - V_T)V_1 \cos \omega_1 t \\ + \frac{\beta}{2} V_1 V_2[\cos (\omega_1 + \omega_2)t + \cos (\omega_1 - \omega_2)t] \\ + \frac{\beta}{2} V_1 V_2 m_2 \cos \omega_{m2} t[\cos (\omega_1 + \omega_2)t + \cos (\omega_1 - \omega_2)t] \end{aligned} \tag{9-24}$$

Now if the narrow-band tuned amplifier can exclude $\cos (\omega_1 \pm \omega_2)t$ signals, then Eq. (9-24) reduces to

$$i_D = \beta(V_{GSQ} - V_T)V_1 \cos \omega_1 t \tag{9-25}$$

and zero cross modulation occurs! Only if third and higher order cross-product terms occur will cross-modulation distortion be finite in the narrow-band amplifier. These terms do not exist for the amplifier employing an ideal square-law device.

Intermodulation distortion arises from nonlinearity of the amplifier characteristics, and depends on the input signal level. Suppose that an amplitude modulated signal

$V_1(1 + m_1 \cos \omega_{m1}t) \cos \omega_1 t$ is applied to the gate of a MOSFET amplifier stage tuned to frequency ω_1, i.e.,

$$v_{GS} = V_{GSQ} + V_1(1 + m_1 \cos \omega_{m1}t) \cos \omega_1 t \tag{9-26}$$

Substituting Eq. (9-26) into Eq. (9-15) yields

$$i_D = \frac{\beta}{2}[V_{GSQ} + V_1(1 + m_1 \cos \omega_{m1}t) \cos \omega_1 t - V_T]^2 \tag{9-27}$$

Simplifying Eq. (9-27) by collecting terms in $\cos \omega_1 t$ and using the identity $\cos^2 \omega t = \dfrac{1 + \cos 2\omega t}{2}$ yields

$$i_D = \beta(V_{GSQ} - V_T)V_1(1 + m_1 \cos \omega_{m1}t) \cos \omega_1 t \tag{9-28}$$

and no harmonics are introduced; i.e., the output signal is equivalent in harmonic content to the input signal. Again only through third and higher order cross-product terms will intermodulation distortion be finite in the narrow-band RF amplifier.

Source parasitic resistance and gate-voltage dependent mobility plus other effects which modify the ideal square-law device equation of the MOSFET can lead to third and higher order terms in the transfer characteristic. Thus in practice the MOSFET narrow-band (tuned) RF amplifier will exhibit small but nonetheless finite cross-modulation and intermodulation distortion.[8]

9.2.4 Noise Characteristics

Components of the output signal of an amplifier which are not in some manner related to the input signal are called *amplifier noise*. This noise remains when the signal is removed entirely, and thereby sets a lower limit to the strength of signals which can be amplified. We will be concerned in this section only with noise generated by fluctuating electronic phenomena in the MOSFET.

The noise figure is used for rating the performance of a device. It is defined as the available signal-to-noise ratio of the input source divided by the available signal-to-noise ratio at the device output. Although the MOSFET generates large quantities of low-frequency noise, it does have an exceptionally low noise figure for frequencies above the audio range. This characteristic coupled with low cross-modulation and intermodulation distortion makes the device very useful in RF applications.

Noise generation in the MOSFET is mainly attributed to the following three sources:

1. Thermal channel noise (Johnson noise)
2. Generation-recombination noise
3. $1/f$ noise

Let us briefly consider the nature of each of these three noise sources.

Thermal noise in the channel (Johnson noise) is caused by random thermal motion of charge carriers in the conducting channel medium. The thermal noise generated in the conducting channel gives rise to a white-noise spectrum and is approximately the equivalent of the noise generated in an external source resistor equal to $1/g_m$. The thermal-noise voltage generated in such a resistor is

$$\overline{e^2}_{th} = 4KT\frac{1}{g_m}\Delta f \tag{9-29}$$

where Δf is the bandwidth and other terms have been previously defined. The thermal noise contribution is dominant at high frequencies. For the MOSFET in saturation, the product of the equivalent input thermal noise resistance R_n* and the transconductance g_m is equal to $\frac{2}{3}$ for the special case of near-intrinsic substrate.[9] The product increases as the substrate doping level increases.[10] The substrate effect thus becomes more significant as the background doping increases and the equivalent noise resistance then becomes increasingly dependent on the effective gate-oxide voltage $(V_G - V_T)$. A thin gate oxide and large effective gate voltage make the $R_n g_m$ product approach the limit of $\frac{2}{3}$ under saturation conditions.[11]

The generation-recombination noise[12] is caused by the fluctuation of the charge carriers at the recombination and defect centers in the depletion region between the channel and the semiconductor substrate. The random generation of carriers in the channel-substrate depletion layer causes a fluctuation in the channel charge, thereby leading to noise components in the drain current. This noise source is important at intermediate-frequency values.

The $1/f$ noise[13] in the MOSFET is caused by the random fluctuations of carriers in the surface states at the silicon–silicon dioxide interface.[14] The $1/f$ noise is the dominant noise source at low frequencies and in fact dominates the noise characteristics of the MOSFET at all frequencies. Abowitz et al.[15] have shown that the equivalent $1/f$ input noise resistance is proportional to the surface-state density at the Fermi level. They point out that temperature and substrate orientation do not have a direct influence on the output noise. It is only through the influence of temperature and substrate orientation on surface-state density and position of the Fermi level at the surface that the noise is affected.

The noise-equivalent resistance for the MOSFET as a function of frequency is shown in Fig. 9.14.[10] The $1/f$ noise component for the MOSFET is considerably larger than for the JFET. In the JFET, $1/f$ noise arises from generation noise in the gate-substrate depletion region. On the other hand, the MOSFET not only has this noise component but also suffers from the component generated by charge fluctuation in the surface states at the silicon–silicon dioxide interface. It is advisable, therefore, not to utilize the MOSFET in audio applications where noise considerations are an important factor. Conversely, the shot-noise current in the MOSFET is essentially negligible in comparison to the JFET, since the gate-substrate current of the MOSFET is several orders of magnitude less than that of the JFET. The high-frequency noise performance of the MOSFET is therefore attractive and the device can compete favorably at high frequencies with low-noise bipolar transistors and JFETs.

*It is convenient to think of the noise contributed from all sources within the device as due to thermal noise generation by a resistance designated as the *noise-equivalent resistance* R_n, placed in series with the gate of the MOSFET. The thermal noise voltage generated by the equivalent noise resistance is $e^2 = 4KTR_n\Delta f$.

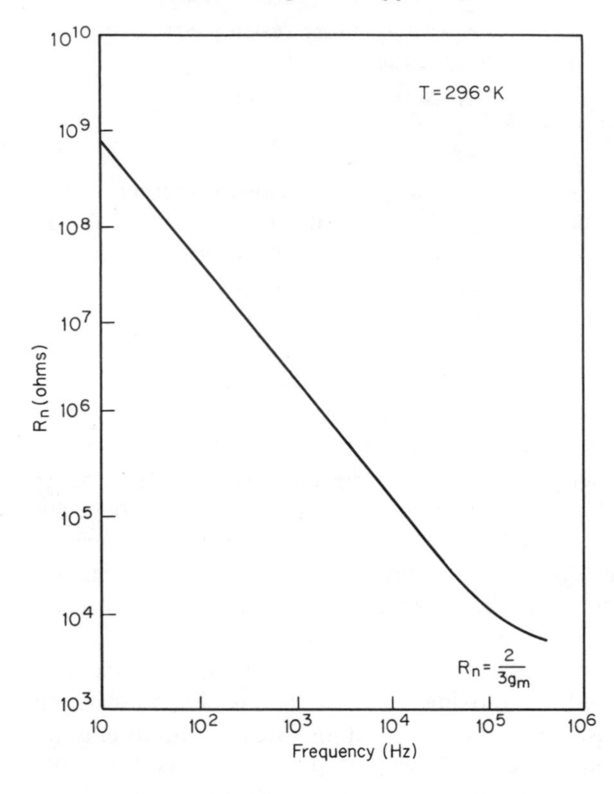

Fig. 9.14. Input noise resistance for the MOSFET as a function of frequency (p-channel device).[10]

9.2.5 The Integrable Cascode Configuration

The basic cascode configuration for the MOSFET is defined as one in which the load on a common-source stage is a common-gate stage as shown in Fig. 9.15. The cascode arrangement of Fig. 9.15 lends itself nicely to integrated-circuit technology. A cross-section of an n-channel depletion-mode cascode stage is shown in Fig. 9.16. A common n^+ region serves as both drain for T_1 and source for T_2.

The cascode configuration shown in Fig. 9.16 is also referred to as a dual-gate

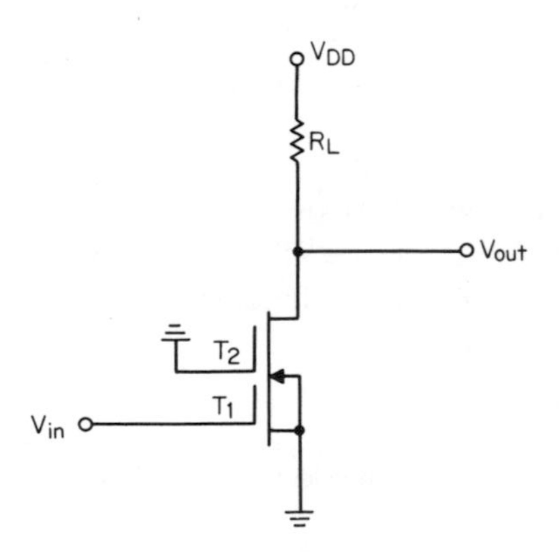

Fig. 9.15. MOSFET cascode stage featuring the dual gate.

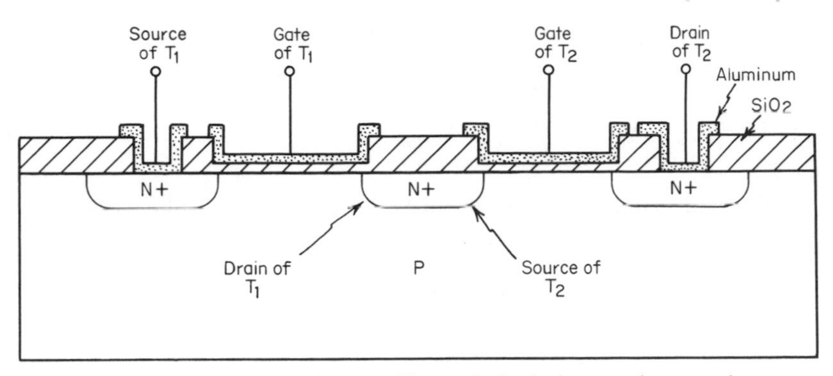

Fig. 9.16. Cross section of an n-Channel depletion-mode cascode stage.

transistor and as such exhibits all the features of a single-gate MOSFET, such as high input impedance, low noise at high frequency, and wide dynamic range. In addition, the dual-gate MOSFET exhibits better AGC capability, better cross modulation, and lower feedback capacitance than does the single-gate MOSFET.[16]

We shall defer the analysis of the cascode array to Sec. 9.3.5, and here merely note that the dual-gate MOSFET is yet another configuration which is available for MOS/LSI analog application.

9.3 MOS ANALOG BUILDING BLOCKS

9.3.1 The Common-source Amplifier

The common-source amplifier configuration is shown in Fig. 9.17. The linear, low-frequency, small-signal equivalent circuit can be obtained by expressing the drain current i_D as a function of the gate voltage v_{GS} and the drain voltage v_{DS}; i.e.,

$$i_D = f(v_{GS}, v_{DS}) \tag{9-30}$$

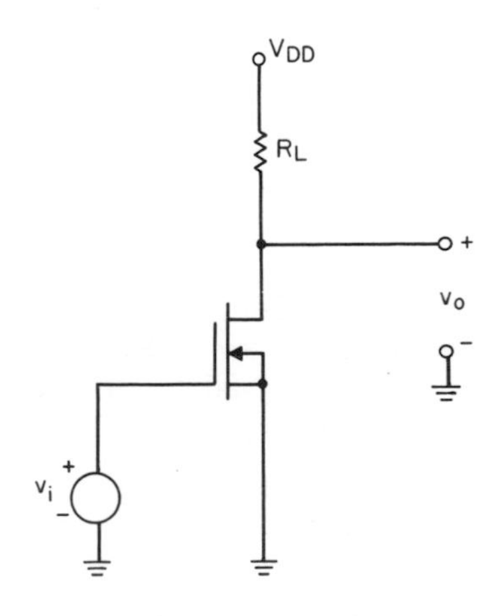

Fig. 9.17. Common-source MOSFET configuration.

If both the gate and drain voltage are varied independently through small incremental values, the change in the drain current is given by

$$\Delta i_D = \frac{\partial i_D}{\partial v_{GS}}\bigg|_{v_{DS}} \Delta v_{GS} + \frac{\partial i_D}{\partial v_{DS}}\bigg|_{v_{GS}} \Delta v_{DS} \tag{9-31}$$

where the standard notation of partial derivatives (chain rule) has been employed. In small signal notation we can write

$$\Delta i_D \approx i_d$$
$$\Delta v_{GS} \approx v_{gs}$$
$$\Delta v_{DS} \approx v_{ds}$$

and Eq. (9-31) becomes

$$i_d = \frac{\partial i_D}{\partial v_{GS}}\bigg|_{v_{DS}} v_{gs} + \frac{\partial i_D}{\partial v_{DS}}\bigg|_{v_{GS}} v_{ds} \tag{9-32}$$

Then employing the following definitions:

$$\text{Transconductance} \equiv g_m \equiv \frac{\partial i_D}{\partial v_{GS}}\bigg|_{v_{DS}} \approx \frac{i_d}{v_{gs}}\bigg|_{v_{DS}} \tag{9-33}$$

$$\text{Channel resistance} \equiv r_{ds} \equiv \frac{\partial v_{DS}}{\partial i_D}\bigg|_{v_{GS}} \approx \frac{v_{ds}}{i_d}\bigg|_{v_{GS}} \tag{9-34}$$

Equation (9-32) becomes

$$i_d = g_m v_{gs} + \frac{v_{ds}}{r_{ds}} \tag{9-35}$$

Furthermore, if the amplification factor u for the MOSFET is defined as

$$u \equiv -\frac{\partial v_{DS}}{\partial v_{GS}}\bigg|_{i_D} \approx -\frac{v_{ds}}{v_{gs}}\bigg|_{i_D} \tag{9-36}$$

then from Eq. (9-35) with $i_d = 0$ and hence $i_D =$ constant and the definition of Eq. (9-36), we can write

$$u = g_m r_{ds} \tag{9-37}$$

We are now in a position to draw the small-signal equivalent circuit for the common-source configuration. Equation (9-35) can be rewritten as

$$-i_d r_{ds} + g_m v_{gs} r_{ds} + v_{ds} = 0 \tag{9-38}$$

and Eq. (9-38) can be expressed in a circuit as shown in Fig. 9.18*a*. Combining the results shown in Fig. 9.18*a* with the circuit of Fig. 9.17 gives the desired small-signal equivalent circuit as shown in Fig. 9.18*b*. Infinite gate input resistance has been assumed.

A small-signal equivalent circuit which uses a voltage generator rather than the current generator of Fig. 9.18*b* can be developed by combining Eqs. (9-37) and (9-38) to obtain

$$-i_d r_{ds} + u v_{gs} + v_{ds} = 0 \tag{9-39}$$

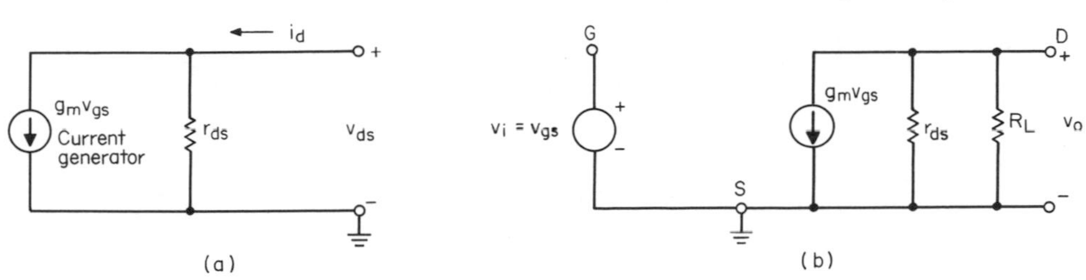

Fig. 9.18. (*a*) Small-signal equivalent circuit of Eq. (10.38); (*b*) small-signal equivalent circuit for the common-source configuration employing a current generator.

Eq. (9-39) can then be expressed in circuit form as shown in Fig. 9.19*a*. Combining the results shown in Fig. 9.19*a* with the circuit of Fig. 9.17 thus gives a second version of the small-signal equivalent circuit of the MOSFET common-source amplifier with infinite gate-input resistance (Fig. 9.19*b*). Either of the two equivalent circuits can be adopted for use in the following derivations.

The voltage gain of the configuration under discussion can thus be derived by referring to Fig. 9.18*b*. We can write by inspection that

$$A_v = \frac{v_o}{v_i} = \frac{-1}{v_{gs}} \cdot \frac{g_m v_{gs} r_{ds}}{r_{ds} + R_L} \cdot R_L \tag{9-40}$$

since $v_i = v_{gs}$. Therefore

$$A_v = -g_m(R_L \| r_{ds}) \tag{9-41}$$

As an example consider

$$\begin{aligned} g_m &= 20 \cdot 10^{-3} \text{ mhos} \\ r_{ds} &= 10^4 \text{ Ohms} \\ R_L &= 10^4 \text{ Ohms} \end{aligned}$$

Then from Eq. (9-41),

$$A_v = -\frac{20 \cdot 10^{-3} \cdot 10^8}{10^4 + 10^4} = -100$$

The negative sign indicates that the output signal has been inverted and is out of phase by 180° with respect to the input signal. Voltage gain for the configuration

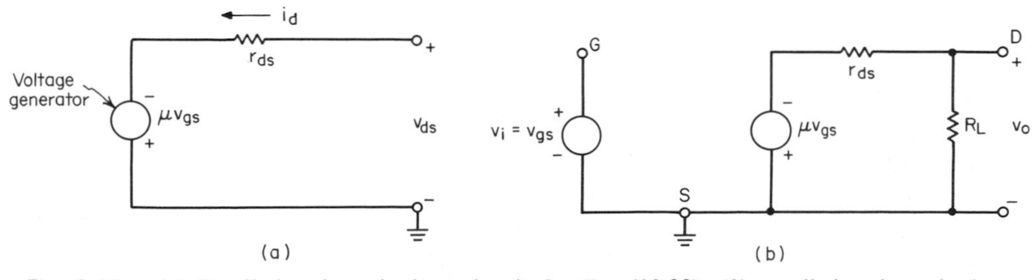

Fig. 9.19. (*a*) Small-signal equivalent circuit for Eq. (10.39); (*b*) small-signal equivalent circuit for the common-source configuration employing a voltage generator.

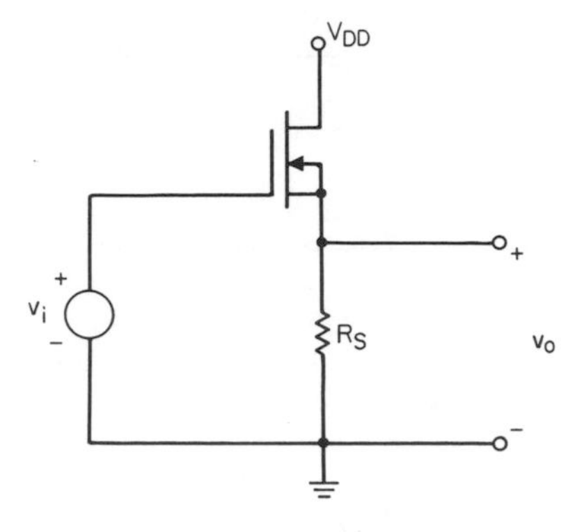

Fig. 9.20. Common-drain (source-follower) MOSFET configuration.

is appreciable. (The power gain approaches infinity since we have assumed that the input impedance is infinite.)

9.3.2 The Common-drain Amplifier (Source Follower)

The common-drain amplifier configuration is shown in Fig. 9.20. Let us derive the amplification of this configuration. From Figs. 9.18*a* and 9.20 the complete small-signal equivalent circuit can be derived as shown in Fig. 9.21*a*. Transforming from a Norton equivalent to a Thevenin equivalent, Fig. 9.21*a* becomes Fig. 9.21*b*, and we can write

$$i_s = \frac{g_m v_{gs} r_{ds}}{R_S + r_{ds}} \tag{9-42}$$

Since

$$v_{gs} = v_i - R_S i_s \tag{9-43}$$

then combining Eqs. (9-42) and (9-43) yields

$$v_o = i_s R_S = \frac{R_s g_m r_{ds}(v_i - R_S i_s)}{R_S + r_{ds}} \tag{9-44}$$

Equation (9-44) yields the expression for voltage gain

$$A_v = \frac{v_o}{v_i} = \frac{g_m R_S r_{ds}}{R_S + r_{ds} + R_S g_m r_{ds}} \tag{9-45}$$

For the practical case where $r_{ds} \gg R_S$, Eq. (9-45) becomes

$$A_v \approx \frac{g_m R_S}{1 + g_m R_S} \tag{9-46}$$

As an example consider

$$g_m = 20 \cdot 10^{-3} \text{ mhos}$$
$$R_S = 200 \text{ Ohms}$$

Then from Eq. (9-46),

$$A_v \approx \frac{20 \cdot 10^{-3} \cdot 200}{1 + 20 \cdot 10^{-3} \cdot 200} = 0.8$$

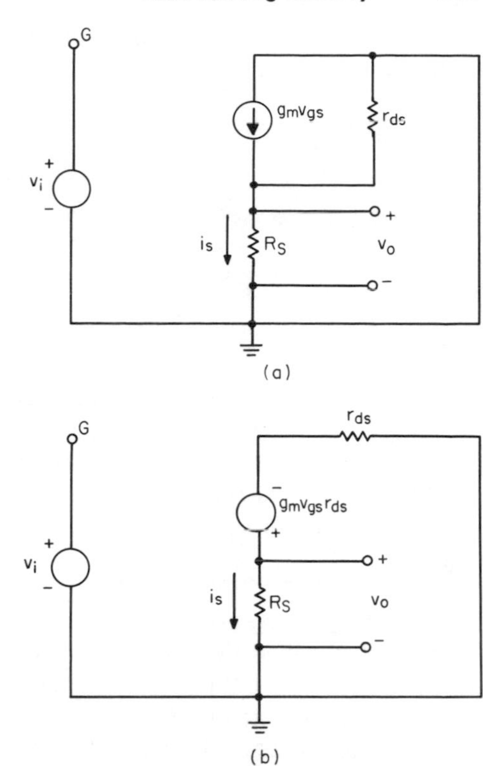

Fig. 9.21. (*a*) Small-signal equivalent circuit for the common-drain configuration; (*b*) transformed small-signal equivalent circuit for the common-drain configuration.

The output impedance of the source follower can be calculated with the gate grounded by applying a signal to the output terminals of Fig. 9.21*a* and forming the resulting voltage-current ratio. This operation is readily performed by rearranging the circuit of Fig. 9.21*b* to that shown in Fig. 9.22 and noting that $v_{gs} = -v_o$. Then writing the Kirchoff loop equations for Fig. 9.22 as

$$i_1r_{ds} + v_og_mr_{ds} + (i_1 - i_2)R_S = 0 \tag{9-47}$$

$$v_o + R_S(i_2 - i_1) = 0 \tag{9-48}$$

and solving Eqs. (9-47) and (9-48) simultaneously, yields the output impedance Z_o. Thus

$$Z_o = \frac{v_o}{-i_2} = \frac{R_Sr_{ds}}{r_{ds} + g_mr_{ds}R_S + R_S} = R_S\|r_{ds}\|\frac{1}{g_m} \tag{9-49}$$

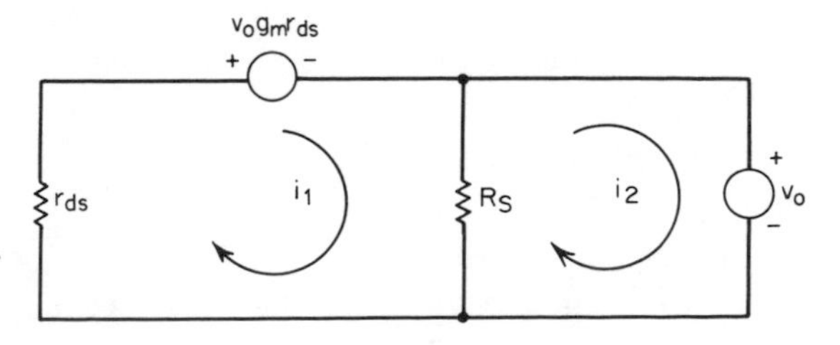

Fig. 9.22. Circuit for analysis of the common-drain configuration.

As an example of a typical output impedance value for the common-drain configuration, consider

$$R_S = 100 \text{ Ohms}$$
$$r_{ds} = 10^4 \text{ Ohms}$$
$$g_m = 20 \cdot 10^{-3} \text{ mhos}$$

Then from Eq. (9-49),

$$Z_o = \frac{10^2 \cdot 10^4}{10^4 + 2 \cdot 10^{-2} \cdot 10^6 + 100} \approx 30 \text{ Ohms}$$

and

$$Z_o < R_S$$

In summary of the common-drain configuration, note that the output signal is in phase with respect to the input signal, input impedance is high, and output impedance is low, being less than the source resistance. Although the voltage gain will always be less than unity, the configuration finds extensive use as an active impedance transformer.

9.3.3 The Common-gate Amplifier

The circuit configuration for the common-gate amplifier is shown in Fig. 9.23. The small-signal equivalent circuit for the configuration is obtained by combining the circuitry of Figs. 9.18*a* and 9.23, and is shown in Fig. 9.24*a*. Transforming from a Norton equivalent to a Thevenin equivalent, Fig. 9.24*a* becomes Fig. 9.24*b*. We can write the Kirchoff loop equation for the circuit of Fig. 9.24*b* with current flow i indicated as

$$iR_L - v_i + iR_S + g_m v_{gs} r_{ds} + ir_{ds} = 0 \tag{9-50}$$

Also note that

$$v_{gs} = iR_S - v_i \tag{9-51}$$

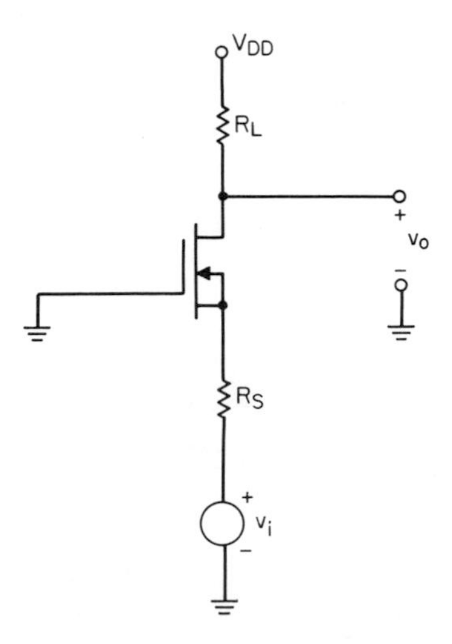

Fig. 9.23. Common-gate MOSFET configuration.

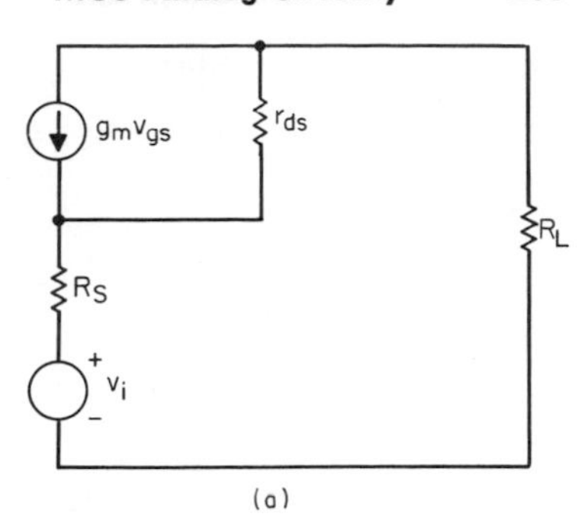

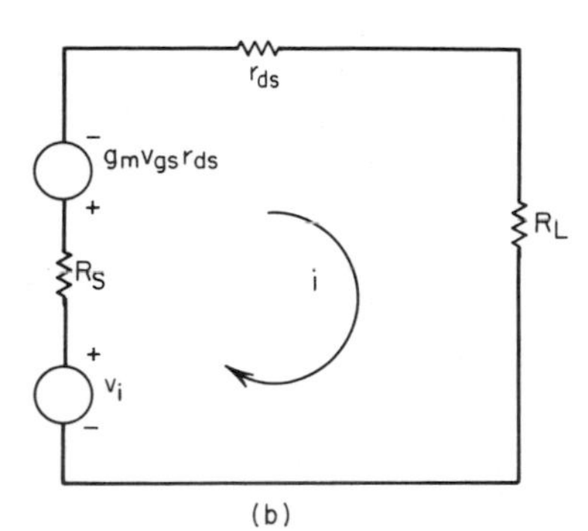

Fig. 9.24. (*a*) Small-signal equivalent circuit for the common-gate configuration; (*b*) transformed small-signal equivalent circuit for the common-gate configuration.

Then combining Eqs. (9-50) and (9-51) we obtain

$$i = \frac{v_i(1 + g_m r_{ds})}{R_L + R_S + r_{ds} + g_m r_{ds} R_S} \tag{9-52}$$

The input impedance for the configuration then becomes

$$Z_i = \frac{v_i}{i} = \frac{R_L + R_S + r_{ds} + g_m r_{ds} R_S}{1 + g_m r_{ds}} \tag{9-53}$$

The voltage gain of the configuration is obtained by utilizing Eq. (9-52) and writing

$$A_v = \frac{v_o}{v_i} = \frac{iR_L}{v_i} = \frac{R_L(1 + g_m r_{ds})}{R_L + r_{ds} + (1 + g_m r_{ds})R_S} \tag{9-54}$$

As an example calculation for this configuration consider

$$R_L = 10^4 \text{ Ohms}$$
$$g_m = 20 \cdot 10^{-3} \text{ mhos}$$
$$r_{ds} = 10^4 \text{ Ohms}$$
$$R_S = 100 \text{ Ohms}$$

Then the voltage gain is given from Eq. (9-54) as

$$A_v = \frac{10^4(1 + 20 \cdot 10^{-3} \cdot 10^4)}{10^4 + 10^4 + (1 + 20 \cdot 10^{-3} \cdot 10^4)100} \approx 50$$

and the input impedance is given from Eq. (9-53) as

$$Z_i = \frac{10^4 + 10^2 + 10^4 + 20 \cdot 10^{-3} \cdot 10^4 \cdot 100}{1 + 20 \cdot 10^{-3} \cdot 10^4} \approx 200 \text{ Ohms}$$

The common-gate configuration is thus characterized by low input impedance, high voltage gain, and zero phase change between input and output signals.

9.3.4 The Phase Splitter

A phase-splitting circuit ideally performs the function of transforming a given input signal to two signals 180° apart in phase but of equal amplitude. The MOSFET circuit for performing this function is shown in Fig. 9.25. The voltage gain at the source terminal can be calculated by referring to the small-signal equivalent circuit for the configuration as shown in Fig. 9.26*a*. Transformation of the Norton equivalent portion of the circuit to a Thevenin equivalent has been performed in Fig. 9.26*b*. Then writing the Kirchoff loop equation for the circuit of Fig. 9.26*b* yields

$$iR_L + iR_S + g_m v_{gs} r_{ds} + i r_{ds} = 0 \tag{9-55}$$

The small-signal gate to source voltage v_{gs} is given by

$$v_{gs} = v_i + iR_S \tag{9-56}$$

Combining Eqs. (9-55) and (9-56) and forming the ratio of v_{o1} to v_i yields an expression for the voltage gain at the drain terminal A_{v1}, i.e.,

$$A_{v1} = \frac{v_{o1}}{v_i} = \frac{iR_L}{v_i} = -\frac{R_L g_m r_{ds}}{R_L + R_S + r_{ds}(1 + g_m R_S)} \tag{9-57}$$

Combining Eqs. (9-55) and (9-56) and forming the ratio of v_{o2} to v_i yields an expression for the voltage gain at the source terminal A_{v2}, i.e.,

$$A_{v2} = \frac{v_{o2}}{v_i} = \frac{-iR_S}{v_i} = \frac{R_S g_m r_{ds}}{R_L + R_S + r_{ds}(1 + g_m R_S)} \tag{9-58}$$

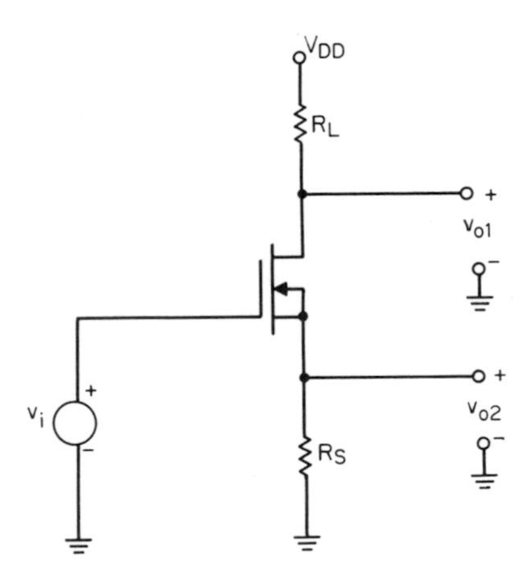

Fig. 9.25. The MOSFET phase-splitting circuit.

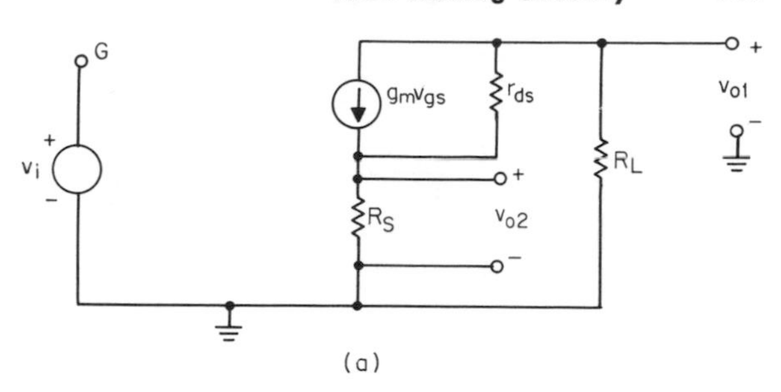

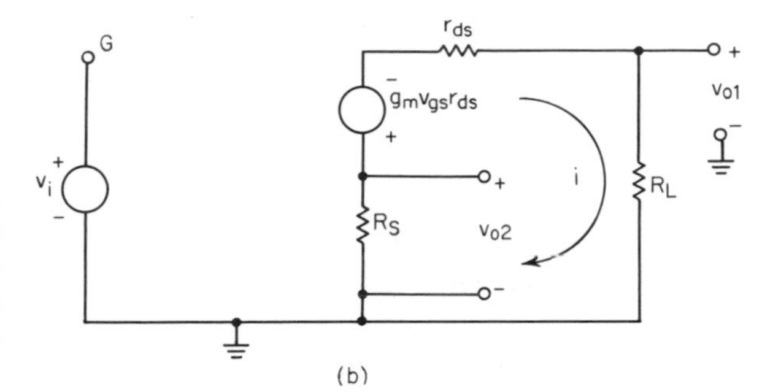

Fig. 9.26. (*a*) Small-signal equivalent circuit for the phase-splitting circuit; (*b*) transformed small-signal equivalent circuit for the phase-splitting circuit.

Let us consider the following example for calculation of A_{v1} and A_{v2} with

$$R_S = R_L = 10^3 \text{ Ohms}$$
$$g_m = 20 \cdot 10^{-3} \text{ mhos}$$
$$r_{ds} = 10^4 \text{ Ohms}$$

then from Eq. (9-57),

$$A_{v1} = -\frac{10^3 \cdot 20 \cdot 10^{-3} \cdot 10^4}{10^3 + 10^3 + 10^4(1 + 20 \cdot 10^{-3} \cdot 10^3)} \approx -0.95$$

and from Eq. (9-58),

$$A_{v2} = \frac{10^3 \cdot 20 \cdot 10^{-3} \cdot 10^4}{10^3 + 10^3 + 10^4(1 + 20 \cdot 10^{-3} \cdot 10^3)} \approx +0.95$$

We thus conclude that for the phase-splitting circuit to accomplish its desired function, R_S must equal R_L and this constraint in turn results in a voltage gain at either output terminal being less than unity.

9.3.5 The Cascode Pair

(a) High-frequency Behavior. A compound type of MOSFET connection referred to as a cascode configuration is shown in Fig. 9.27. In the configuration a common-source stage is followed by a common-gate stage. The configuration offers desirable high-frequency characteristics which will be discussed upon completion of a frequency-independent small-signal analysis of the voltage gain.

The small-signal equivalent circuit for the cascode configuration is obtained by combining the small-signal equivalent circuits for the common-source (Fig. 9.18*b*)

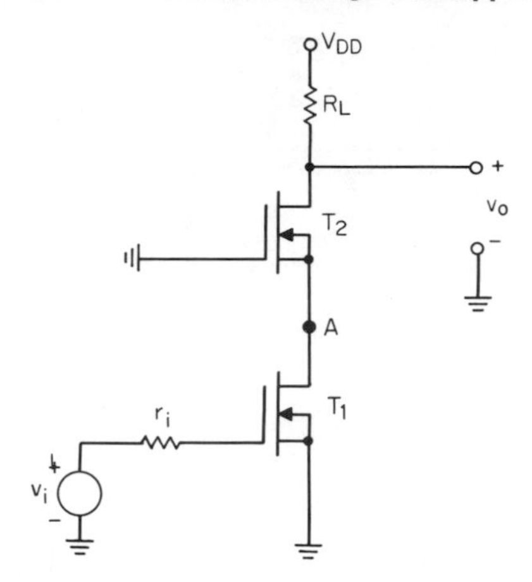

Fig. 9.27. The MOSFET cascode configuration (discrete version).

and the common-gate (Fig. 9.24*a*) configurations. The results are shown in Fig. 9.28*a*. Note reference point *A* in Figs. 9.27 and 9.28*a*. The Norton equivalent portions of the circuit have been converted to Thevenin equivalents as shown in Fig. 9.28*b*. Writing the Kirchoff loop equation for the circuit of Fig. 9.28*b* yields

$$iR_L + ir_{ds2} - g_{m2}r_{ds2}(g_{m1}r_{ds1}v_i - ir_{ds1}) + ir_{ds1} - g_{m1}r_{ds1}v_i = 0 \qquad (9\text{-}59)$$

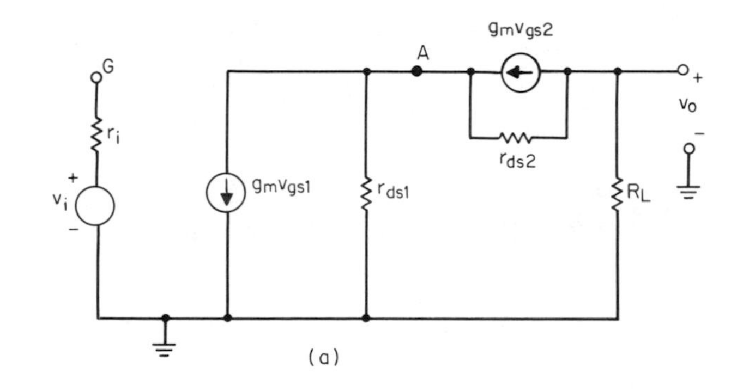

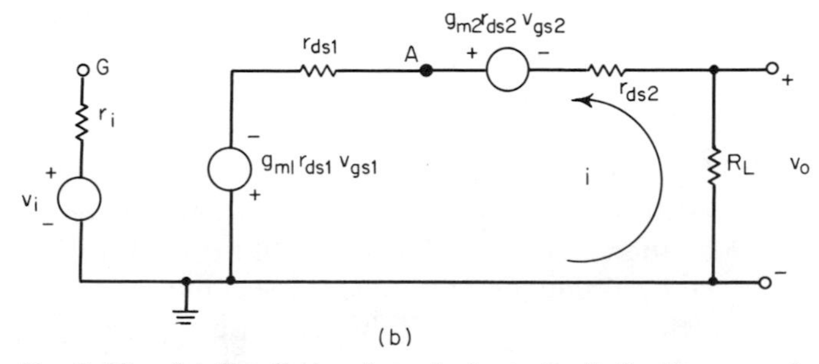

Fig. 9.28. (*a*) Small-signal equivalent circuit for the cascode configuration; (*b*) transformed small-signal equivalent circuit for the cascode configuration.

where the expressions

$$v_{gs1} = v_i \tag{9-60}$$

and

$$v_{gs2} = g_{m1} r_{ds1} v_i - i r_{ds1} \tag{9-61}$$

have been employed. The small-signal gain for the configuration can be obtained by combining Eq. (9-59) with the expression for the output voltage:

$$v_o = -iR_L \tag{9-62}$$

and thus

$$A_v = \frac{v_o}{v_i} = -R_L \frac{g_{m1} r_{ds1} + g_{m1} g_{m2} r_{ds1} r_{ds2}}{R_L + r_{ds1} + r_{ds2} + g_{m2} r_{ds1} r_{ds2}} \tag{9-63}$$

As an example of a voltage gain calculation for the cascode stage, consider the following:

$$R_L = 10^3 \text{ Ohms}$$
$$g_{m1} = g_{m2} = 20 \cdot 10^{-3} \text{ mhos}$$
$$r_{ds1} = r_{ds2} = 10^4 \text{ Ohms}$$

and therefore from Eq. (9-63),

$$A_v = -10^3 \frac{20 \cdot 10^{-3} \cdot 10^4 + 20 \cdot 10^{-3} \cdot 20 \cdot 10^{-3} \cdot 10^4 \cdot 10^4}{10^3 + 10^4 + 10^4 + 20 \cdot 10^{-3} \cdot 10^4 \cdot 10^4} \approx -20$$

It is instructive to calculate the voltage gain of the cascode stage by separately calculating the gain of the common-source section and multiplying the obtained value by the calculated gain of the common-gate section. We will perform the calculation utilizing the above circuit parameters. We first note that the combination of T_2 and R_L presents an effective load impedance to T_1 given by Eq. (9-53) as

$$Z_L' = \frac{10^3 + 0 + 10^4 + 0}{1 + 20 \cdot 10^{-3} \cdot 10^4} \approx 50 \text{ Ohms}$$

and hence the gain of the common-source section is from Eq. (9-41):

$$A_v|_{\text{common source}} = -20 \cdot 10^{-3}(50 \| 10^4) \approx -1$$

The gain of the common-gate section is obtained from Eq. (9-54) as

$$A_v|_{\text{common gate}} = \frac{10^3(1 + 20 \cdot 10^{-3} \cdot 10^4)}{10^3 + 10^4 + 0} \approx 20$$

The overall gain for the cascode stage is:

$$A_v = A_v|_{\text{common source}} \times A_v|_{\text{common gate}} = (-1)(20) = -20$$

The result is of course in agreement with our previously calculated example of the voltage gain for the cascode stage using a small-signal equivalent circuit for the entire stage.

We note from these calculations that a voltage gain of -20 was obtained for a specific cascode pair; whereas, if only a single common-source stage had been

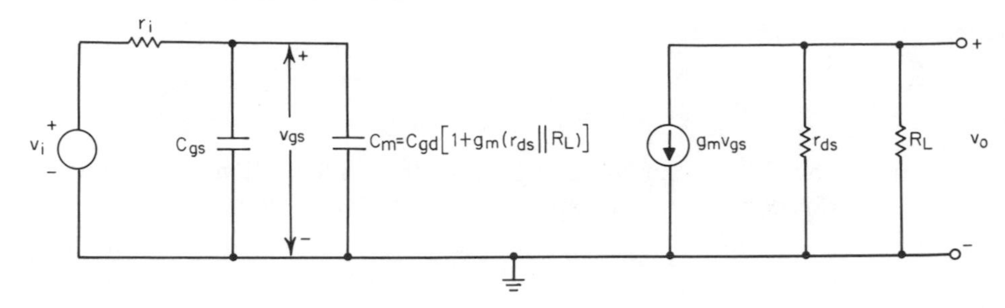

Fig. 9.29. The small-signal, high-frequency model for the common-source configuration.

employed with a 10^3-Ohm load resistor and $g_m = 20 \cdot 10^{-3}$ mhos, a voltage gain of ≈ -20 would also have been developed. Thus the cascode array has required two active devices to provide the gain achievable with one active device in common-source configuration—at least as far as frequency-independent gain is concerned. It is, however, at high frequencies that the cascode configuration finds important application.

To understand the major characteristics of the cascode array at high frequencies, recall that the high-frequency small-signal model for the common-source configuration is given as shown in Fig. 9.29.[17] Note in Fig. 9.29 how the Miller capacitance C_M reflects into the input circuit and hence causes a deleterious input voltage division of v_i. The Miller capacitance C_M can be made small if the R_L of Fig. 9.29 is small. R_L appears small to the T_1 section of the cascode pair (Fig. 9.27) since T_1 "looks" into a common-gate configuration T_2. Although T_1 provides only unity gain, it does offer the desired high input impedance to the incoming signal. The gate of T_2 is grounded, thereby eliminating the negative feedback (Miller effect) caused by the gate-drain capacitance of T_2. T_2 thus provides voltage gain in the absence of the Miller effect for the cascode stage. In essence, the cascode stage enjoys all the attributes of the common-source stage with the additional feature that Miller capacitance is greatly reduced, thereby extending the voltage gain to high frequencies.

It is to be understood, of course, that the above calculations have been performed with the substrate of T_2 (Fig. 9.27) connected to the source of T_2. In conventional MOS/LSI configuration, the substrate of T_2 is grounded, and appropriate back-gate bias effects must be taken into account throughout the derivations.

(b) Dual Gate Utilization. If the gate lead of T_2 (Fig. 9.27) is lifted from ground, the cascode stage becomes an active *tetrode* device. Signals can then be "mixed" under dc isolation conditions by employing the dual gates of the structure. Practical cases of high-frequency mixers and RF amplifier-AGC configurations will be presented in Sec. 9.4.6.

9.3.6 The Differential Pair.

(a) The Basic Squaring Circuit. As an introduction to the MOSFET differential amplifier, consider the pairing arrangement shown in Fig. 9.30. If we assume that T_1 and T_2 are matched and are set at identical Q point values, and that the output impedance of the devices is infinite, then for operation in the saturation region the current flow through R_L is given by

$$i_L = i_{D1} + i_{D2} \tag{9-64}$$

and therefore

$$i_L = \frac{\beta(v_{GS1} - V_T)^2}{2} + \frac{\beta(v_{GS2} - V_T)^2}{2} \tag{9-65}$$

If the input signals are of equal amplitude but 180° different in phase, Eq. (9-65) reduces to

$$i_L = \beta(V_Q^2 - 2V_QV_T + v_i^2 + V_T^2) \tag{9-66}$$

where $v_{GS1} = V_{Q1} + v_i$
$v_{GS2} = V_{Q2} - v_i$
$V_{Q1} = V_{Q2} = V_Q$

The dc components are removed from the signal appearing at the drain of T_1 and T_2 by taking the output signal through capacitor C. The output voltage thus becomes

$$v_o = -\beta R_L v_i^2 \tag{9-67}$$

The configuration of Fig. 9.30 can therefore be used as a squaring circuit and finds application in numerical calculations by analog methods, measurement of power in complex waveforms, and in noise measurements.

(b) The MOSFET Differential Amplifier. The circuit of Fig. 9.30 can be modified to extract the signal in true differential form as shown in Fig. 9.31*a*. The low-frequency small-signal equivalent circuit for Fig. 9.31*a* is shown in Fig. 9.31*b*. The Norton equivalent portions of the circuit of Fig. 9.31*b* appear as Thevenin equivalent configurations in Fig. 9.31*c*. From Fig. 9.31*c* we can write

$$v_{gs1} = v_{i1} - R_S(i_1 - i_2) \tag{9-68}$$
$$v_{gs2} = v_{i2} - R_S(i_1 - i_2) \tag{9-69}$$

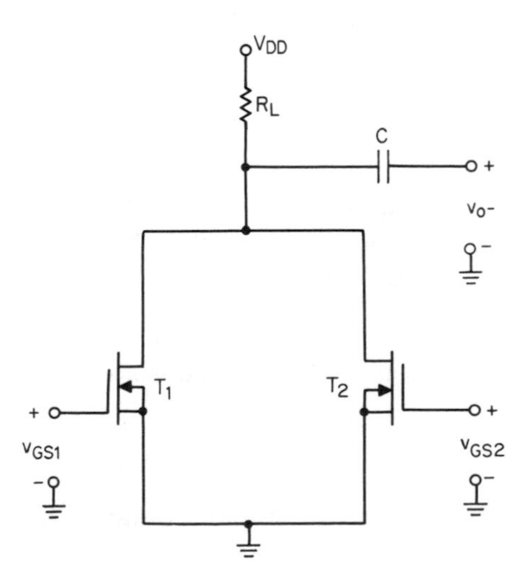

Fig. 9.30. MOSFET squaring circuit.

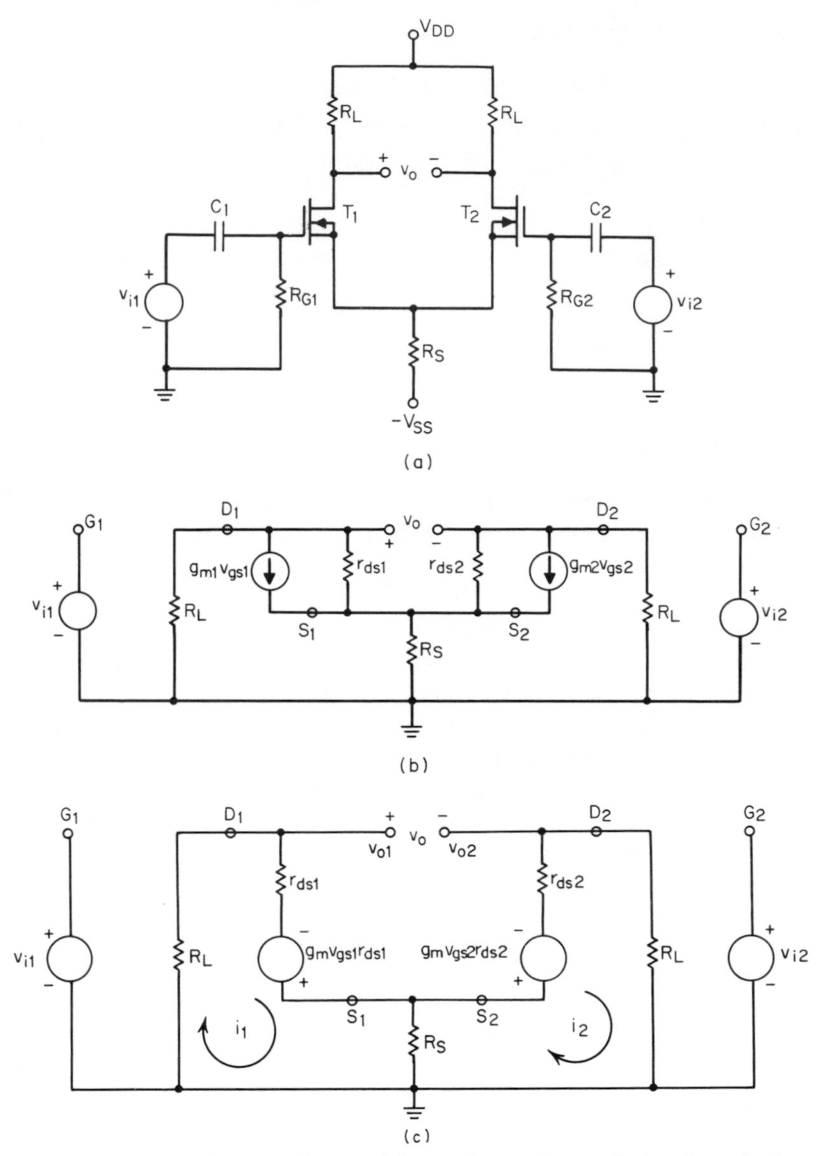

Fig. 9.31. (*a*) MOSFET differential amplifier; (*b*) small-signal equivalent circuit for MOSFET differential amplifier; (*c*) transformed small-signal equivalent circuit for MOSFET differential amplifier.

The Kirchoff loop equations for Fig. 9.31*c* can be written as

$$i_1R_L + i_1r_{ds1} - g_m v_{gs1} r_{ds1} + R_S(i_1 - i_2) = 0 \tag{9-70}$$

$$i_2R_L + i_2r_{ds2} + g_m v_{gs2} r_{ds2} + R_S(i_2 - i_1) = 0 \tag{9-71}$$

The differential output voltage is given by

$$v_o = v_{o1} - v_{o2} = -i_1R_L - i_2R_L \tag{9-72}$$

Then driving the circuit of Fig. 9.31*a* differentially, i.e.,

$$v_{i1} = v_i \qquad \text{and} \qquad v_{i2} = -v_i \tag{9-73}$$

and solving Eqs. (9-70) and (9-71) simultaneously with the aid of Eqs. (9-68), (9-69), and (9-73) with

$$r_{ds1} = r_{ds2}$$

$$g_{m1} = g_{m2}$$

$$\frac{v_{o1}}{v_{i1}} = \frac{v_{o1}}{v_i} = -g_m(r_{ds}\|R_L) \tag{9-74}$$

and

$$\frac{v_{o2}}{-v_{i2}} = \frac{v_{o2}}{v_i} = g_m(r_{ds}\|R_L) \tag{9-75}$$

Therefore the differential voltage gain A_{vd} is given by

$$A_{vd} = 2\frac{v_o}{v_i} = \frac{v_{o1} - v_{o2}}{2v_i} = -g_m(r_{ds}\|R_L) \tag{9-76}$$

where v_i and v_o are defined by Eqs. (9-73) and (9-72), respectively.

The differential output voltage varies linearly with the input signal v_i as shown in the following derivation. Since

$$v_o = v_{o1} - v_{o2} = -g_m(r_{ds}\|R_L)v_{i1} + g_m(r_{ds}\|R_L)v_{i2}$$

From Eq. (1-117)

$$v_o = (r_{ds}\|R_L)\beta[-(V_{Q1} + v_{i1} - V_T)v_{i1} + (V_{Q2} + v_{i2} - V_T)v_{i2}]$$

Thus with $V_{Q1} = V_{Q2} = V_Q$ and the aid of Eq. (9-73),

$$v_o = (r_{ds}\|R_L)\beta[-(V_Q + v_i - V_T)v_i + (V_Q - v_i - V_T)(-v_i)]$$

and therefore,

$$v_o = -2(r_{ds}\|R_L)(V_Q - V_T)v_i \tag{9-77}$$

The differential amplifier configuration of Fig. 9.31*a* finds its greatest utility in applications where the input signal source is required to drive differentially into an essentially infinite input impedance.

9.4 MOS ANALOG SYSTEM APPLICATIONS

9.4.1 Introduction

To date MOS/LSI has experienced its greatest impact in digital applications; very few MOS analog circuits have reached the marketplace. It is the purpose of this section to summarize some of the analog applications for which MOS/LSI may find future use. Most of the analog circuits to be discussed are not presently available from MOS/LSI manufacturers. Indeed, considerable design effort and technology development must be completed before most of these examples become practical in an engineering sense. It is hoped that the reader will find some of the concepts attractive enough so that he will personally provide the impetus which will make these and other MOS analog circuits a commercial reality.

9.4.2 The Analog Shift Register

The basic concepts for an electronic delay line of variable delay (now referred to as an analog shift register) were initially defined by Janssen.[18] Janssen pointed out that the then-conventional low-frequency delay lines built from capacitors and iron-cored inductors can exhibit nonlinear distortion. The distortion gives rise to dispersion, since the time delay depends upon frequency. It is therefore possible to realize a time delay that is only approximately constant over a limited frequency range.[19] The nonlinear distortion is caused by the iron-cored inductors. Moreover, if a very long time delay is wanted, then either the value of the inductances or the value of the characteristic impedance of the line becomes impractical.

To overcome these difficulties, the delay line shown in Fig. 9.32*a* was suggested by Janssen. The amplifiers are buffer amplifiers with unity gain and infinite input impedance. The transfer of analog information along the electronic delay line is brought about by alternately actuating all even-numbered and all odd-numbered switches. For example, all even-numbered switches can be activated at times 0, T, $2T$ . . . and all odd-numbered switches at times $T/2$, $3T/2$, $5T/2$ . . . . Figure 9.32*b* shows the voltage at the input of the first section as a solid line and the output of the first section as a staircase voltage at C_1. The voltage at the output of the second section at C_2 has the same shape as the output of the first section, but has a time

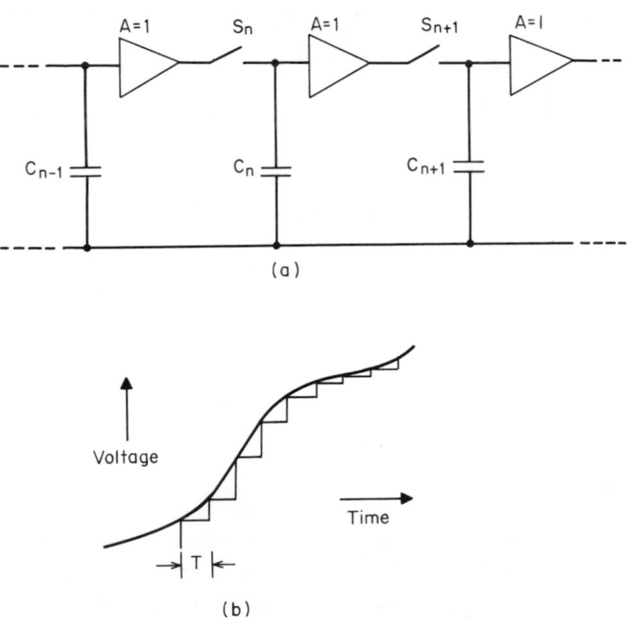

Fig. 9.32. (*a*) Idealized electronic delay line[18]; (*b*) voltage versus time at the input and output of first section of the electronic delay line.[18]

delay of $T/2$. The time delay per section can be varied continuously by varying the period of the switching pulse.

An MOS/LSI version of Janssen's electronic delay line has been fabricated by Sangster.[20] Its basis for operation is a chain of storage capacitors and charge transfer circuits functioning as an analog shift register (Fig. 9.33*a*). Sangster refers to the circuit as having "bucket brigade" capacitor storage, a name which derives from the resemblance of the circuit to the historical fire brigade. The analog shift register uses two complementary clocks with a frequency equal to the sampling frequency applied to the input signal. Signal delay can thus be accurately controlled, or if required, can be changed electronically. Because there are no dc gate currents, signal attenuation is negligible even after hundreds of stages, and no amplifiers are necessary. A cross section of the circuit configuration is shown in Fig. 9.33*b*. The storage capacitors are formed by the enlarged gate areas and accompanying p regions. No interconnection pattern between adjacent stages is necessary, as the drain of one stage also forms the source of the following stage. Sangster has fabricated a p-channel enhancement-mode analog shift register comprising 72 stages with 8-pF storage capacitors on a 60 × 95-mil chip. The circuit uses two clock signals of 5 V amplitude, and has been operated at clock frequencies between 100 Hz and 3 MHz. The circuit has been used experimentally for speech processing and audio delay. Presently its speed is not high enough for most video applications, but required improvement should be forthcoming.

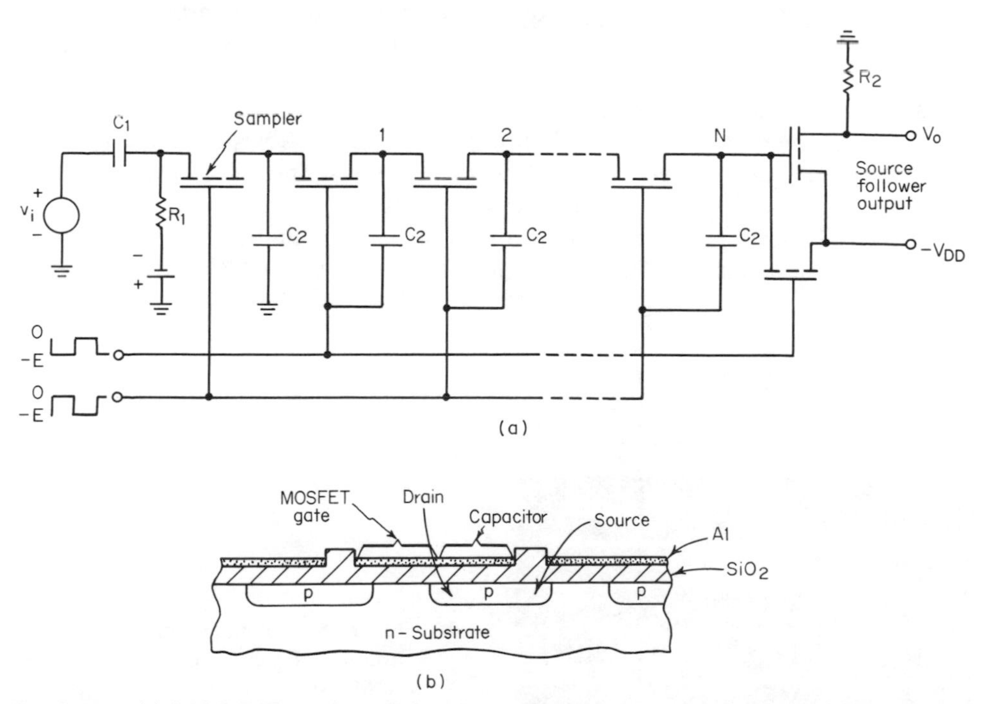

Fig. 9.33. (*a*) MOS/LSI version of the electronic delay line (bucket brigade)[20]; (*b*) cross section of a portion of the MOS/LSI electronic delay line.[20]

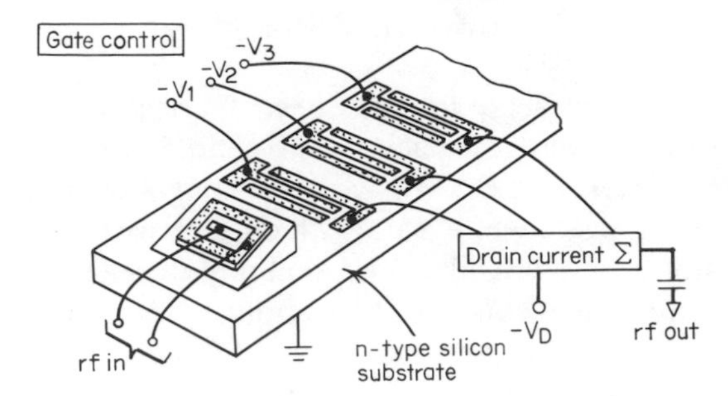

Fig. 9.34. Schematic of the MOSFET tapped delay line.[22]

The applications for analog shift registers in analog data processing include signal correlation, sensor scanning, programed analog signal sources of arbitrary waveform, and variable delay line functions in general. Previous design attempts to fabricate these circuits have met with only limited success primarily for economic reasons. MOS/LSI techniques may, however, make the analog shift register feasible both technically and economically.

9.4.3 Surface-wave Arrays

Surface-wave technology has been shown to be capable of performing complex operations in signal processing and communications.[21] To date, the utilization of surface waves has involved hybrid structures which employ piezoelectric materials such as lithium niobate ($LiNbO_3$), or quartz, for wave generation, propagation, and detection. It has, however, been demonstrated that arrays of silicon MOSFETs (Fig. 9.34) can also be used for detection of surface acoustic waves on silicon.[22] In the investigation, surface waves were launched on silicon substrates by utilizing conventional wedge (mode-conversion) techniques employing $LiNbO_3$ transducers. The MOSFET is a piezoresistive device[23] and therefore responds electrically to strain components in the surface waves as they propagate along the silicon.

Arrays of MOSFETs can thus be arranged to perform the function of a tapped delay line as shown in Fig. 9.35. Note that signals from the MOSFET delay-line

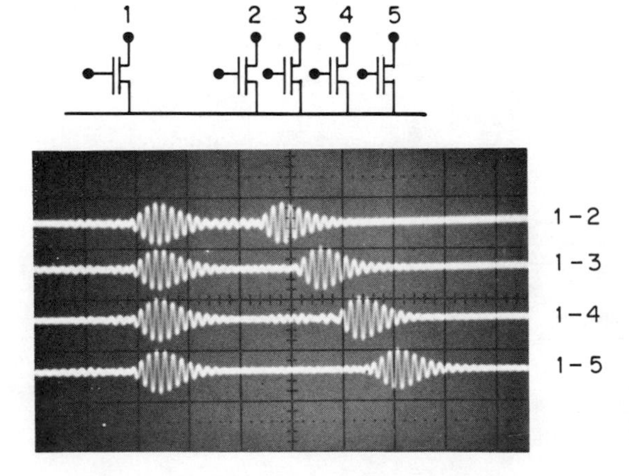

Fig. 9.35. Demonstration of the variable delay-line characteristics of the MOSFET detector array. Horizontal scale is 0.5 second per division.[22]

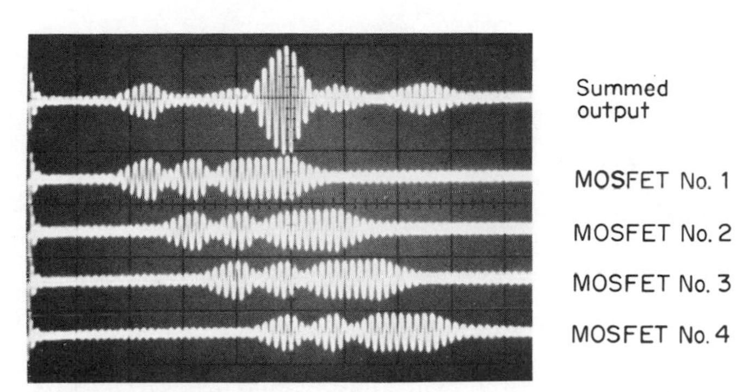

Fig. 9.36. Correlation of a 4-bit Barker sequence with MOSFET surface-wave array.[22]

taps are capable of being summed in a parallel sense. The MOSFET arrays can thus be positioned along the chip at the proper wavelength spacings to function as programable matched filters for decoding Barker coded sequences,[24] signal compression, etc. The programable feature is realized by application of activating gate voltage to the appropriate enhancement-mode devices in the array.

An example of decoding a 4-bit Barker sequence (+ − + +) with MOSFET arrays is shown in Fig. 9.36. Each bit has six cycles of 14.5-MHz center frequency. The MOSFETs are separated by multiples of one-half wavelength at this frequency and can be sampled for 180° phase shifts. The lower traces of Fig. 9.36 show the output from each of the four taps used for transverse summation of the signal. The upper trace is the analog correlation for the 4-bit Barker sequence.

The MOSFET has a geometry ideally suited to surface-wave detection since (1) current flow is limited to the surface inversion-layer region of thickness <100 Å whereby effective interaction with the energy content of the surface wave is realized, (2) the device can be fabricated with channel length less than $\lambda/2$ for surface waves in silicon less than 100 MHz, and (3) a linear gauge factor is exhibited and a mechanical biasing stress is not required to activate the device. In addition, the fabrication of the MOSFET as a surface-wave detector is fully compatible with existing MOS/LSI. The development of a batch-processed, relatively inexpensive programable matched filter offered by MOS/LSI could, in conjunction with surface-wave interaction, have a major impact on radar and communication system technologies.

9.4.4 Camera Control

The high input impedance of the MOSFET makes it ideally suited for sensing photocurrents from silicon p-n junctions operated in short-circuit mode. The short-circuit mode of p-n junction photosensing operation is defined in Fig. 9.37 by points A_1, A_2, and A_3 which represent illumination intensities I_1, I_2, and I_3, respectively. Figure 9.37 has been drawn with $I_3 > I_2 > I_1$. MOS circuitry can offer an input impedance of $\approx 10^{14}$ Ohms to the photodiode. This very high input impedance implies essentially zero input bias current and hence the photodiode light-generated

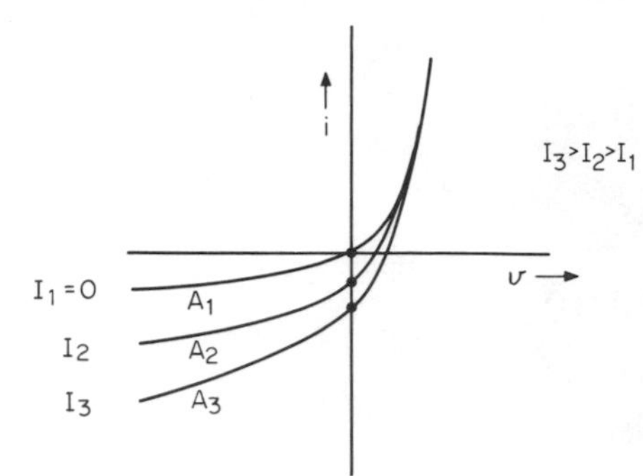

Fig. 9.37. Photo-response of the p-n junction.

current can be accurately processed by feedback elements. In contrast, bipolar input circuitry will exhibit a temperature-dependent input bias current. If, for example, bipolar input bias current were several nanoamperes, then 10-nanoampere signals are probably as low as can be processed with that form of circuitry. The latter considerations then relate to the area requirements and low-level sensing limit in the design of the photodiode.

Three examples of MOSFET amplifier front-end camera control circuitry are shown in Fig. 9.38*a*, *b*, and *c*. The examples are self-explanatory in view of previous discussions in this chapter. They all employ the short-circuit mode of photodiode operation and are intended to point out some of the possibilities available from the MOS technology arsenal for design of this particular system.

The MOS camera control system would be designed to automatically determine and set *f* stop and exposure time. In addition to the unique features of the MOS front end, the system would probably be comprised of 2 MOS op-amps, each having a gain of approximately 1,000. RC elements in the feedback loops coupled with MOS level-sensing circuitry would also be included. Integrated grounded-collector bipolar transistors could be used for driving the camera solenoids. Included on the chip would be the diode for photosensing. In all, some 50 MOSFETs would be used in this analog MOS/LSI system. The system should in turn provide an accurate yet economical camera control unit.

9.4.5 The Analog Switch

The MOSFET serves as a very useful component in analog switching. Other devices which perform the function include mechanical choppers, bipolar transistors, JFETs and photoelectric devices. Particular advantages offered by the MOSFET in this application are: (1) bilateral symmetry characteristics, (2) essentially infinite input impedance, (3) zero offset voltage, and (4) high source-to-drain resistance in the *off* state. Since a small drain-to-source resistance is required in the *on* state for analog switching, the device should be operated in the triode region rather than in the saturation region. In the triode region, the *on* resistance of the switch is given by

$$R_{ON} = \frac{1}{\dfrac{\mu\epsilon_{\text{Si}}\epsilon_o W}{t_{\text{Ox}} L}|V_{GS} - V_T - V_D|} \qquad (1\text{-}113)$$

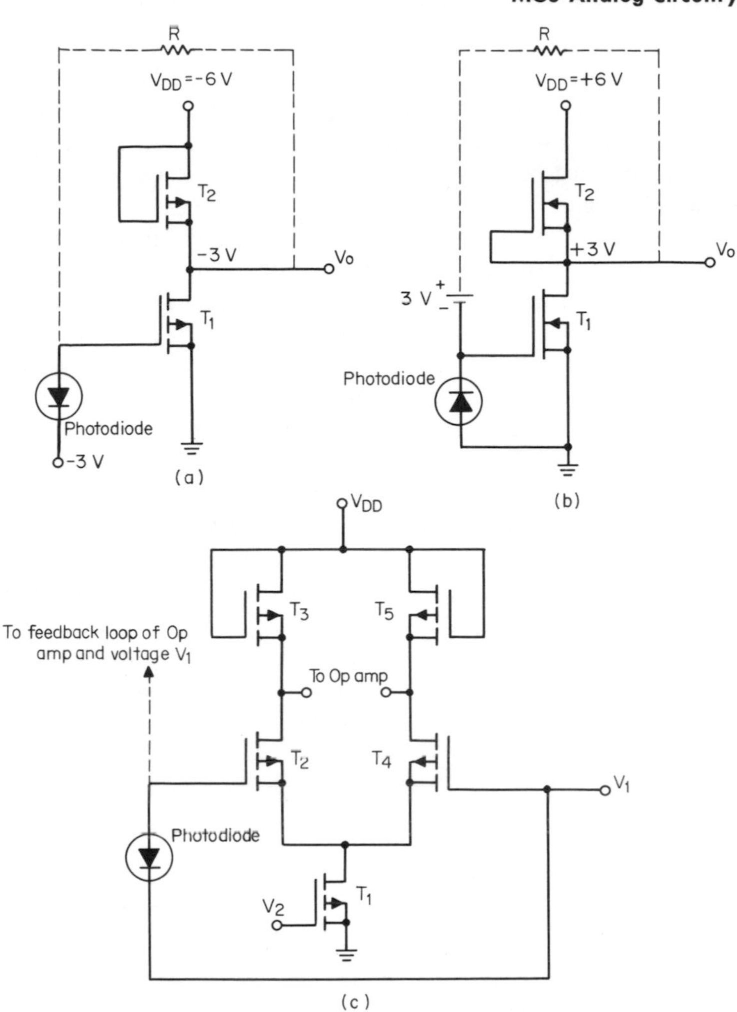

Fig. 9.38. (MOSFET amplifier front end for camera control. (*a*) p-channel enhancement-mode input pair; (*b*) n-channel depletion-mode input pair; (*c*) p-channel enhancement-mode differential input pair.

Note from Eq. (1-113) that to make the *on* resistance as low as possible, a device of large gate width W is required. W thus becomes a device design constraint since with its increase the device parasitic capacitances increase. In addition, p-n junction leakage currents increase as junction size increases. Practical values for *on* resistance of MOS analog switches are thus found to be approximately 200 Ohms as limited by the channel width W constraint and other pertinent factors in Eq. (1-113).

Another important device design consideration arises from the condition that since it is often desired to transmit an alternating voltage through the analog switch, the source and substrate must be disconnected from each other and back-gate biasing must be applied to the substrate. This in turn increases the absolute value of threshold voltage for the p-channel enhancement-mode device, which in turn increases the *on* resistance of the MOSFET [Eq. (1-113)]. The *on* resistance of the device is also temperature dependent through mobility and threshold voltage, and will increase with increasing temperature in a predictable manner (cf. Chap. 3). All of these factors must be taken into account in the design of the analog switch for a given application.

MOSFETs are often grouped together to form a useful analog switch. They can be grouped in common-source or common-gate fashion. The common-source arrangement is pictured in Fig. 9.39*a*. The common-source analog switch could also be called a common-drain analog switch because of the bilateral symmetry characteristics of the device. The common-source (common-drain) grouping is used in multiplexing operations. In a common-gate analog switch grouping, all gates are tied together with the drain and sources separated as shown in Fig. 9.39*b*. This arrangement is often used for gating and logic implementation.

Since MOS analog switches present a very low *on* resistance, they are compatible with gating applications; one example of this is the transfer of output data from a ROM to a serial shift register for data storage (Fig. 9.40). Other digital applications include the use of MOS analog switches in common-gate groupings for timing a feedback loop in ROM sequential logic implementation. Still another use for MOSFET analog switches is that of realizing simple gating functions: e.g., 2-input NAND gates, 3-input NOR gates, etc. (The latter is an example of MOS/LSI "in reverse" and should be resorted to only when the designer needs a simple MOS gate!)

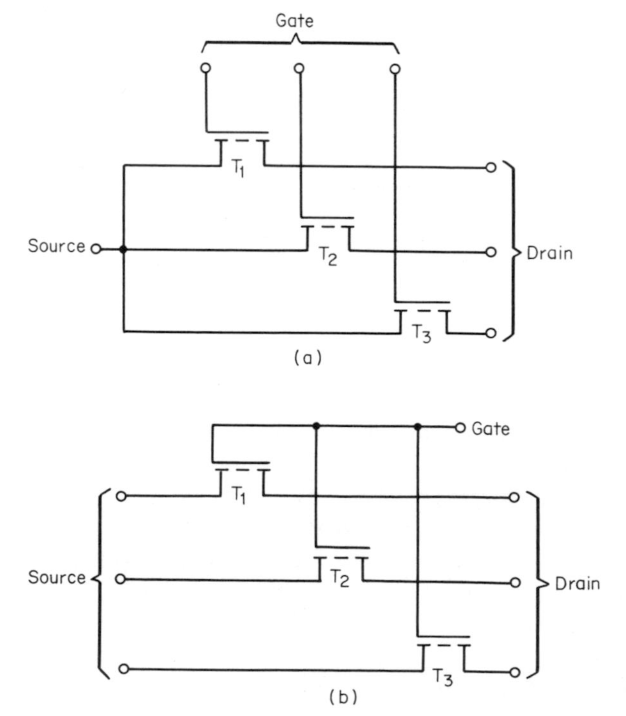

Fig. 9.39. MOSFET analog switch. (*a*) Common-source grouping; (*b*) common-gate grouping.

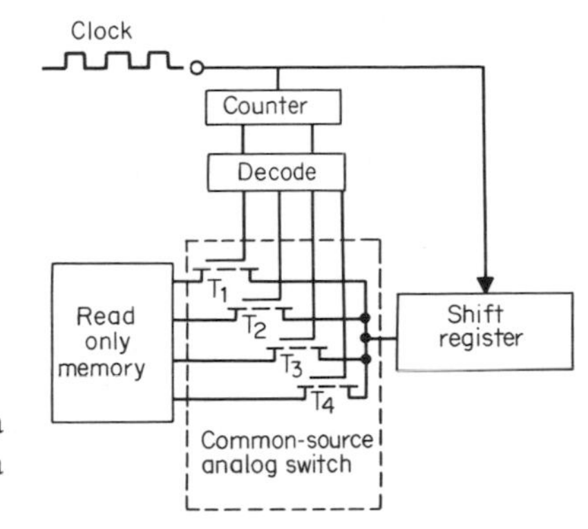

Fig. 9.40. MOS analog switching of output data from a ROM to a serial shift register for data storage.

Probably the major application for large arrays of MOS analog switches is found in performing multiplexing functions. For this application, common-source (common-drain) groupings of analog switches are utilized as shown in Fig. 9.41. (Note that the clocks on the multiplex and demultiplex ends are kept synchronous.) The potential applications for multiplexing are graphically demonstrated in the transmission of signals between cockpit, cabin, engine activators, and sensors, of a jet transport. It is estimated that up to 1 ton of wiring per aircraft can be eliminated by these multiplexing schemes!

9.4.6 FM and TV Receivers

The use of dual gate MOSFETs in FM tuners offers several advantages over bipolar, JFET and single-gate MOSFETs. RF amplification and mixing are aided by the use of this low-feedback-capacitance transistor, which also exhibits a low noise figure and large dynamic range. A second gate is available on the device for either AGC or local oscillation injection. In addition, high and stable RF and conversion gains are easily obtained in conjunction with inexpensive commercially available coils and without the need for neutralization. As an amplifier, the dual-gate MOSFET provides high gain, low noise, and large AGC range without overload. As a mixer, the device has large conversion gain and high spurious rejection.

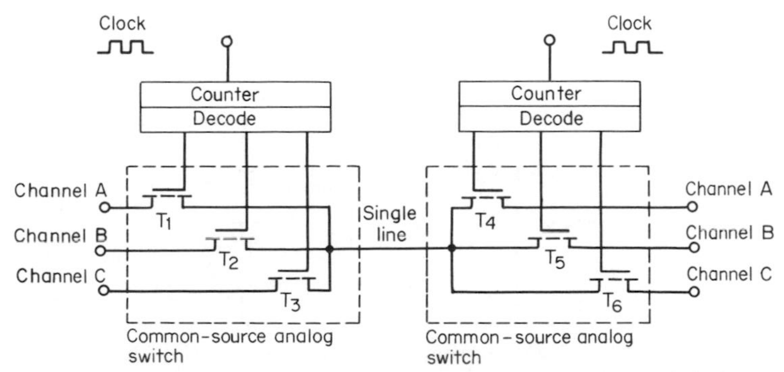

Fig. 9.41. Common-source analog switch used for multiplexing.

A complete FM tuner[25] with the following performance specifications:

Power gain	40 dB
3 dB limiting	0.9 μV
30 dB quieting	1.5 μV
Image rejection	50 dB
f_o + 1/2 IF	80 dB
All other spurious	>100 dB
VSWR	1.2:1

is shown in Fig. 9.42. The tuner uses SFB 8970 dual-gate MOSFETs fabricated by Texas Instruments.

The SFB 8970 dual-gate MOSFET has also been employed in TV IF Amplifiers. One design has been discussed by Weaver.[26] The circuit is shown in Fig. 9.43. Measured parameters for the amplifier are:

IF response

39.75 MHz	30 dB down
41.25 MHz	40 dB down
47.25 MHz	30 dB down
42.17 MHz	8 dB down
45.75 MHz	6 dB down

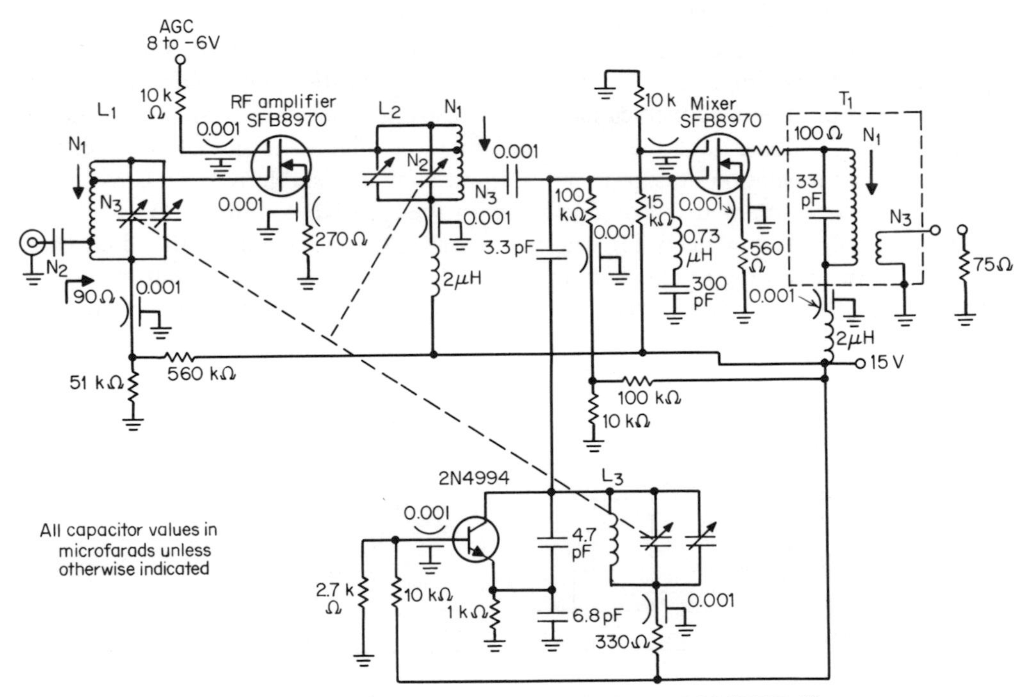

Fig. 9.42. FM tuner employing dual-gate MOSFETs.[25]

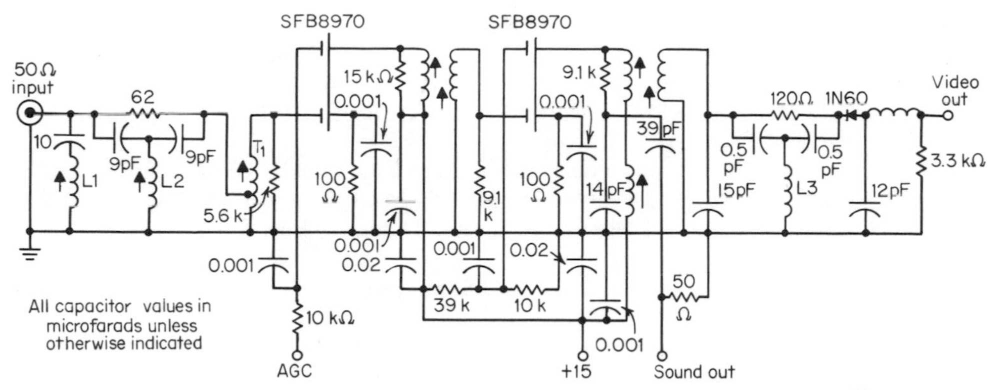

Fig. 9.43. TV receiver IF amplifier employing dual-gate MOSFETs.[26]

Sensitivity = 700 μV for 1 V of video (carrier 100 percent Mod.)
AGC range = 43 dB
AGC voltage = +4.0 V to −6.0 V
Sound output = 570 μV of 4.5 MHz signal

Complete color processing circuitry for TV receivers using MOS ICs has been reported by Mitchell and Sheets.[27] Their design employs four MOS chips performing the following functions:

1. Chroma- burst separator- Amp- keyer- color killer
2. Color demodulator
3. Color matrix
4. Color synchronization

The authors include a component and connection count comparison for the system circuitry rendered in vacuum tube versus discrete bipolar versus MOS IC. Their analysis points up the advantages enjoyed by MOS circuitry and indicates future potential for the technology in this particular application.

9.4.7 Conclusions

In our survey of MOS analog applications we have detailed only a few of the many possibilities that are "known and unknown and remembered and forgotten!" Mention should be made of the use of the MOSFET for such widely diverse circuitry as A/D and D/A converters, electrometers, oscillators, and in automotive sensing and control systems. Conversely, it is apparent that MOS/LSI will not compete with bipolar in present-day operational amplifiers. Furthermore, of the analog applications we have discussed, some may not come to fruition while others may eventually be performed with devices which differ from the MOSFET. If, however, the unique characteristics of the MOSFET are matched to the given application in the light of economic considerations, we can proceed with meaningful and creative design efforts in the still relatively undeveloped field of analog MOS/LSI.

REFERENCES

1. A. B. Phillips, "Transistor Engineering and Introduction to Integrated Semiconductor Circuits," Chap. 9, McGraw-Hill Book Company, New York, 1962.
2. J. M. Early, PNIP and NPIN Junction Transistor Triodes, *Bell System Tech. J.,* **33:** 517–533, May 1954.
3. S. M. Sze, "Physics of Semiconductor Devices," Chap. 6, Wiley-Interscience, 1969.
4. L. J. Sevin, U. S. Patent 3,434,068: Integrated Circuit Amplifier Utilizing Field-effect Transistors Having Parallel Reverse Connected Diodes as Bias Circuits Therefor.
5. R. H. Crawford, private communication.
6. J. L. Chalfan and J. C. Looney, Linear MOS Integrated Circuits, *Solid-state Technology,* **12:** 31–39, May 1969.
7. "RCA COS/MOS Integrated Circuits Manual," 134–137, 1971.
8. D. M. Miller and R. G. Meyer, Nonlinearity and Cross-modulation in Field-effect Transistors, *IEEE J. Solid-state Circuits,* **SC-6:** 244, August 1971.
9. A. G. Jordan and N. A. Jordan, Theory of Noise in Metal Oxide Semiconductor Devices, *IEEE Trans. Electron Devices,* **ED-12:** 148–156, March 1965.
10. C. T. Sah, S. Y. Wu, and F. H. Hielscher, The Effects of Fixed Bulk Charges on the Thermal Noise in Metal-Oxide-Semiconductor Transistors, *IEEE Trans. Electron Devices,* **ED-13:** 410–414, April 1966.
11. R. S. C. Cobbold, "Theory and Applications of Field-effect Transistors," p. 341, Wiley-Interscience, 1970.
12. S. Y. Wu, Theory of Generation-recombination Noise in MOS Transistors, *Solid State Electron.,* **11:** 25 (1968).
13. I. Flinn, G. Bew, and F. Berg, Low-frequency Noise in MOS Field-effect Transistors, *Solid State Electron.,* **10:** 833 (1967).
14. C. T. Sah and F. H. Hielscher, Evidence of the Surface Origin of the $1/f$ Noise, *Phys. Rev. Letters,* **17:** 956 (1966).
15. G. Abowitz, E. Arnold, and E. Leventhal, Surface States and $1/f$ Noise in MOS Transistors, *IEEE Trans. Electron Devices,* **ED-14:** 775, November 1967.
16. H. M. Kleinman, Application of Dual-gate MOS Field-effect Transistors in Practical Radio Receivers, *IEEE Trans. on Broadcast and TV Receivers,* **BTR 13:** 72, July 1967.
17. D. L. Schilling and C. Belove, "Electronic Circuits: Discrete and Integrated," Chap. 13, McGraw-Hill Book Company, New York, 1968.
18. J. M. L. Janssen, Discontinuous Low-frequency Delay Line with Continuously Variable Delay, *Nature,* **148:** January 1952.
19. H. W. Bode and R. C. Dietzold, *Bell System Tech. J.,* **14:** 215 (1935).
20. F. L. J. Sangster, Integrated MOS and Bipolar Analog Delay Lines Using Bucket-brigade Capacitor Storage, *Digest of Technical Papers. Int'l. Solid-State Circuits Conf.,* **9:** 74 (1970).
21. R. M. White, *Proc. IEEE* **58:** 1238 (1970).
22. L. T. Claiborne, E. J. Staples, J. L. Harris, and J. P. Mize, MOSFET Ultrasonic Surface-wave Detectors for Programmable Matched Filters, *Appl. Phys. Letters,* **19:** 58 (1971).
23. D. Colman, R. T. Bate, and J. P. Mize, *J. Appl. Phys.* **39:** 1923 (1968).
24. R. H. Barker, in *Communication Theory,* edited by W. Jackson, p. 273, Academic Press, Inc., New York, 1953.
25. R. Klein, MOSFET FM Tuner Design, *IEEE Trans.* **BTR 16:** 67, May 1970.
26. S. Weaver, Dual-gate MOSFETs in TV Amplifiers, *IEEE Trans.* **BTR 16:** 96, May 1970.
27. M. M. Mitchell and W. Sheets, Integrated MTOS Circuits for Color TV Applications, *IEEE Trans.,* **BTR 14:** 28, July 1968.

10

The Economics of MOS/LSI

10.1 INTRODUCTION

The purpose of this chapter is to provide information for the system designer which will assist him in making economic decisions pertaining to the use of MOS/LSI. By way of introduction it should be pointed out that historically we have experienced a price decline in the bipolar transistor from $\approx$\$100 per device in mid-1950 to less than a few cents per device in present-day integrated-circuit form. No other product or commodity in the history of man, with the possible exception of salt and certain ethical drug products, has experienced the four-order-of-magnitude price decline witnessed by the transistor. This economic trend has certainly stimulated the use of silicon bipolar integrated circuits in electronic systems. Now we stand amidst an entirely new integrated-circuit technology—that of MOS/LSI. This new technology promises, through its extremely high circuit density and favorable economics, to have an even greater impact on electronic systems than did the bipolar integrated circuit.

An empirical relationship has evolved which will assist us in our understanding of the price behavior of semiconductor products with time. The relationship is referred to as the *learning curve*. The concept can be stated as: "the price of a semiconductor product decreases by 30 percent for every consecutive doubling of *cumulated* production of that given product." A hypothetical example of the learning curve is shown in Fig. 10.1. The learning curve has its origin in precisely the source from which it derives its name: As the semiconductor manufacturer becomes familiar with the product, his manufacturing skills improve (he learns!), and this is reflected in the diminishing cost of his product. As a product matures through high-volume production, it thus eventually becomes quite difficult to maintain dramatic price reductions because of the sheer volume required to double cumulative production. The product price as a function of time can, of course, be obtained from Fig. 10.1 by relating cumulative production to time and then preparing a graph of unit price versus time from these data.

The concept of the learning curve gives us an immediate understanding of the apparent leveling off of prices in the now-mature TTL small-scale integrated-circuits

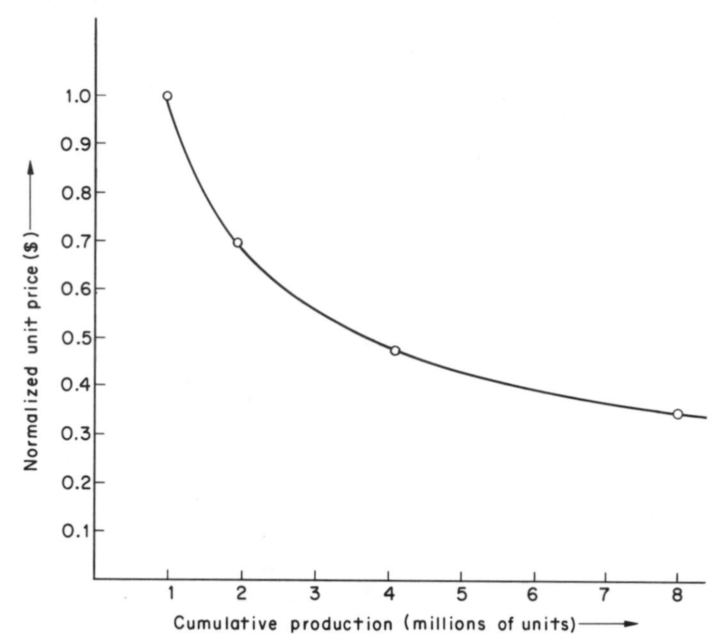

Fig. 10.1. A representation of the learning curve.

market. Conversely, production of the new MOS/LSI technology has only just begun to start us down the slope of the learning curve. The potential for dramatic price reduction for this latter circuitry should, therefore, be realized in the 1970s.

10.2 NATURE OF THE MOS/LSI MARKET

Presently the MOS market is divided almost equally between catalog and custom circuitry. Catalog circuitry has taken the form of structured logic implementation, and the products that are available in catalog form consist mainly of shift registers, read-only memories (ROMs) and character generators, and random access memories (RAMs). The custom circuitry has taken the form of random logic implementation and specialized versions of catalog circuitry.

The MOS market is in a rapid state of growth as shown in Fig. 10.2. Why is MOS technology in such a growth pattern? Basically the answer is that *MOS/LSI offers the highest logic function density of any presently existing technology*. This central feature of the technology will be discussed in detail because it will in turn provide definition and insight concerning the impact that MOS/LSI will have on electronic systems.

10.3 COST FACTORS AT THE SYSTEM LEVEL

Additional understanding of the impact of MOS/LSI on electronic systems is realized through definition of the various cost factors at the system level. These cost factors are simply the sum total of the various costs that accrue during the creating and manufacturing of an electronic system. These factors are:

1. Component cost. This cost includes electronic elements such as resistors, diodes, TTL, MOS/LSI, etc. Mechanical components are also part of these expenditures.
2. Tooling cost. This item encompasses engineering labor costs and materials expenses such as printed circuit board layout and procurement. Basically it is a one-time engineering charge.
3. Assembly cost. This expense is incurred in the physical structuring of the system.
4. System testing. Included in this item are debugging of system operation and establishment of system reliability.
5. System repair and maintenance.
6. Inventory. This expense arises from the requirement for stocking component spare parts for the system.

Each of the above six factors must be considered for any electrical system that is to be mass produced. The relative importance of each of the system cost factors depends, of course, upon the particular system to be manufactured. It is important to ask, "What is the overriding factor that can make the greatest impact on reducing the expenses associated with the above six factors?" The answer is: *Reduce the number of packaged electronic components in the system.* To accomplish this, we must strive for the greatest circuit complexity within each package and we must also obtain the greatest logic density per package pin. This is precisely where and how MOS will make its impact.

10.4 THE CONCEPT OF SILICON REAL ESTATE

10.4.1 Introduction

We have been using the terms "circuit function density" and "circuit complexity." Let us clarify these terms by examining the following photographs of MOS/LSI

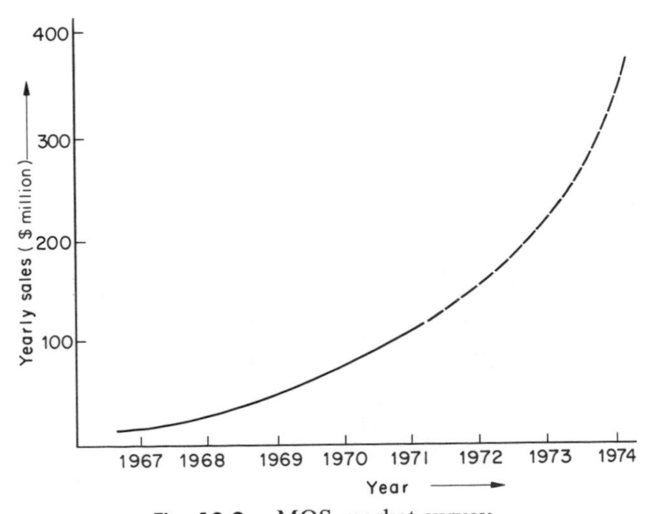

Fig. 10.2. MOS market survey.

Fig. 10.3. Quad 256-bit dynamic shift register.

chips. Figure 10.3 shows a quad 256-bit dynamic shift register. The chip occupies an area which is 127 × 130 mils and employs approximately 6,200 transistors. A 2,560-bit dynamic ROM occupying an area of 150 by 120 mils is shown in Fig. 10.4, and, finally, a 1,024-bit RAM of dimensions 146 × 163 mils employing approximately 4,000 transistors is pictured in Fig. 10.5.

Note for these three typical MOS/LSI chips of high complexity that considerable engineering effort has been expended in compacting the circuitry onto as small a

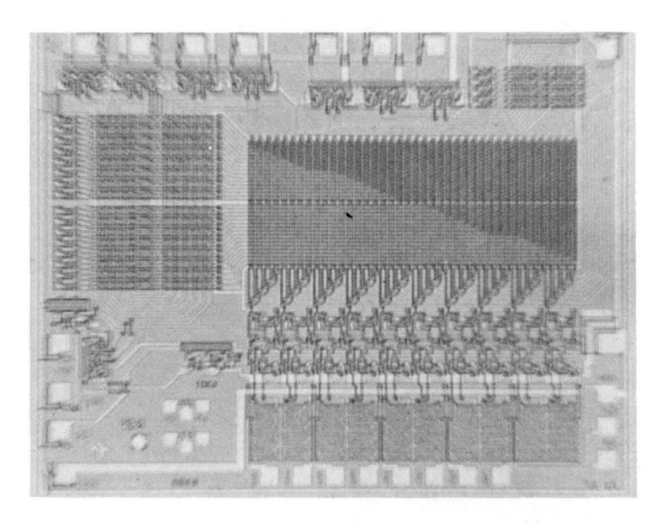

Fig. 10.4. 2,560-bit dynamic ROM.

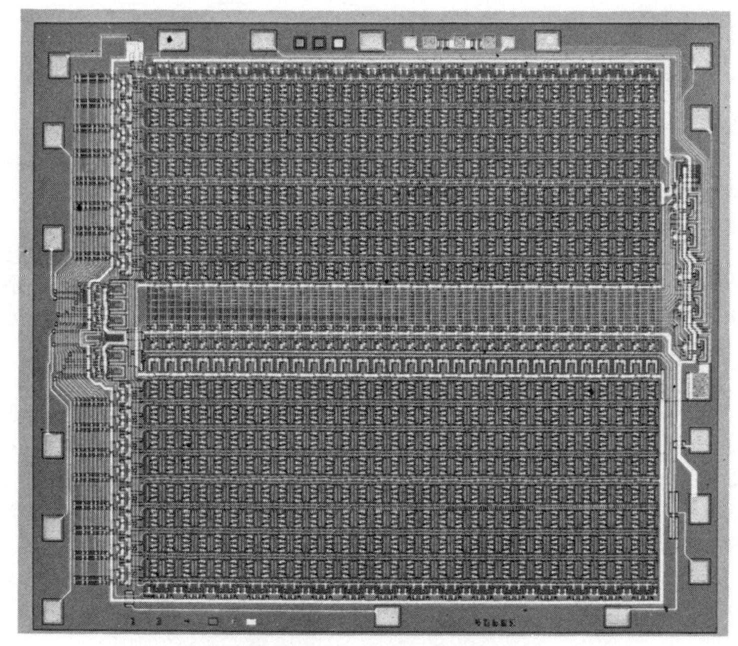

Fig. 10.5. 1,024-bit RAM.

silicon area as possible. We refer to the area utilized for the circuitry as *silicon real estate.* In analogy to conventional real estate, the guiding principle is to develop the land to its highest and best use.[1] The highest and best use of silicon MOS real estate is realized by obtaining the greatest circuit density possible on a given silicon area in accordance with good engineering practices and overall product reliability.

10.4.2 Batch Processing

Development of silicon real estate involves *batch processing.* In MOS/LSI fabrication by batch processing, we start with a 2-inch diameter silicon wafer and simultaneously form, by means of a photolithographic process, subsequent oxidation, diffusion, and metalization regions on the wafer. Thus, approximately 138 structures, each containing a 2,560-bit dynamic ROM (Fig. 10.4), are simultaneously formed on a 2-inch diameter silicon wafer. The wafer is then scribed and broken into 138 chips, each containing a potential ROM.

10.4.3 The Concept of Yield

Note in the above statement that we said *potential* ROM. This implies that a manufacturing yield must be considered. The yield is less than 100 percent mainly because of:

1. Defects in the photomasks
2. Pinholes in the oxide

3. Mask misalignment
4. Defects in photoresist
5. Processing errors (over-etching, etc.)

Yield, then, is the percentage of acceptable integrated circuits (chips) produced by the process. Assuming that there exist no circuit design problems that might place demands on components within the circuit which the process cannot consistently maintain (e.g., g_m, leakage current, etc.), IC yield depends on the *area* of the given circuit. The number of potential flaws a circuit may encounter is a function of the area which the circuit occupies on the silicon wafer. The larger the required area, the more defects likely to be encountered and the lower will be the expected yield.

The following numerical example points up some of the factors that will affect MOS/LSI manufacturing yield. Consider a 2-inch diameter silicon wafer on which four large circuits are formed by batch processing to be represented by Fig. 10.6*a*. The circuit yield will depend on the number of random defects on the wafer, and if the defects which are noted by crosses are distributed as shown in Fig. 10.6*a*, the resulting yield will be zero. If through circuit design and mask layout the area of each chip is reduced by a factor of 4, then eight circuits can be placed on the 2-inch diameter wafer as shown in Fig. 10.6*b*. With the same distribution of random defects as experienced in Fig. 10.6*a*, four circuits would now have defects and four would be defect-free, or in other words, the yield would be 50 percent.

To consider a somewhat more practical example, recall that the 2,560-bit dynamic ROM shown in Fig. 10.4 has dimensions of 150 $\times$ 120 mils and that approximately 138 of these chips can be obtained from a 2-inch diameter silicon wafer. If we experience 69 defects which are distributed on a one-to-one basis per chip, then the resulting yield would be 50 percent. If through circuit and mask layout, just 10 mils were removed from each side of the 2,560-bit dynamic ROM chip, then approximately 150 potentially good chips would be obtained per wafer. If the same random distribution of 69 defects exists as previously cited and is again distributed on a one-to-one basis per chip, then 81 defect-free ROMs would be produced, giving a yield of 54 percent. By merely taking 10 mils off each side of the ROM chip, we have thus effected an 8 percent gain in yield. The improvement in yield will in turn be reflected in reduced circuit costs.

10.4.4 Quantitative Description of Yield

Since manufacturing yield will influence circuit cost, it is necessary to make quantitative statements concerning this important factor. The all-important rela-

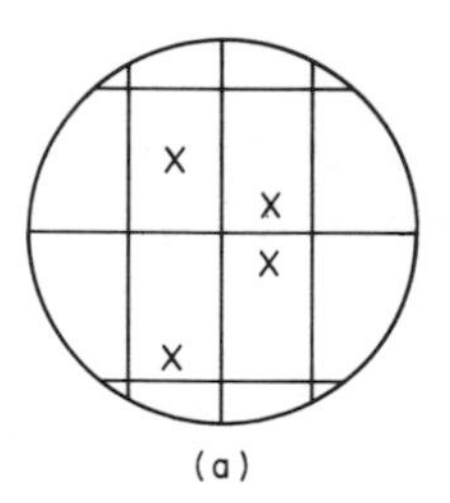

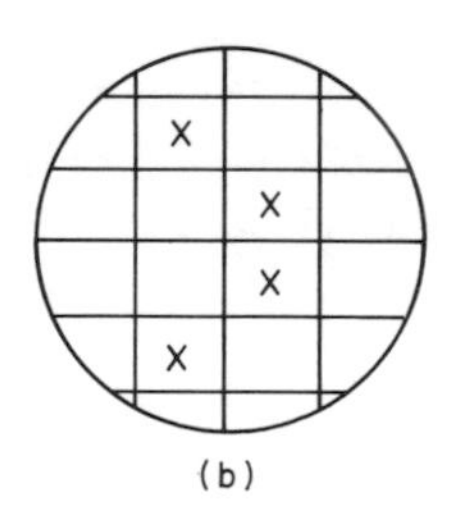

Fig. 10.6. (*a*) Yield consideration with four chips and four random defects. Crosses represent defects; (*b*) yield consideration with eight chips and four random defects. Crosses represent defects.

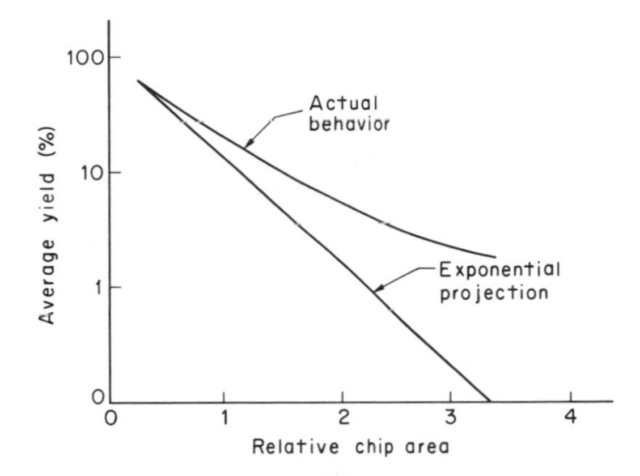

Fig. 10.7. Yield versus relative chip area.[5]

tionship between chip area and yield has been discussed in the literature.[2,3,4] If defects were randomly distributed over the wafer surface, yield would drop off exponentially as active area increases. Moore[5] points out that in actuality the yield of most integrated circuits falls off at a less than exponential rate as active area increases, indicating that defects are clustered on the wafer and not distributed at random. Circuit costs, of course, will be closely related to this distribution. The situation is summarized very approximately in Fig. 10.7.

In spite of the mathematical difficulties involved in predicting yields through modeling, it is possible for a semiconductor manufacturer to arrive at an empirical projection of the average defect density by observing his production-line output as a function of time. By noting the long-term variation in average defect density, the manufacturer can establish a trend for yield as process technology matures. He can use these trends to predict yields and costs in advance, which might typically develop as shown in Fig. 10.8. A survey of the present situation indicates that a 20 to 30 percent yield improvement per year can be expected for a given chip area as the MOS manufacturer gains experience in producing complex MOS/LSI circuitry. As a consequence of this, the chip area for a given yield will increase with time. The circuit complexity will therefore also enjoy a corresponding increase, and this in turn will beneficially affect the factors which constitute the overall system cost as discussed in Sec. 10.3.

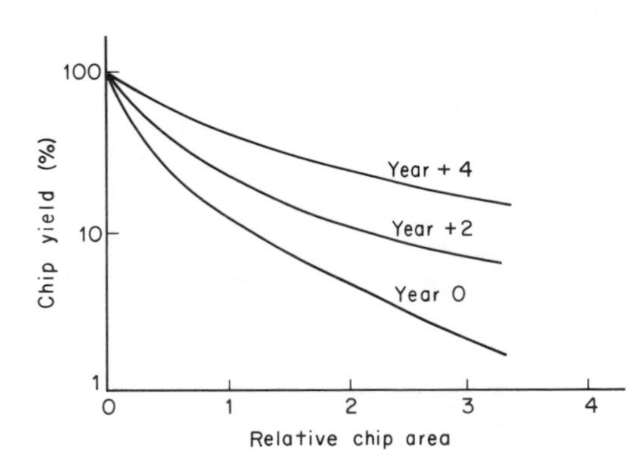

Fig. 10.8. Yield improvement projection.[5]

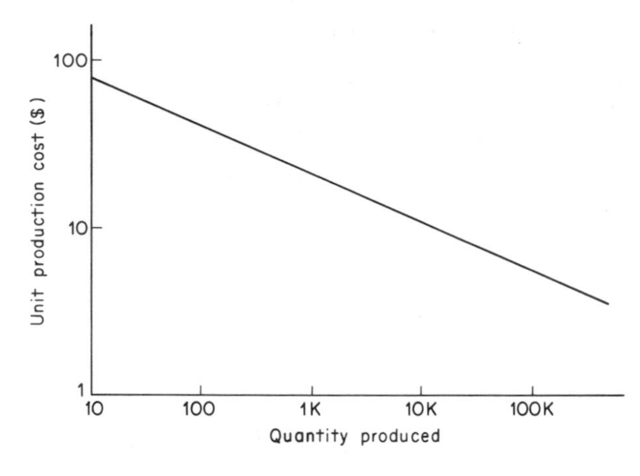

Fig. 10.9. Hypothetical unit production cost versus quantity produced.

10.5 CATALOG MOS/LSI

Having discussed the concept of silicon real estate and factors which contribute to its cost, let us now turn to the specifics of MOS/LSI catalog circuit costs. The cost situation here is governed by supply and demand for a given circuit. If the manufacturer is to offer a low-cost item, he must have a large manufacturing base, i.e., volume production. Figure 10.9 indicates how hypothetical unit production cost may vary with quantity produced. Note that Fig. 10.9 is an admixture of the learning curve and a reflection of economies realized through mass production of a semiconductor product. A dichotomy may therefore arise at this point in that the customer may not want to buy a particular circuit since his volume requirements are not high enough to create a price structure which is acceptable to him. The MOS market has thus developed in part around a set of standard or catalog items designed to be acceptable to many customers. High-volume production is thereby achieved and favorable price structures result.

To date, catalog products have taken the form of (1) cost-competitive direct replacements for previously existing electronic functions and (2) highly structured logic forms wherein the circuitry is of a repetitive nature. Examples of catalog circuitry such as shift registers, ROMs, and RAMs even exhibit their repetitive circuit form in the physical layout of the circuits (Figs. 10.3, 10.4, and 10.5). The repetitive layout contributes to the realization of high circuit density per chip.

The standard or catalog parts are playing an important role in developing a semiconductor memory market in which multiple-1,000-bit dynamic RAMs are being produced in volume. The complexity of these catalog items is continually increasing as further applications for semiconductor memories evolve.[6] Table 10.1 summarizes the economic trends in catalog MOS/LSI. The fundamental reasons for the price declines shown in that table are that more and more logic is being placed on a chip and manufacturing yields are continually being improved.

Trends in catalog circuit complexity and density during the past several years are summarized in Fig. 10.10. A possible future candidate for classification as a catalog item is the experimental 18K-bit MOS RAM super memory circuit shown in Fig. 10.11. The circuit employs approximately 90,000 MOSFETs! From these recent

Table 10.1. Past and Present MOS/LSI Product Economics

Year and circuit		Bit count	Price per bit, cents
1968	RAMs	256	20
1971	RAMs	1,000	1
1971	ROMs	2,560	0.5
1969	Shift registers	200	12
1971	Shift registers	1,024	0.7

developments and trends it takes no great imagination to predict that remarkable advances in circuit complexity and density per package should be realized with MOS/LSI during the next few years.

10.6 CUSTOM MOS/LSI

Although catalog MOS/LSI circuits give one a basic capability in structuring the electronic portion of a system, the system designer may not be able to use the items available in the catalog. If he cannot build his system with standard off-the-shelf catalog items, he should investigate system implementation with custom MOS/LSI.

Perhaps the most important item for consideration in working with custom MOS/LSI is the influence of nonrecurring engineering (NRE) expense on the circuit and system cost. Recall that tooling cost is one of the six major expenses associated with system realization. NRE costs can become the main contributor to the tooling costs, and they are expended by the semiconductor manufacturer in development of a given MOS circuit. The aspects of NRE costs are summarized in Fig. 10.12, where a hypothetical example of unit costs versus circuit quantity produced is plotted. The slope of the straight-line portion of Fig. 10.12 is again an admixture of the learning curve and a reflection of economies realized through mass production of a semiconductor product. The custom approach often does not reap the economic benefits derivable from batch processing because the batches are not large enough.

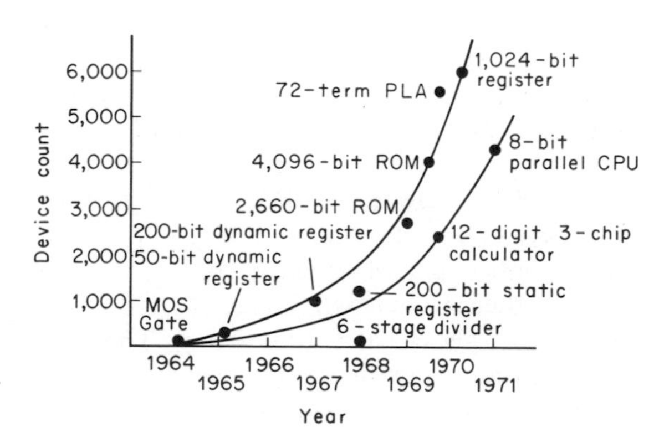

Fig. 10.10. MOS/LSI complexity versus time.

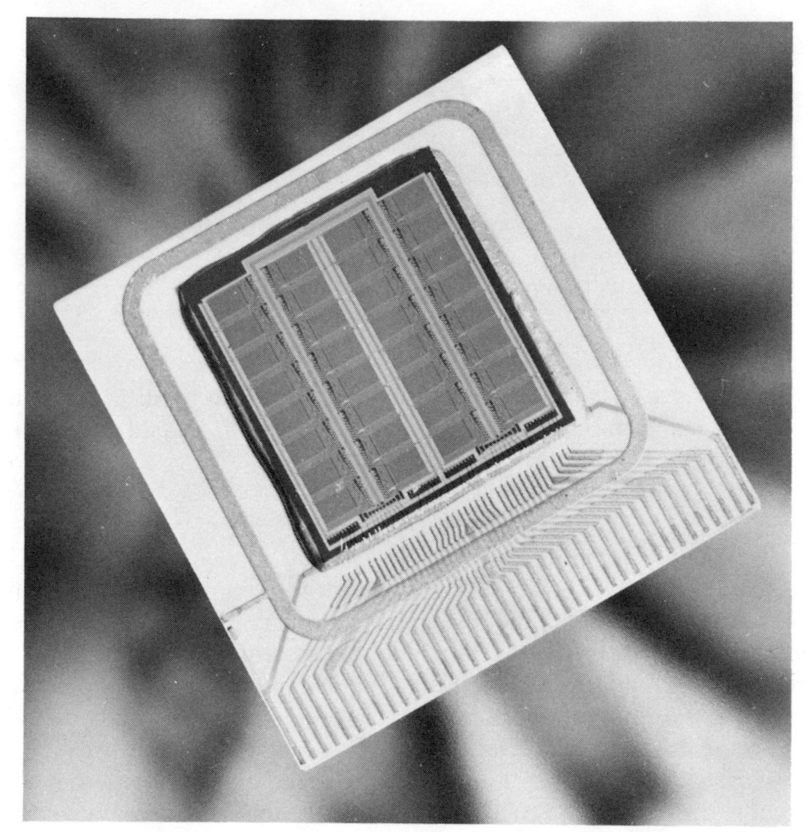

Fig. 10.11. Experimental 18K-bit MOS super RAM.

The location of the straight-line portion of Fig. 10.12 with respect to the unit cost axis is determined by the area (and hence yield) of the circuit.

Superimposed on the cost information of Fig. 10.12 are the NRE costs, which in this example are taken to be \$20,000. Thus if the system quantity buy for this case is 100 circuits, the chips will have a cost of $\approx$\$250, of which $\approx$\$50 is chip cost and $\approx$\$200 is amortized NRE cost. As shown in Fig. 10.12, the NRE cost is diluted by increasing the quantity purchased. For example, if 1,000 circuits were purchased, the NRE cost would contribute only \$2.00 to each chip. The two lines essentially merge for this example at the 5,000 unit production volume, and above that volume the NRE cost is insignificant. It should be further noted that for a costly large chip, the NRE cost is amortized quickly in contrast to a less expensive smaller chip where the NRE cost remains a large percentage of the total cost. We have here of course assumed identical NRE cost for both the large and small chip.

The contribution of NRE cost to the entire system cost must be carefully evaluated to justify the custom MOS/LSI approach. NRE costs can often obscure the apparent cost advantages of batch processing. The semiconductor manufacturer is aware of the importance of this tooling cost to his customer and strives to keep the NRE

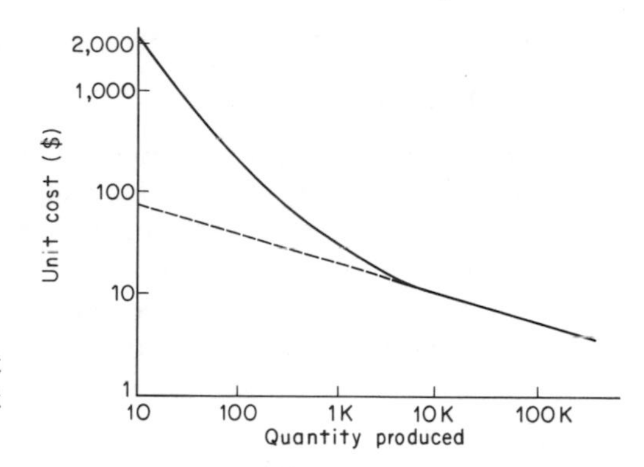

Fig. 10.12. Hypothetical unit cost versus quantity produced. NRE cost factor is indicated.

expense minimal by employing efficient computer aided design (CAD) and skillful circuit implementation. In most cases the system engineer will be able to justify the NRE cost of custom circuitry, but if not, he may have to design his system to utilize MOS catalog items or bipolar medium- or small-scale integrated TTL, DTL, or ECL circuits.

10.7 TRENDS IN CHIP SIZE AND COMPLEXITY

The recent general trends in MOS chip size and complexity are summarized in Fig. 10.13, where chip size versus system cost is plotted for various years. Note that whereas a few years ago a 130 × 130-mil chip was an optimum size for minimization of system costs, presently a chip of 200 × 200 mils offers the user the greatest cost savings. The reduction in system cost as the chip area, and hence the circuit size, increases is generated by the impact of high circuit complexity on the six system cost factors described in Sec. 10.3. In addition, the semiconductor manufacturer is continually improving his yields through improved process and mask-generation capabilities; hence, he achieves larger and larger chip sizes at lower cost. Finally, the techniques of system implementation are continually being improved, resulting in reduced system cost. An example of this improvement is that of the recent departure from the gate-reduction approach to that of the programable logic array (PLA), or matrix, method.

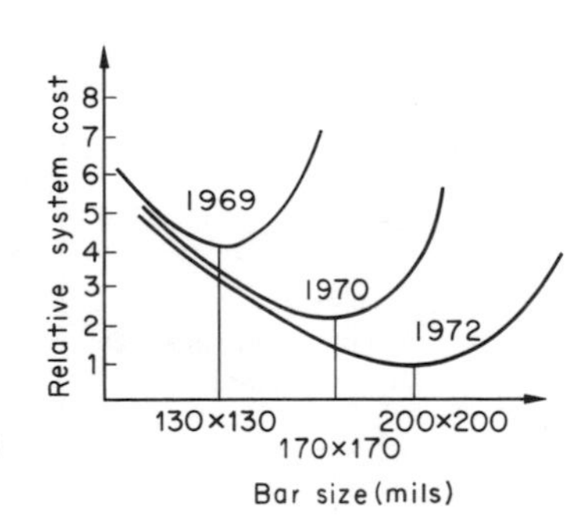

Fig. 10.13. Relative system cost as a function of chip size during recent years.

10.8 COMPARISON OF SYSTEM IMPLEMENTATION TECHNIQUES

10.8.1 Statement of the Problem

To illustrate possible approaches to system implementation, consider the following example. At a certain phase of system design, a logic diagram will be created. Consider for illustration a diagram of the form shown in Fig. 10.14. This logic diagram is an arithmetic subsystem for a large arithmetic unit. The diagram is implemented by TTL SN 7400 series gates. The parts count indicates a requirement of 16 TTL packages to implement the 51 gates. The design represents what appears to be a cost-effective solution to system implementation using off-the-shelf TTL components.

Now let us consider the conversion of the logic diagram shown in Fig. 10.14 to implementation with MOS. We will immediately be concerned with how much logic can be placed on a given MOS chip and the general manner in which to proceed with the conversion. The gate-reduction method and PLA technique will be considered separately in the conversion of the logic diagram to MOS.

10.8.2 The Gate-reduction Method

In the gate-reduction method we initially replace, on a one-to-one basis, the 51 TTL gates with their MOS equivalents. The semiconductor manufacturer would thus refer to his library of MOS standard cells for forming the required gates. His

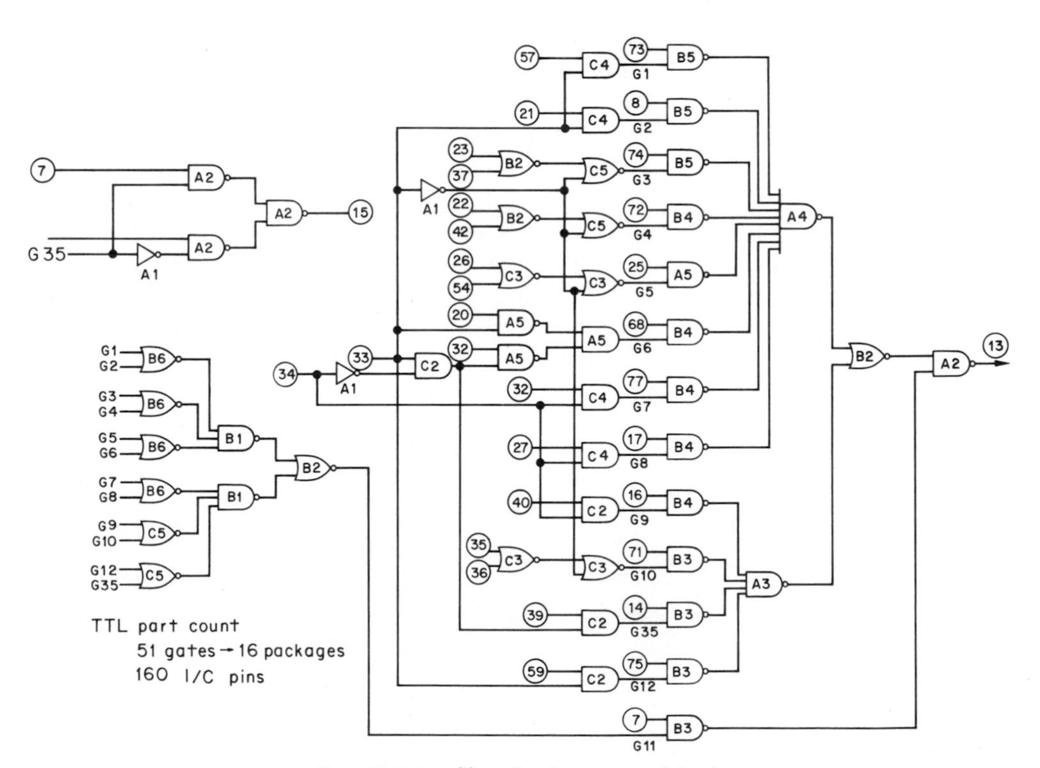

Fig. 10.14. Signal select control logic.

standard logic cells typically might have a height of 12 mils. If an average of 3 pins per gate (2 input and 1 output) with a spacing of 1.2 mils between pins is required, then the corresponding chip size would be calculated as

$$12 \text{ mils/gate} \times 1.2 \text{ mils/pin} \times 3 \text{ pins} \times 51 \text{ gates} = 12 \text{ mils} \times 184 \text{ mils}$$

The above calculation includes the 51 gates and not their intraconnections. To calculate the area consumed by the gates and their intraconnections, assume that for random logic implementation by means of standard cells, approximately 20 percent of the area will be occupied by cells and 80 percent will be occupied by intraconnections and bonding pads. Thus for our example the chip size would finally be given by

$$\frac{12 \times 184}{0.20} = 11{,}040 \text{ mils}^2$$

In other words, a chip of silicon real estate approximately 100 × 100 mils will be required to implement 51 logic gates.

The above chip size could be effectively reduced by combining various gates into one cell so as to reduce the routing at the points where fan-out equals unity. Figure 10.15 demonstrates this approach, which results in 28 standard cells with a total of 107 pins. Using the same rules as employed in our initial gate reduction, we

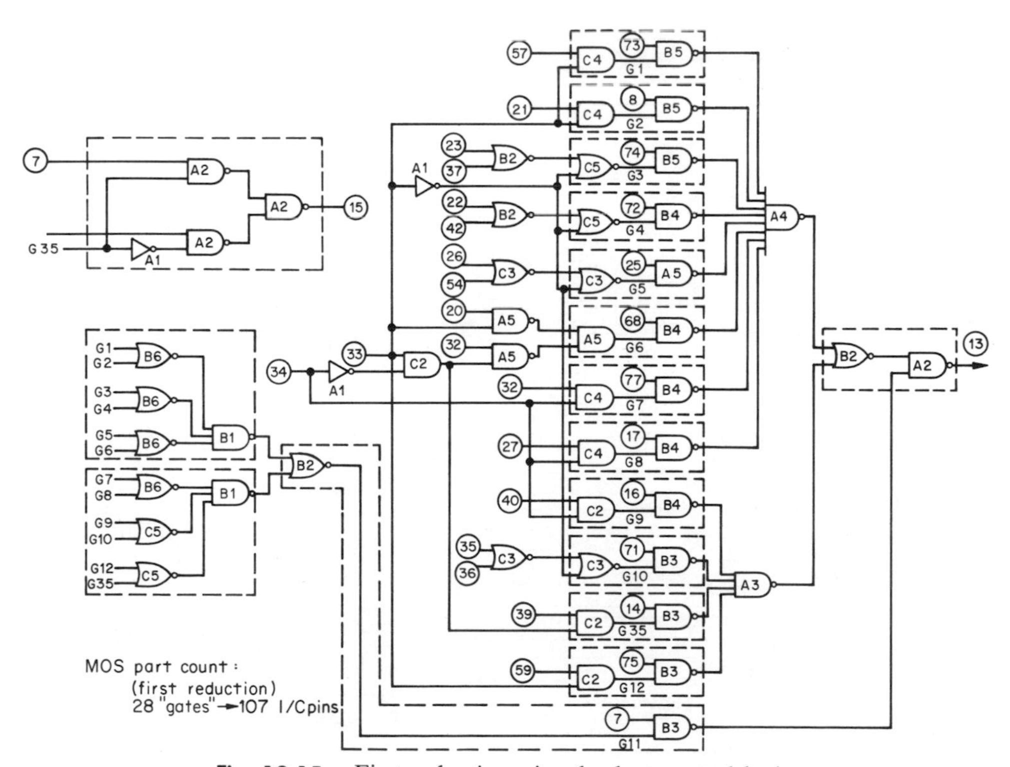

Fig. 10.15. First reduction, signal-select control logic.

will find that this latter approach yields a silicon area of 7,704 mils^2 for the 51 gates, which is a reduction in required silicon area of 30 percent.

Additional customizing of the cells could possibly lead to a chip where the active and passive areas are equally divided and the final chip size would then be 60 × 60 mils. This represents a valuable reduction in area, made possible by a thorough optimization of the logic cells and their intraconnections.

The optimizing of cells, or gate-reduction technique, is a commonly used approach for converting TTL systems to MOS/LSI. Unfortunately it is a dead-end approach because we are optimizing, for circuit layout purposes, perhaps only 5 percent of the entire system at a time. In the case considered, the logic diagram of Fig. 10.15 was only part of an arithmetic unit shown in Fig. 10.16. Just the one small triangle of Fig. 10.16 represents *all* of Fig. 10.15!

10.8.3 PLA or Matrix Technique

(a) Introduction. Modern system implementation with MOS/LSI demands a more sophisticated technique than that offered by the gate-reduction method. To realize this improvement, we must first examine the overall logic of the system and break it down into functional blocks. We can then apply the PLA or matrix technique. This implementation technique is demonstrated in detail in Chap. 8. A PLA consists of a unique combination of master-slave *J-K* flip-flops and static ROMs. It is used to perform sequential and combinational logic on a single

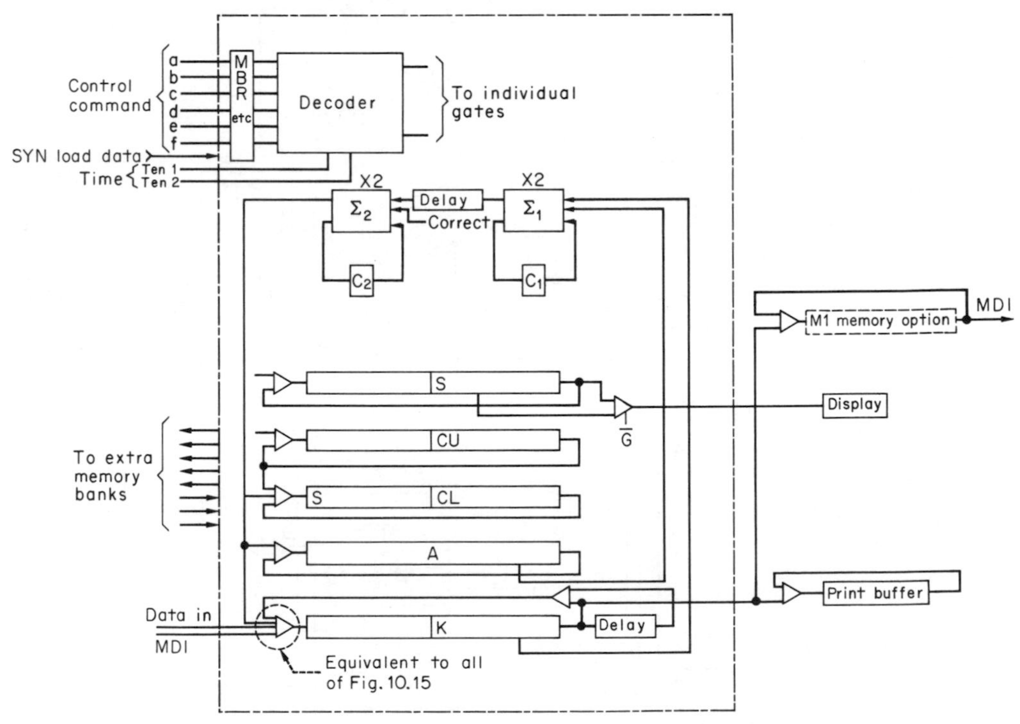

Fig. 10.16. Arithmetic unit.

Table 10.2. Quantification of Silicon Real Estate Requirements for Selected MOS/LSI Circuits

Symbol	Meaning	Silicon area, mil²/unit
RL	Random logic	60-200/gate
J-K	J-K flip-flop	300-600/flip-flop
SSR	Static shift register	40-60 bit
FSR	Feedback shift register	40-60/bit
DSR	Dynamic shift register	20-30/bit
RAM	Random-access memory	8-16/bit
ROM	Read-only memory	1-2/bit
ALU	Structured parallel adder selector logic	
⇨	Parallel interconnection	1-2/mil length
→	Serial interconnection	1-2/mil length
Pin	Bonding terminal	0-10/mil periphery
Scribe	Peripheral wasted space	10/mil periphery

MOS/LSI chip. The PLA represents an economical and efficient way to implement random logic using programable techniques at the gate-oxide mask step. When a random logic circuit is implemented by conventional methods in MOS, a large part of the chip area is used for intraconnection of cells. The PLA corrects this waste of silicon real estate.

Using the PLA technique we can adjust our thinking toward the direction of building a hardware framework for an electronic system with a software program to drive it. If this concept is coupled with identification of parallel operational sequences in the system logic, we can realize functional complexities in MOS chips that were formerly considered impossible to achieve. To illustrate these concepts, let us trace the technical developments which led to the one-chip calculator at Texas Instruments. Table 10.2 will serve as a useful reference for the discussion since it quantifies typical silicon real estate requirements for various circuit entities.

(b) Evolution of the One-chip Calculator. In 1969, TI implemented its first MOS calculator (Fig. 10.17). All the logic for this 13-digit, fixed-point machine was contained in four chips. This represented a considerable step forward in technology since the number of electronic components for this type calculator had been reduced by an order of magnitude from the bipolar version. Economically speaking, the greater percentage of the cost of integrated-circuit packages had not been reduced because of the higher cost-per-package of MOS over TTL. But drastic reductions in printed circuit board area and associated interconnection complexity had been achieved. Several PC boards now became a single board which served all the electronic packages. The cabinet size and cost for the calculator were also reduced in the initial MOS version. Interconnect reliability of the system was also improved over the TTL counterpart by the mere reduction of the electronic package count.

In 1970 a new design effort on the MOS calculator was initiated with the objective of reducing the system cost further by decreasing both the MOS component count within the circuits as well as the peripheral circuit requirements. The first step taken to realize these improvements was that of simplifying the external clocking circuitry

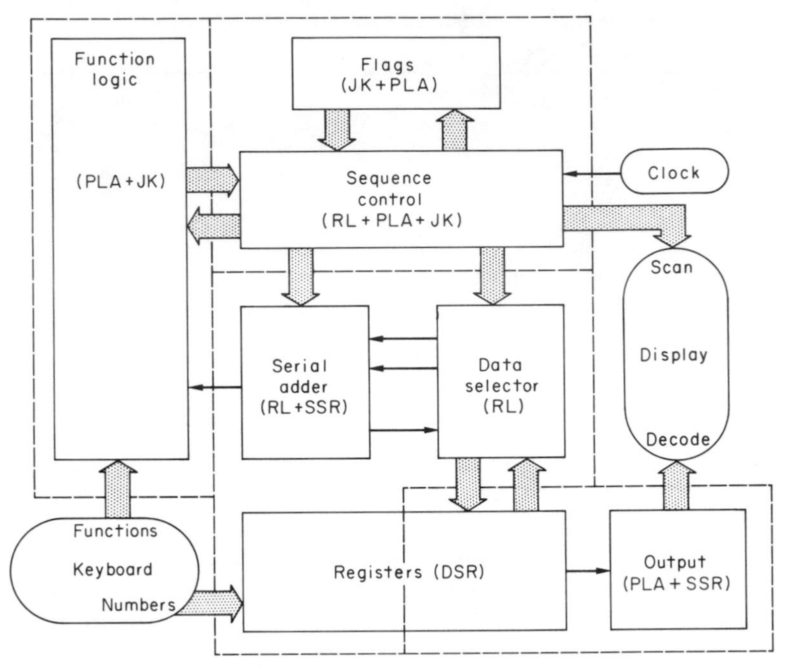

Fig. 10.17. Block diagram of the 1969 MOS calculator.

by requiring the use of only one clock instead of the previous two. Thus not only was one package pin eliminated, but more importantly a factor of 4 reduction in cost of the clock generator circuitry was achieved. Another major problem remaining was the requirement for external encoding of numbers at the keyboard. Numbers which were entered from the keyboard entered the data path directly, whereas the functions entered the function PLA. As a consequence the sequence control logic circuitry was very cumbersome. PLA was being utilized but a relatively large amount of random logic circuitry was still required.

Now consider the 1970 TI version of the MOS calculator in somewhat more detail as shown in Fig. 10.18. As mentioned above, the external clock requirements were reduced from two clock signals to one. A major change was next realized by combining both functions and numbers from the keyboard and feeding them directly to the function and sequence control block with the aid of PLAs and *J-K* flip-flops. Each key thus became a function key. The problem of data handling was solved by placing all the main program sequences in a ROM. The ROM functions in this case as a look-up table, and it can be implemented in MOS/LSI more economically than can random logic. So instead of using fixed random logic to control the logic flow through the system, programs stored in the ROM were utilized. Accomplishment of these modifications thus represented an advance toward realization of a "hardware framework driven by a software program."

The remaining logic required in the 1970 MOS calculator was also simplified by taking the numbers out of the data path. It was then possible to implement both the serial adder and the data selector logic with the PLA approach instead of by

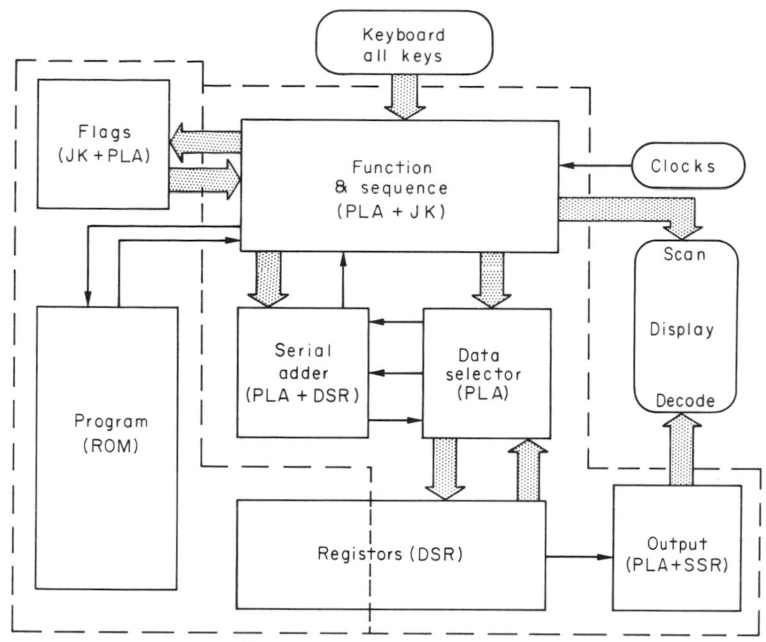

Fig. 10.18. Block diagram of the 1970 MOS calculator.

fixed random logic paths. Other refinements made possible by the new approach included a greater use of dynamic shift registers instead of the costlier static shift registers.

The combination of all these changes resulted in a two-chip MOS calculator for 1970 which featured a twofold reduction in MOS part count over the 1969 version, plus simplified external circuitry.

Development was continued in 1970 and 1971 with the goal of realizing a one-chip MOS calculator. The "hardware framework driven by a software program" approach was again employed. To this point in the development, bit serial arithmetic was used for performing computations, and this was reflected in the MOS circuitry. To modify this method, a structured parallel arithmetic unit operation at the digit level instead of at the bit level was employed. Thus the arithmetic could be handled in a parallel mode rather than in a serial mode. The ramifications of these modifications are evident by examination of the 1971 one-chip calculator block diagram shown in Fig. 10.19. In the calculator a separate block for control flags has been dispensed with and dynamic RAMs are employed for data storage. The RAM cell is considerably smaller than the dynamic shift register previously used for data storage. The RAM also enjoys great flexibility when used with the adder and selector logic since the RAM can also be used for storing data and control flags. The ROM has been retained for program storage and the keyboard interface has been simplified by taking the scan signal used for display and using it for encoding keyboard entries. The silicon real estate used for the 1971 MOS one-chip calculator is shown in Fig. 10.20.

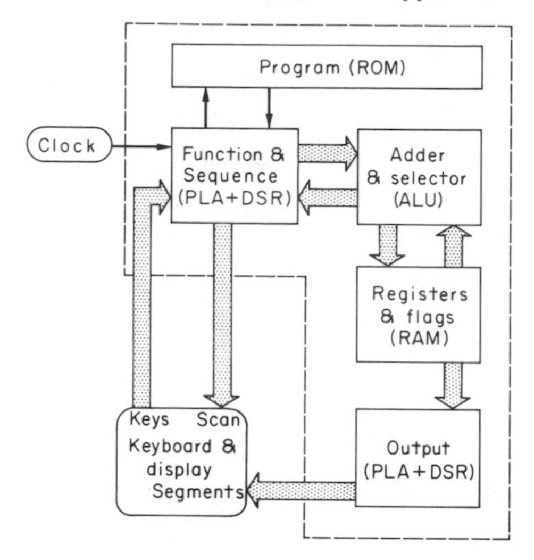

Fig. 10.19. Block diagram of the 1971 MOS calculator.

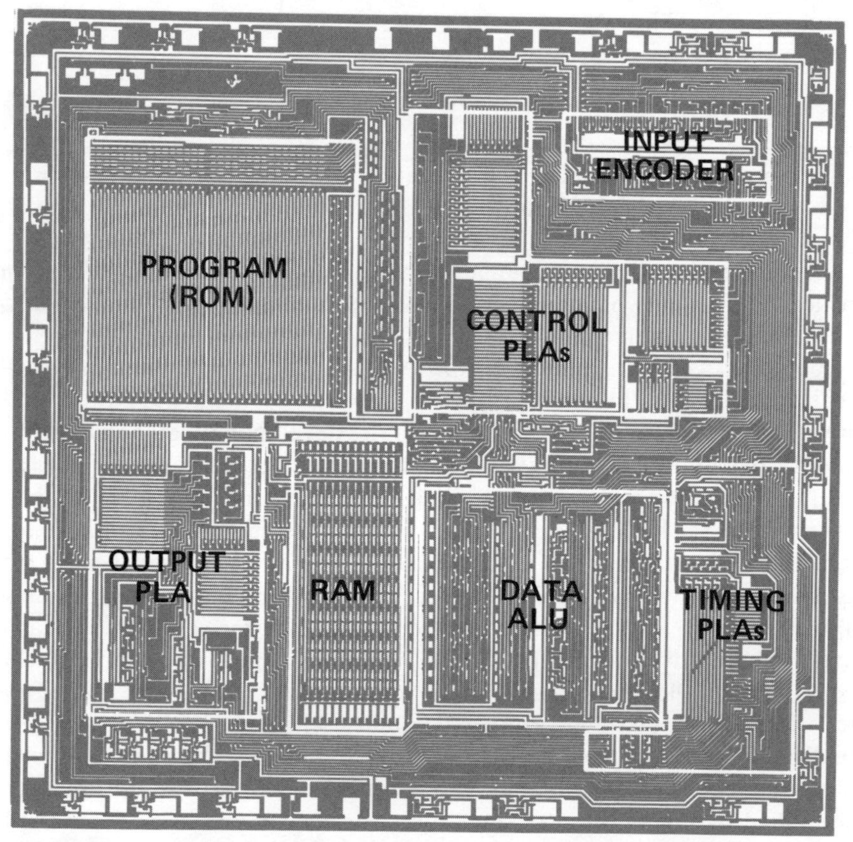

Fig. 10.20. Photograph of the silicon real estate used for 1971 MOS one-chip calculator.

10.9 SUMMARY AND CONCLUSIONS

The MOS/LSI market is in a state of dynamic growth. This is brought about by the economic characteristics of the MOSFET and its integrated-circuit technology. MOS exhibits its most favorable economic aspects in circuits of high complexity. Simple MOS logic gates are not economical. The price per integrated MOSFET has been reduced below a fraction of a cent and the technology produces the highest logic density available. The true evaluation of the cost of MOS circuitry can, however, be made properly only by considering its economic impact at the system level.

The system designer may choose between catalog and custom MOS/LSI. The nonrecurring engineering (NRE) expense and its contribution to system cost must be carefully evaluated when working with custom circuitry. The semiconductor manufacturer will be continually improving his technology capability and will offer ever-increasing logic density per MOS/LSI package. The circuit-and-logic designer is also presently advancing his methods for logic implementation. The recently developed PLA technique is one example of this advance, and it has resulted in at least a factor of 2 reduction in the cost of systems implemented with MOS/LSI.

REFERENCES

1. N. L. North and A. A. Ring, "Real Estate Principles and Practices," chap. 1, Prentice-Hall, Inc., Englewood Cliffs, N.J., 1960.
2. B. T. Murphy, Cost-size Optima of Monolithic Integrated Circuits, *Proc. IEEE,* **52:** 1537 (1964).
3. R. B. Seeds, Yield and Cost Analysis of Bipolar LSI, *IEEE International Electron Devices Meeting,* Washington, October 1967.
4. A. G. F. Dingwall, High-yield-processed Bipolar LSI Arrays, *IEEE International Electron Devices Meeting,* Washington, October 1968.
5. G. E. Moore, What Level of LSI is Best for You?, *Electronics,* **43:** 126, February 16, 1970.
6. S. Thompson, A Way of Thinking About Memories, *The Electronic Engineer,* **29:** 30, December 1970.

Index

Index